Harley-Davidson Big Twins Owners Workshop Manual

by Curt Choate, Tom Schauwecker and John H Haynes
Member of the Guild of Motoring Writers

Models covered:
1200 FL models: FL, FLH, FLHS. 1970 to 1980
1200 FX models: FX, FXE, FXS, FXEF. 1970 to 1980
1340 FL models: FLT, FLTC/U, FLH, FLHS, FLHT, FLHF, FLHTC/U. 1978 to 1993
1340 FXR models: FXR, FXRS, FXRS-SP, FXRS-CONV, FXLR, FXRD, FXRT. 1982 to 1993
1340 FX/Softails: FXB, FXE, FXDG, FXEF, FXS, FXSB, FXWG, FXST, FXSTC, FXSTS, FLST, FLST-SP, FLSTC, FLSTF, FLSTN. 1979 to 1993
1340 Dynas: FXDB, FXDC, FXDL, FXDWG. 1991 to 1993

(9W3 – 703)

Haynes Publishing
Sparkford Nr Yeovil
Somerset BA22 7JJ England

Haynes Publications, Inc
861 Lawrence Drive
Newbury Park
California 91320 USA

Acknowledgements

Our thanks to NGK Spark Plugs (UK) Ltd for contributing the spark plug condition photos used in Chapter 1. Thanks also to Jon LaCourse for supplying several tools and the motorcycle used for photography of the Evolution engine repair procedures and the rear cover illustration. Riders of Bridgwater supplied the Heritage Softail Classic shown on the front cover.

A book in the **Haynes Owners Workshop Manual Series**

Printed in the USA

ISBN 1 56392 081 6

Library of Congress Catalog Card Number 93-081024

Contents

Harley-Davidson Glide (FLH shown)

Right-side view of the engine and transmission

About this manual

Its purpose

The purpose of this manual is to help you maintain and repair your motorcycle. It can do so in several ways. It can help you decide what work must be done, even if you choose to have it done by a dealer service department or a repair shop, it provides information and procedures for routine maintenance and it offers diagnostic and repair procedures to follow when trouble occurs.

We hope you will use the manual to tackle the work yourself. For many simple jobs, doing it yourself may be quicker than arranging an appointment to get the machine into a shop and making the trips to leave it and pick it up. More importantly, a lot of money can be saved by avoiding the expense the shop must pass on to you to cover its labor and overhead costs. An added benefit is the sense of satisfaction and accomplishment you feel after doing the job yourself.

Using the manual

The manual is divided into several Chapters. Each Chapter is divided into numbered Sections which are headed in bold type between horizontal lines. Each Section consists of consecutively numbered Paragraphs (often referred to as "Steps" in the text).

The two types of illustrations used (line drawings and photographs) are all referenced by a number preceding each caption. The number denotes the Section and Paragraph the illustration is intended to clarify (i.e. illustration 3.4 means Section 3, Paragraph [or Step] 4). A "Refer to illustration . . ." entry under the Section head (or in some cases, a sub-head) will alert you to the illustrations which apply to the procedure you're following.

Procedures, once described in the text, are not normally repeated. When it's necessary to refer to another Chapter, the reference will be given as Chapter and Section number (i.e. Chapter 1, Section 16). Cross-references given without use of the word "Chapter" apply to Sections and/or Paragraphs in the same Chapter. For example, "see Section 8" means in the same Chapter.

Reference to the left or right side of the motorcycle is based on the assumption you're sitting on the seat, facing forward.

Even though extreme care has been taken during the preparation of this manual, neither the publisher nor the author can accept responsibility for any errors in, or omissions from, the information given.

NOTE

A **Note** provides information necessary to properly complete a procedure or information which will make the procedure easier to understand.

CAUTION

A **Caution** provides a special procedure or special steps which must be taken while completing the procedure where the **Caution** is found. Not heeding a **Caution** can result in damage to the assembly being worked on.

WARNING

A **Warning** provides a special procedure or special steps which must be taken while completing the procedure where the **Warning** is found. Not heeding a **Warning** can result in personal injury.

Introduction to the Harley-Davidson Big Twins

William S. Harley and Arthur Davidson jointly designed and built their first motorcycle during 1903, constructed along the lines of a powered bicycle. The single cylinder, air-cooled engine had a bore and stroke of 2-1/8 x 2-7/8 inches and is alleged to have been one of several built for experimental purposes. The weaknesses of the powered bicycle approach soon became apparent and a larger capacity engine followed, with much heavier flywheels. With the new, more powerful engine, the bicycle was able to climb hills more easily but, in turn, weaknesses in the frame design became apparent. And so the design progressed until something usable and marketable was available. Land was purchased in Milwaukee for the erection of a simple, two story factory and the Harley-Davidson joined the ranks of the world's motorcycles. By the end of 1906, fifty machines had been assembled at Chestnut Street and the order book was full, portending well for the year to follow.

The year 1909 proved to be a milestone in the history of the company, for it was during that year the company launched its first V-twin. It was this model, and others that followed, that helped create the legend of the large capacity V-twin, a field in which Harley-Davidson ruled supreme for many years. In private hands it gave a standard of comfort, performance and reliability second to none, while on the racetrack it became a snarling, powerful beast, with the ability to perform incredible feats of speed and endurance that made the headlines.

Contrary to expectations, the American motorcycle boom didn't last long and by the 1930s, it had all but petered out. It was largely the more sporting side that kept things going, although motorcycles were also used by police for law enforcement purposes. The American "speed cop' was very much a part of life and a big, powerful V-twin gave him a certain amount of flexibility that wouldn't have been possible in a patrol car.

The modern Harley-Davidson Electra Glide was never introduced as a model, as such, but is a development of such machines as the model 74K, which was introduced in 1941, and more recently the Duo-Glide of the late 50's. This can be seen when it's noted that many parts installed on the latest models are not only similar, but in many cases identical, to components installed on the wartime machines. The policies of the company, reflecting the motorcycling needs of the home market, have produced a machine ideally suited to the task of traveling vast distances in comfort, and which has evolved a personality and charisma which has made Harley-Davidson a household name. The Super Glide model was introduced in 1971 to cater to the rider requiring a sportier and lighter machine, yet not lacking in the glamor of its more heavily dressed stablemate.

The development of the Evolution engine, which was introduced in 1984 models, ensures the big twin mystique will continue, since the new engine is as reliable, oil-tight and serviceable as the engines in competitive motorcycles, without discarding the unique characteristics of previous Harley engines. Because it's more powerful across the board, lighter, cooler running and cleaner, in terms of exhaust emissions, than its predecessor, it launched a new era for the big V-twin engine and continues the trend toward refinement that should produce many more generations of satisfied Harley customers.

Identification numbers

Harley-Davidson motorcycles have a Vehicle Identification Number (VIN) stamped into the frame on the right side of the steering head, as well as on the right side of the engine crankcase **(see illustrations)**. All 1978 and later models also have an adhesive-backed label, attached to the right front frame downtube, with the VIN on it **(see illustration)**. The VIN is made up of a model code, a serial code, a serial number, the model year and the manufacturer's identification.

The VIN number should be recorded and kept in a safe place so it can be furnished to law enforcement officials in the event of theft. The VIN should also be available when purchasing or ordering parts for the machine. It's a good idea to write it on a card and keep it tucked away with your driver's license, then it'll be handy when you need it.

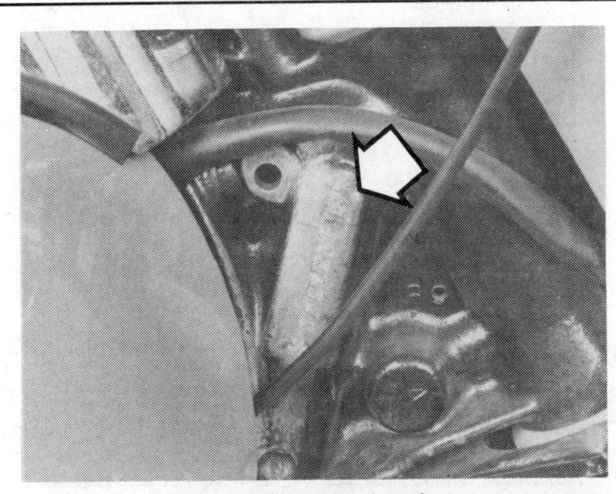

Frame VIN number location

Engine VIN number location

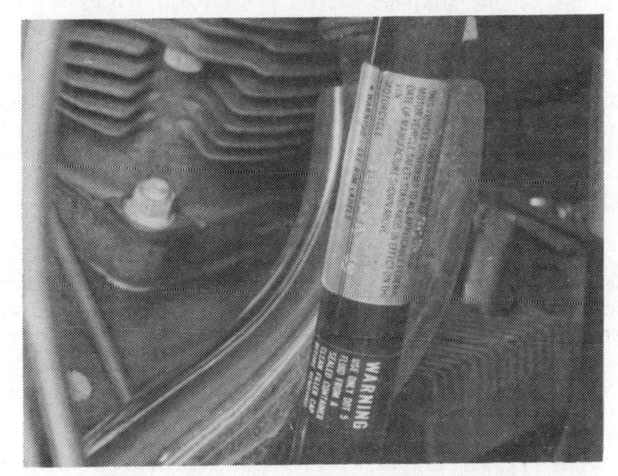

Frame tube label on 1978 and later models

Model	First Two Digits (Model)	Next Five Digits (Sequential Number)	Second Last Digit (Manufacturer)	Last Digit (Model) (Season)
FL or FLP	1A	10000	Harley-Davidson	8
FLH	2A	and up		(1978)
FX	2C	(5 digits)	H	
FXE	9D			
FXS	2F			
FLH-1200	2A			
FX-1200	2C			8
FXE-1200	9D	60000	H	(1978½)
FXEF-1200	5E	and		
FXEF-80	6E	up	H	9
FXS-1200	2F	10000		(1979)
FLH-80	3G	and		
FXE-80	6G	up		
FXS-80	7G			
FXWG-80	9G			
FLH-80				
Classic	3H			
FLHS-80	5H			
FL-80	6H	10000	J	0
FLH-1200		and		(1980)
Police	7H	up		
FLH-1200				
Shrine	8H			
FLH-80				
Police	9H			
FLH-80				
Shrine	1K			
FLT	5G	00001	J	0
		and up		(1980)

Examples: 1979 FLH-1200, 2A12141H9
1980 FLH-1200, 2A12141JO

VIN key for pre-1980 models

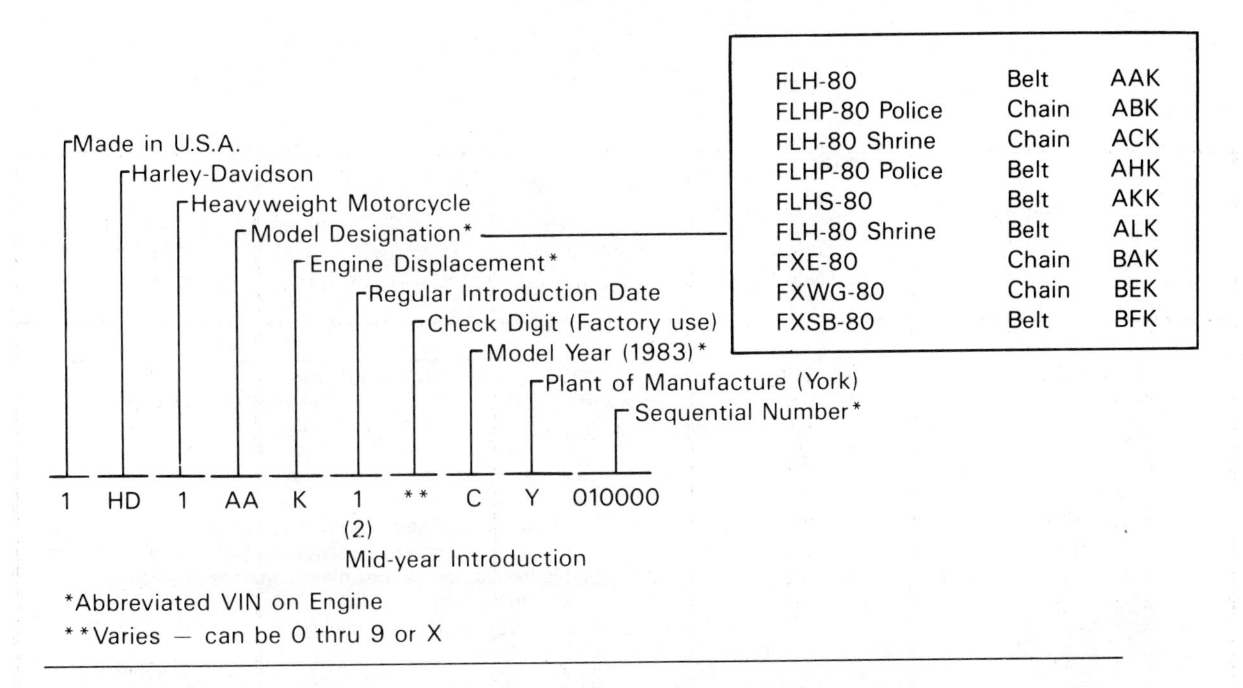

FLH-80	Belt	AAK
FLHP-80 Police	Chain	ABK
FLH-80 Shrine	Chain	ACK
FLHP-80 Police	Belt	AHK
FLHS-80	Belt	AKK
FLH-80 Shrine	Belt	ALK
FXE-80	Chain	BAK
FXWG-80	Chain	BEK
FXSB-80	Belt	BFK

*Abbreviated VIN on Engine
**Varies – can be 0 thru 9 or X

Sample V.I.N. as it appears on the steering head – 1 HD 1AAK1 1 CY010000

Sample abbreviated V.I.N. as it appears on the engine – AAKC 010000

VIN key for 1980 and later four-speed models (typical)

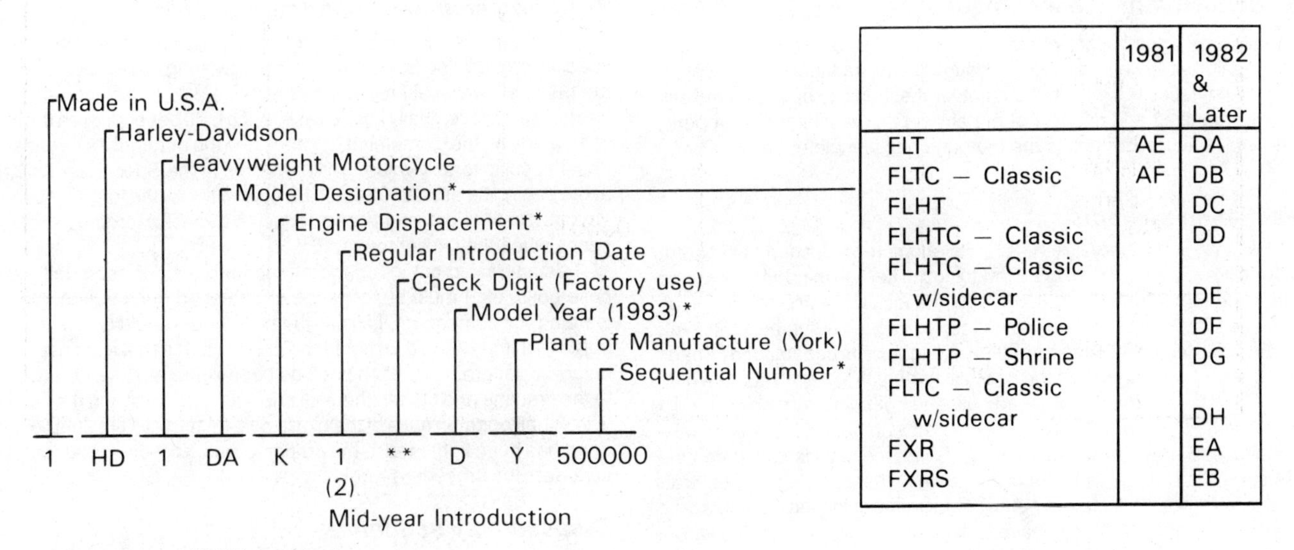

	1981	1982 & Later
FLT	AE	DA
FLTC – Classic	AF	DB
FLHT		DC
FLHTC – Classic		DD
FLHTC – Classic w/sidecar		DE
FLHTP – Police		DF
FLHTP – Shrine		DG
FLTC – Classic w/sidecar		DH
FXR		EA
FXRS		EB

*Abbreviated VIN on Engine
**Varies – can be 0 thru 9 or X

Sample V.I.N. as it appears on the steering head – 1 HD 1DAK1 1 DY500000

Sample abbreviated V.I.N. as it appears on the engine – DAKD 500000

VIN key for 1980 and later five-speed models (typical)

Buying parts

Once you've located all the identification numbers, record them for reference when buying parts. Since manufacturers change specifications, parts and vendors (companies that manufacture various components on the machine), providing the ID numbers is the only way to be reasonably sure you're buying the correct parts.

Whenever possible, take the worn part to the dealer so direct comparison with the new component can be made. Along the trail from the manufacturer to the parts shelf, there are numerous places the part can end up with the wrong number or be listed incorrectly.

The two places to purchase new motorcycle parts – franchised dealers and independent accessory stores – differ in the type of parts they carry. While a dealer can obtain virtually every stock part on the motorcycle, as well as aftermarket items, the accessory dealer is usually – not always – limited to items such as shock absorbers, tune-up parts, engine gaskets, cables, brake parts, etc. Often, however, an accessory outlet will sell aftermarket suspension components, cylinders, transmission gears and other major components.

Used parts can be obtained for roughly half the price of new ones, but you can't always be sure of what you're getting. Once again, take the worn part to the salvage yard for direct comparison.

Whether buying new, used or rebuilt parts, it's a good idea to deal directly with someone who specializes in parts for Harley-Davidson motorcycles.

Maintenance techniques, tools and working facilities

Basic maintenance techniques

There are a number of techniques involved in motorcycle maintenance and repair that will be referred to throughout this manual. Application of these techniques will enable the amateur mechanic to be more efficient, better organized and capable of performing the various tasks properly, which will ensure the repairs are thorough and complete.

Fastening systems

Fasteners, basically, are nuts, bolts and screws used to hold two or more parts together. There are a few things to keep in mind when working with fasteners. Almost all of them require a locking device of some type (either a lock washer, locknut, lockplate, locking tab or thread adhesive). All threaded fasteners should be clean, straight, have undamaged threads and undamaged corners on the hex head where the wrench fits. Develop the habit of replacing all damaged nuts and bolts with new ones.

Rusted nuts and bolts should be treated with penetrating oil to make removal easier and prevent breakage. Some mechanics use turpentine in a spout type oil can, which works quite well. After applying the penetrating oil, let it "work" for a few minutes before trying to loosen the nut or bolt. Badly rusted fasteners may have to be chiseled off or removed with a hacksaw or special nut breaker, available at tool stores.

Flat washers and lock washers, when removed from an assembly, should always be replaced exactly as removed. Discard damaged washers and replace them with new ones. Always use a flat washer between a lock washer and any soft metal surface (such as aluminum), thin sheet metal or plastic. Special locknuts can only be used once or twice before they lose their locking ability and must be replaced.

If a bolt or stud breaks off in an assembly, it can be drilled out and removed with a special tool called an E-Z out. Most dealer service departments and motorcycle repair shops (and automotive machine shops) can perform this task, as well as others (such as the repair of threaded holes that have been stripped out).

Torquing sequences and procedures

When threaded fasteners are tightened, they are often tightened to a specific torque value (torque is basically a twisting force). Over-tightening the fastener can weaken it and cause it to break, while under-tightening can cause it to eventually come loose. Each bolt, depending on the material it's made of, the diameter of its shank and the material it's threaded into, has a specific torque value, which is noted in the Specifications Section at the beginning of each Chapter. Be sure to follow the torque recommendations closely. For fasteners not requiring a specific torque, use common sense when they're tightened.

Fasteners laid out in a pattern (such as cylinder head bolts, engine case bolts, etc.) must be loosened or tightened in a sequence to avoid warping the component. Initially, the bolts/nuts should go on finger-tight only. Next, they should be tightened one full turn each, in a criss-cross or diagonal pattern. After each one has been tightened one full turn, return to the first one and tighten them all one-half turn, following the same pattern. Finally, tighten each of them one-quarter turn at a time until each fastener has been tightened to the proper torque. To loosen and remove the fasteners the procedure would be reversed.

Disassembly sequence

Component disassembly should be done slowly and carefully to make sure the parts go back together properly during reassembly. Always keep track of the sequence in which parts are removed. Note special characteristics or marks on parts that can be installed more than one way (a good example would be the rocker arm shafts on the Evolution engine – they have a small cutout on only one end for the bolts to pass through). It's a good idea to lay the disassembled parts out on a clean surface in the order they were removed. It may also be helpful to make sketches or take instant photos of components before removal.

When removing fasteners from a component, keep track of their locations. Sometimes threading a bolt back into a part, or putting the washers and nut back on a stud, can prevent mixups later. If nuts and bolts can't be returned to their original locations, they should be kept in a compartmented box or a series of small boxes. A cupcake or muffin tin is ideal for this purpose, since each cavity can hold the bolts and nuts from a particular area (i.e. engine case bolts, rocker arm cover bolts, engine mount bolts, etc.). A pan of this type is especially helpful when working on assemblies with very small parts (such as the carburetor and the valve train). The cavities can be marked with paint or tape to identify the contents.

When wiring looms, harnesses or connectors are separated, it's a good idea to identify the two halves with numbered pieces of masking tape so they can be easily reconnected.

Gasket sealing surfaces

Gaskets are used to seal the mating surfaces between components and keep lubricants, fluids, vacuum or pressure contained in an assembly.

Many times these gaskets are coated with a liquid or paste-type gasket sealing compound before assembly. Age, heat and pressure can sometimes cause the two parts to stick together so tightly they're very difficult to separate. In most cases, the part can be loosened by striking it with a soft-face hammer near the mating surfaces. A regular hammer can be used if a block of wood is placed between the hammer and the part. Do not hammer on cast parts or parts that could be easily damaged. With any particularly stubborn part, always recheck to make sure all fasteners have been removed.

Avoid using a screwdriver or bar to pry components apart, as they can easily mar the gasket sealing surfaces of the parts (which must remain smooth). If prying is absolutely necessary, use a piece of wood, but keep in mind that extra clean-up will be necessary if the wood splinters.

After the parts are separated, the old gasket must be carefully scraped off and the gasket surfaces cleaned. Stubborn gasket material can be soaked with a gasket remover (available in aerosol cans) to soften it so it can be easily scraped off. Gasket scrapers are widely available – just be careful not to gouge the sealing surfaces if you use one. Some gaskets can be removed with a wire brush, but regardless of the method used, the mating surfaces must be left clean and smooth. If the gasket surface is scratched or gouged, then a gasket sealer thick enough to fill scratches will have to be used during reassembly of the components. For most applications, a non-drying (or semi-drying) gasket sealer is best.

Hose removal tips

Hose removal precautions closely parallel gasket removal precautions. Avoid scratching or gouging the hose fitting or the connection may leak. Because of various chemical reactions, the rubber in hoses can bond itself to the metal fitting it's installed on. To remove a hose, first loosen the clamp(s) securing it to the fitting. Next, using slip-joint pliers, grab the hose near the fitting and rotate it back-and-forth until it's completely free, then pull it off (silicone or other lubricants will ease removal if they can be applied between the hose and the fitting). Apply the same lubricant to the inside of the hose and the outside of the fitting before installation.

If a hose clamp is broken or damaged, don't reuse it. Also, don't reuse hoses that are cracked, hardened or chafed.

Tools

A selection of good tools is a basic requirement for anyone who plans to maintain and repair a motorcycle. For the owner who has few tools, if any, the initial investment might seem high, but when compared to the spiraling costs of routine maintenance and repair, it's a wise one.

To help the owner decide which tools are needed to perform the tasks detailed in this manual, the following tool lists are offered: *Maintenance and minor repair, Repair and overhaul* and *Special.* The newcomer to practical mechanics should start off with the *Maintenance and minor repair* tool kit, which is adequate for the simpler jobs. Then, as confidence and experience grow, the owner can tackle more difficult tasks, buying additional tools as they're needed. Eventually the basic kit will be built into the *Repair and overhaul* tool set. Over a period of time, the experienced do-it-yourselfer will assemble a tool set complete enough for most repair and overhaul procedures and will add tools from the *Special* category when it's felt the expense is justified by the frequency of use.

Maintenance and minor repair tool kit

The tools in this list (many of which are included in the tool kit supplied when the motorcycle is purchased new) should be considered the minimum required for performance of routine maintenance, servicing and minor repair work. We recommend the purchase of combination wrenches (box-end and open-end combined in one wrench); while more expensive than open-end ones, they offer the advantages of both types of wrench.

Combination wrench set
Adjustable wrench – 10-inch
Spark plug socket (with rubber insert)
Spark plug gap adjusting tool
Feeler gauge set
Standard screwdriver (5/16-inch x 6-inch)
Phillips screwdriver (no. 2 x 6-inch)
Combination (slip-joint) pliers – 6-inch
Hacksaw and assortment of blades
Tire pressure gauge
Control cable pressure luber
Grease gun
Oil can
Fine emery cloth
Wire brush
Hand impact screwdriver and bits
Funnel (medium size)
Safety goggles

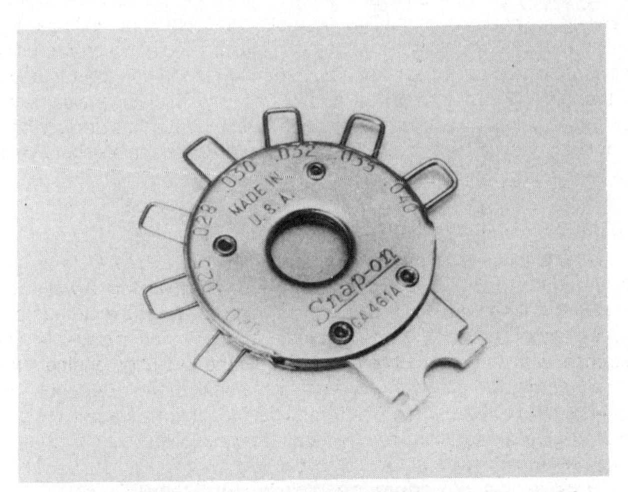

A special wire-type gauge is needed for checking and adjusting spark plug gaps

Feeler gauges are needed for measuring clearances

Pressure lubricating adapters are used with aerosol cable lubricants

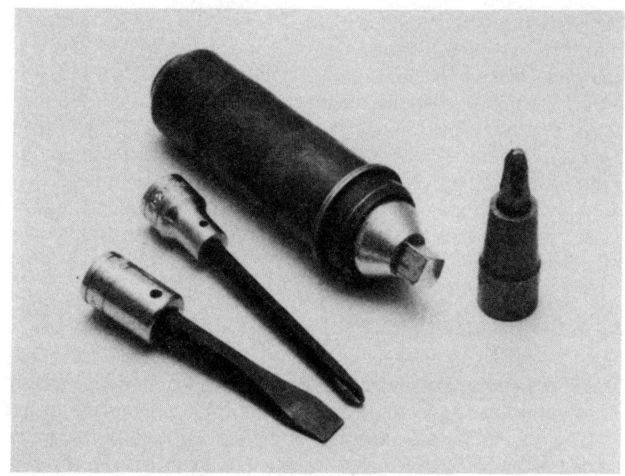

Many factory-tightened screws and bolts are hard to loosen without an impact driver (various size standard, Phillips and Allen-head bits are available)

Buy an oil filter wrench that fits over the end of the filter and accepts a 3/8-inch drive ratchet or breaker bar

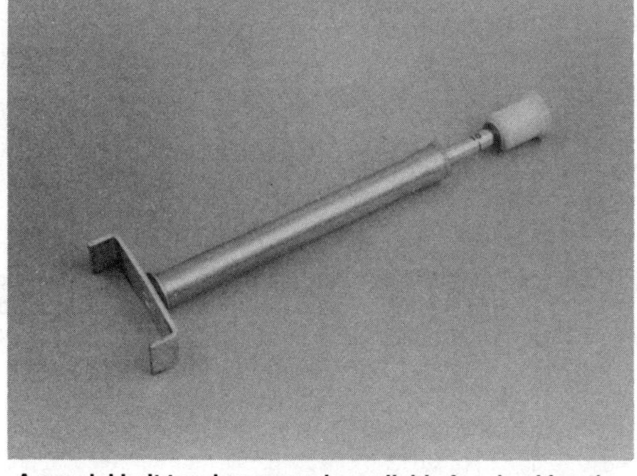

A special belt tension gauge is available for checking the final drive belt used in place of a chain on some models

Oil filter wrench
Drain pan
Belt tension gauge (for belt drive models)
Spoke wrench
Battery hydrometer
Battery charger (1 amp)

Note: *Since basic ignition timing checks are a part of routine maintenance, you'll have to purchase a good quality, inductive pick-up timing light. Although it's included in the list of Special tools, it's mentioned here because ignition timing checks can't be made without it. You'll also need the special threaded clear plastic timing mark view plug (part no. HD-96295-65C) that's installed in the crankcase so the timing marks can be seen without oil spraying out of the view hole.*

Repair and overhaul tool set

These tools are essential for anyone who plans to perform major repairs and are intended to supplement those in the Maintenance and minor repair tool kit. Included is a comprehensive set of sockets which, though expensive, are invaluable because of their versatility (especially when various extensions and drives are available). We recommend 3/8-inch drive over 1/2-inch drive for general motorcycle maintenance and repair (ideally, a mechanic would have both).

Socket set(s)
Reversible ratchet
Extension – 6-inch
Universal joint
Torque wrench (same size drive as sockets)
Ball pein hammer – 12 oz
Soft-face hammer (plastic/rubber)
Standard screwdriver (1/4-inch x 6-inch)
Standard screwdriver (stubby – 5/16-inch)
Phillips screwdriver (no. 3 x 8-inch)
Phillips screwdriver (stubby – no. 2)
Pliers – vise-grip
Pliers – lineman's
Pliers – needle-nose
Pliers – snap-ring (internal and external)
Cold chisel – 1/2-inch
Scribe
Gasket scraper
Center punch
Pin punches (1/16, 1/8, 3/16-inch)
Steel rule/straightedge – 12-inch
Allen wrench set
Pin-type spanner wrench
A selection of files
A selection of brushes for cleaning small passages
Wire brush (large)
Work light with extension cord

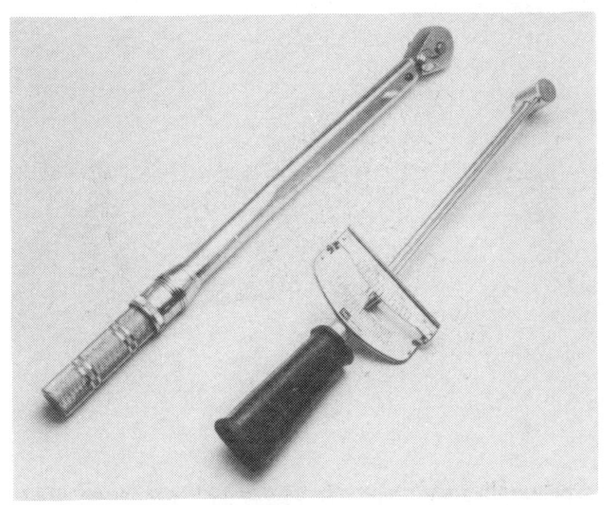

Two types of torque wrenches are commonly available (left – click type; right – deflecting beam type

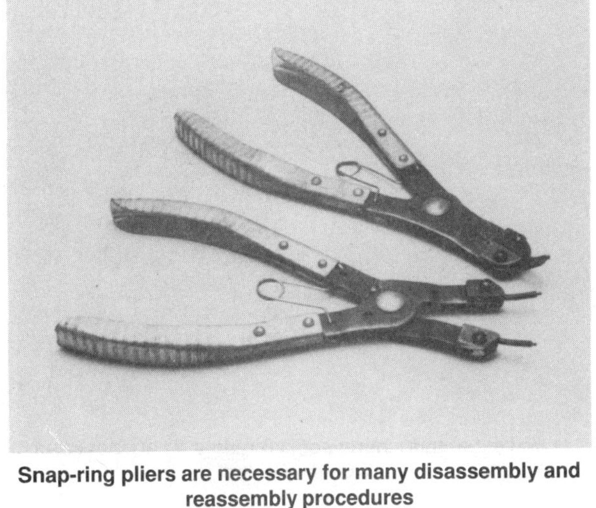

Snap-ring pliers are necessary for many disassembly and reassembly procedures

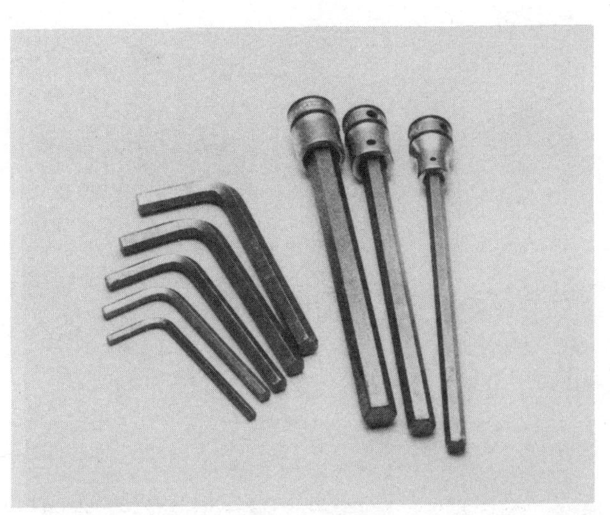

Hex (Allen) wrenches are available in a variety of styles

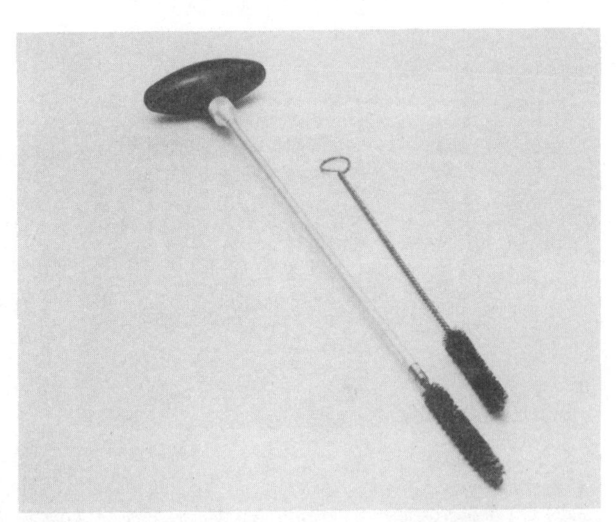

Small nylon or copper brushes can be used to clean internal passages in the engine and other components

Note: *Another tool which is often useful is an electric drill with a chuck capacity of 3/8-inch (and a set of good quality drill bits).*

Special tools

The tools in this list include those which aren't used regularly, are expensive to buy, or which need to be used in accordance with their manufacturer's instructions. Unless these tools will be used frequently, it's not very economical to purchase many of them. A consideration would be to split the cost and use between yourself and a friend or friends (such as members of a motorcycle club).

This list primarily contains tools and instruments widely available to the public, as well as some special tools produced by the vehicle manufacturer for distribution to dealer service departments. As a result, references to the manufacturer's special tools are occasionally included in the text of this manual. Generally, an alternative method of doing the job without the special tool is offered. However, sometimes there is no alternative to their use. Where this is the case, and the tool can't be purchased or borrowed, the work should be turned over to a dealer service department or motorcycle repair shop.

Valve spring compressor
Valve lapping tool
Piston ring removal and installation tool
Piston ring compressor
Telescoping gauges
Micrometer(s) and/or dial/Vernier calipers
Cylinder surfacing hone
Compression gauge
Dial indicator set
Multimeter
Tap and die set
Timing light and timing mark view plug
Small air compressor with blow gun and tire chuck

Buying tools

For the do-it-yourselfer just starting to get involved in motorcycle maintenance and repair, there are a number of options available when purchasing tools. If maintenance and minor repair is the extent of the work to be done, the purchase of individual tools is satisfactory. If, on the other hand, extensive work is planned, it would be a good idea to purchase a modest tool set from one of the large retail chain stores. A set can usually be bought at a substantial savings over the individual tool prices (and they often come with a tool box). As additional tools are needed, add-on sets, individual tools and a larger tool box can be purchased to expand the tool collection. Building a tool set gradually allows the cost of the tools to be spread over a longer period of time and gives the mechanic the freedom to choose only those tools that will actually be used.

Tool stores and motorcycle dealers will often be the only source of some of the special tools that are needed, but regardless of where tools

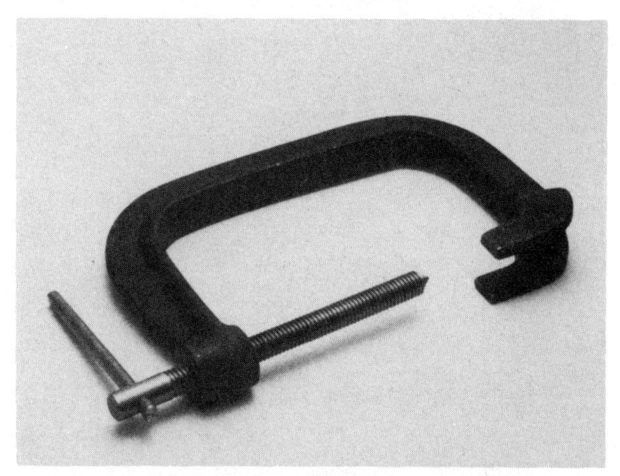

A valve spring compressor is required if cylinder head work is needed

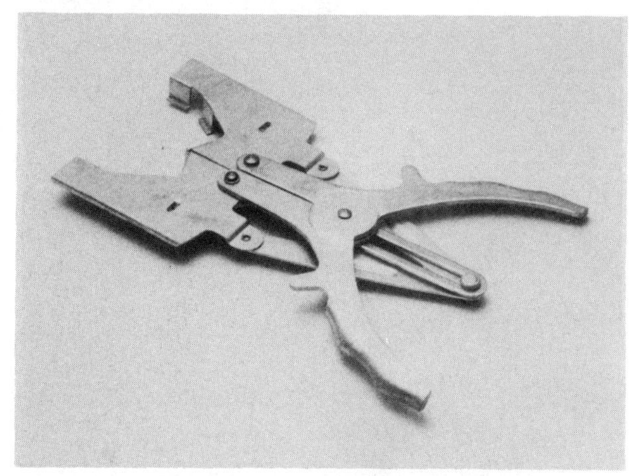

A piston ring removal and installation tool should be used to remove and install rings

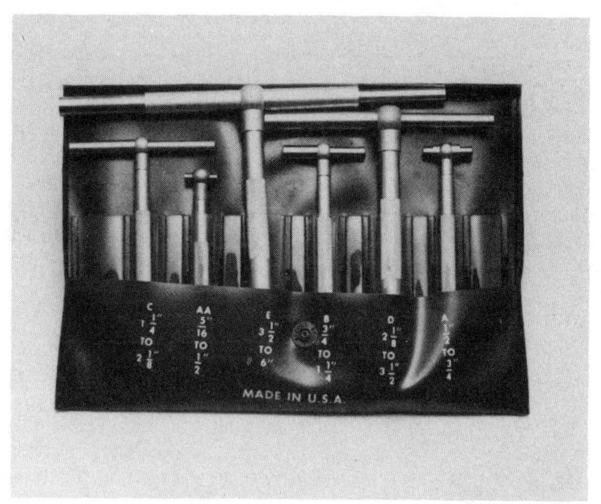

Telescoping gauges are used with micrometers to measure cylinders diameters and other internal dimensions

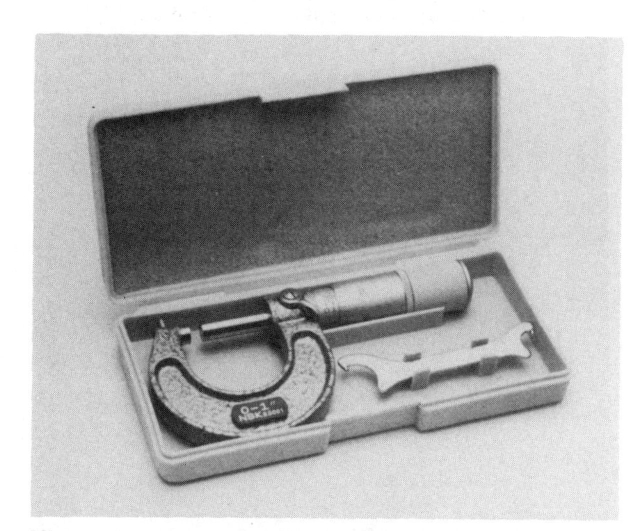

Micrometers are used to accurately measure shaft diameters and other outside dimensions

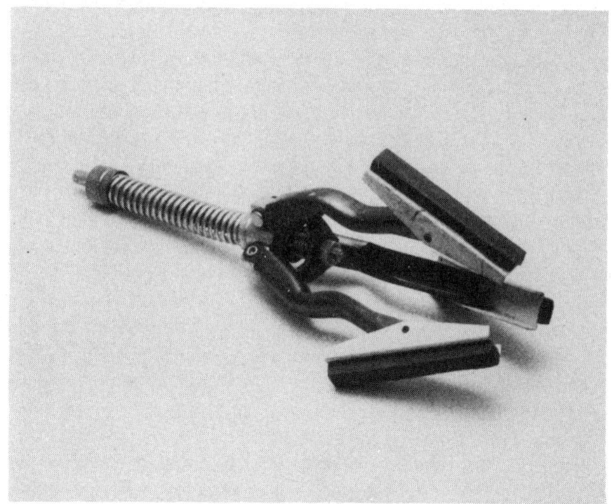

A hone is needed to resurface cylinders so new piston rings will seat properly

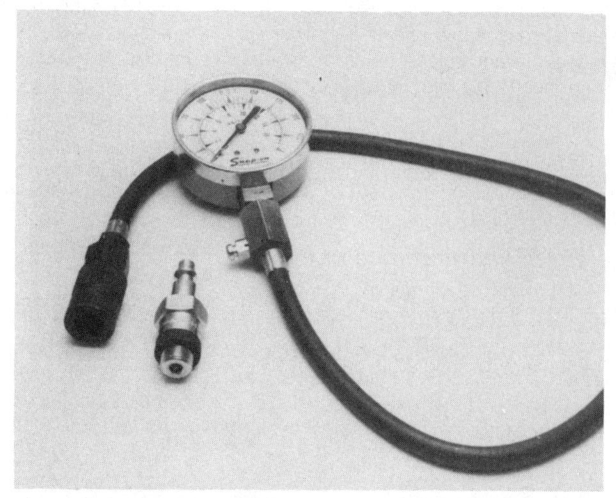

A compression gauge with a threaded fitting for the spark plug hole is preferred over the type that requires hand pressure to maintain the seal at the plug hole

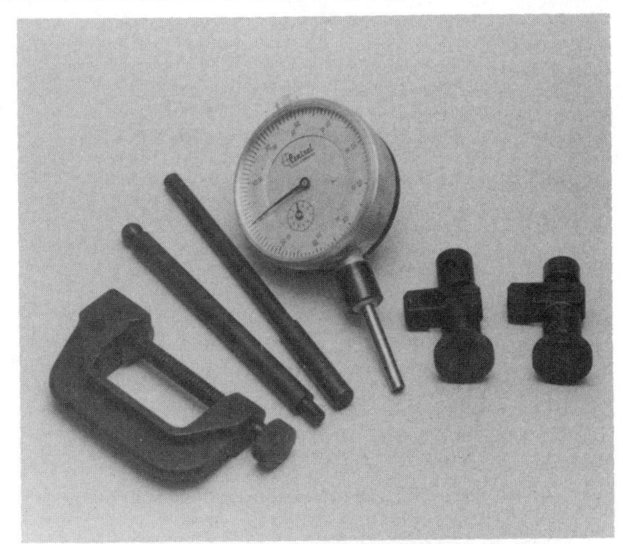

A dial indicator is used to determine shaft runout, gear backlash and other critical specifications

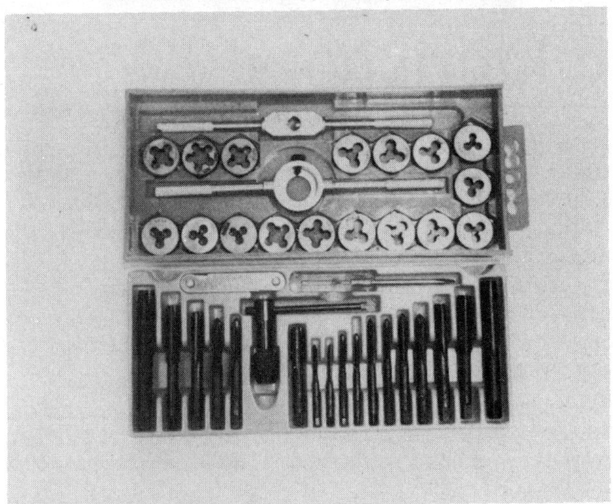

A tap and die set is handy for cleaning and restoring stripped or dirty threads

are bought, try to avoid cheap ones (especially when buying screwdrivers and sockets) because they won't last very long. There are plenty of tools around at reasonable prices, but always aim to purchase items which meet the relevant national safety standards. The expense involved in replacing cheap tools will eventually be greater than the initial cost of quality tools.

It is obviously not possible to cover the subject of tools fully here. For those who wish to learn more about tools and their use, there is a book entitled Motorcycle Workshop Practice Manual (Book no. 1454) available from the publishers of this manual.

Care and maintenance of tools

Good tools are expensive, so it makes sense to treat them with respect. Keep them clean and in usable condition and store them properly. Always wipe off dirt, grease and metal chips before putting them away. Never leave tools lying around in the work area.

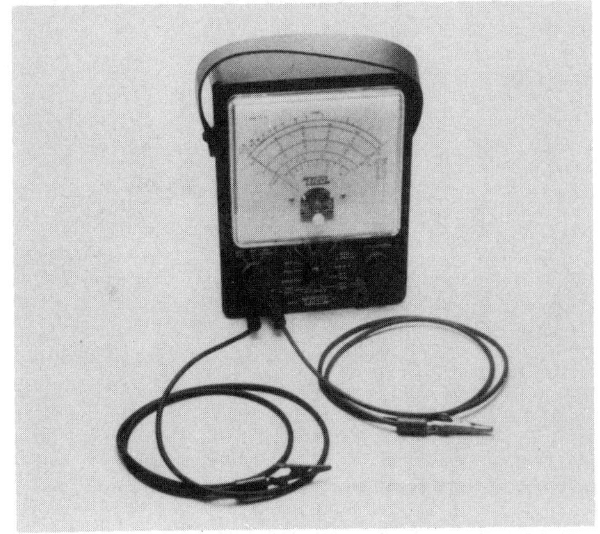

A multimeter (volt/ohm/ammeter) is needed if electrical system diagnosis is attempted

Some tools, such as screwdrivers, pliers, wrenches and sockets, can be hung on a panel mounted on a garage or workshop wall, while others should be kept in a tool box or tray. Measuring instruments, gauges, meters, etc. must be carefully stored where they can't be damaged by weather or impact from other tools.

When tools are used with care and stored properly, they'll last a very long time. Even with the best of care, tools will wear out if used frequently. When a tool is damaged or worn out, replace it; subsequent jobs will be safer and more enjoyable if you do.

Working facilities

Not to be overlooked when discussing tools is the workshop. If anything more than routine maintenance is going to be done, some sort of special work area is essential.

It's understood, and appreciated, that many home mechanics don't have a good workshop or garage available and end up removing an engine or doing major repairs outside (the overhaul or repair should be completed under the cover of a roof).

A clean, flat workbench or table of comfortable working height is an absolute necessity. The workbench should be equipped with a vise that will open at least four inches.

As mentioned previously, some clean, dry storage space is also required for tools, as well as the lubricants, fluids, cleaning solvents, etc. which soon become necessary.

Sometimes waste oil and fluids, drained from the engine or transmission during normal maintenance or repairs, present a disposal problem. To avoid pouring them on the ground or into a sewage system, simply pour the used fluids into large containers, seal them with caps and take them to an authorized disposal site or service station. Plastic jugs (such as old antifreeze containers) are ideal for this purpose.

Always keep a supply of old newspapers and clean rags available. Old towels are excellent for mopping up spills. Many mechanics use rolls of paper towels for most work because they're readily available and disposable. To help keep the area under the motorcycle clean, a large cardboard box can be cut open and flattened to protect the garage or shop floor.

When working over a painted surface (such as the fuel tank) cover it with an old blanket or bedspread to protect the finish.

Motorcycle chemicals and lubricants

A number of chemicals and lubricants are available for use in motorcycle maintenance and repair. They include a wide variety of products ranging from cleaning solvents and degreasers to lubricants and protective sprays for rubber, plastic and vinyl.

Contact point/spark plug cleaner is a solvent used to clean oily film and dirt from points, grime from electrical connectors and oil deposits from spark plugs. It's oil free and leaves no residue. It can also be used to remove gum and varnish from carburetor jets and other orifices.

Carburetor cleaner is similar to contact point/spark plug cleaner but it usually has a stronger solvent and may leave a slight oily reside. It isn't recommended for cleaning electrical components or connections.

Brake system cleaner is used to remove grease or brake fluid from brake system components (where clean surfaces are absolutely necessary and petroleum-based solvents must not be used); it also leaves no residue.

Silicone-based lubricants are used to protect rubber parts such as hoses and grommets, and are used as lubricants for hinges and locks.

Multi-purpose grease is an all purpose lubricant used wherever grease is more practical than a liquid lubricant such as oil. Some multi-purpose grease is colored white and specially formulated to be more resistant to water than ordinary grease.

Gear oil (sometimes called gear lube) is a specially designed oil used in transmissions, as well as other areas where high friction, high-temperature lubrication is required. It's available in a number of viscosities (weights) for various applications.

Motor oil, of course, is the lubricant specially formulated for use in the engine. It normally contains a wide variety of additives to prevent corrosion and reduce foaming and wear. Motor oil comes in various weights (viscosity ratings) of from 5 to 80. The recommended weight of the oil depends on the seasonal temperature and the demands on the engine. Light oil is used in cold climates and under light load conditions; heavy oil is used in hot climates and where high loads are encountered. Multi-viscosity oils are designed to have characteristics of both light and heavy oils and are available in a number of weights from 5W-20 to 20W-50.

Gas additives perform several functions, depending on their chemical makeup. They usually contain solvents that help dissolve gum and varnish that build up on carburetor and intake parts. They also serve to break down carbon deposits that form on the inside surfaces of the combustion chambers. Some types contain upper cylinder lubricants for valves and piston rings.

Brake fluid is a specially formulated hydraulic fluid that can withstand the heat and pressure encountered in brake systems. Care must be taken not to get it on painted surfaces or plastics. An opened container should always be resealed to prevent contamination by water or dirt.

Chain lubricants are formulated especially for use on motorcycle final drive chains. A good chain lube should adhere well and have good penetrating qualities to be effective as a lubricant inside the chain and on the side plates, pins and rollers. Most chain lubes are either the foaming type or quick drying type and are usually marketed as sprays.

Degreasers are heavy duty solvents used to remove grease and grime that may accumulate on engine and frame components. They can be sprayed or brushed on and, depending on the type, are rinsed off with either water or solvent.

Solvents are used alone or in combination with degreasers to clean parts and assemblies during repair and overhaul. The home mechanic should use only solvents that are non-flammable and don't produce irritating fumes.

Gasket sealing compounds may be used in conjunction with gaskets, to improve their sealing capabilities, or alone, to seal metal-to-metal joints. Many gasket sealers can withstand extreme heat, some are impervious to gasoline and lubricants, while others are capable of filling and sealing large cavities. Depending on the intended use, gasket sealers either dry hard or stay relatively soft and pliable. They're usually applied by hand, with a brush, or are sprayed on the gasket sealing surfaces.

Thread adhesive is a liquid chemical locking compound that solidifies after application and prevents threaded fasteners from loosening because of vibration. It's available in a variety of types for different applications.

Moisture dispersants are usually sprays that can be used to dry out electrical components such as the fuse block and wiring connectors. Some types can also be used as treatment for rubber and as a lubricant for hinges, cables and locks.

Waxes and polishes are used to help protect painted and plated surfaces from the weather. Different types of paint may require the use of different types of wax polish. Some polishes utilize a chemical or abrasive cleaner to help remove the top layer of oxidized (dull) paint on older machines. In recent years, many non-wax polishes (that contain a wide variety of chemicals such as polymers and silicones) have been introduced. These non-wax polishes are usually easier to apply and last longer than conventional waxes and polishes.

Safety first

Professional mechanics are trained in safe working procedures. However enthusiastic you may be about getting on with the job at hand, take the time to ensure that your safety is not put at risk. A moment's lack of attention can result in an accident, as can failure to observe simple precautions.

There will always be new ways of having accidents, and the following is not a comprehensive list of all dangers; it is intended rather to make you aware of the risks and to encourage a safe approach to all work you carry out on your bike.

Essential DOs and DON'Ts

DON'T start the engine without first ascertaining that the transmission is in neutral.
DON'T suddenly remove the filler cap from a hot cooling system - cover it with a cloth and release the pressure gradually first, or you may get scalded by escaping coolant.
DON'T attempt to drain oil until you are sure it has cooled sufficiently to avoid scalding you.
DON'T grasp any part of the engine or exhaust system without first ascertaining that it is cool enough not to burn you.
DON'T allow brake fluid or antifreeze to contact the machine's paint work or plastic components.
DON'T siphon toxic liquids such as fuel, hydraulic fluid or antifreeze by mouth, or allow them to remain on your skin.
DON'T inhale dust - it may be injurious to health (see *Asbestos* heading).
DON'T allow any spilled oil or grease to remain on the floor - wipe it up right away, before someone slips on it.
DON'T use ill fitting wrenches or other tools which may slip and cause injury.
DON'T attempt to lift a heavy component which may be beyond your capability - get assistance.
DON'T rush to finish a job or take unverified short cuts.
DON'T allow children or animals in or around an unattended vehicle.
DON'T inflate a tire to a pressure above the recommended maximum. Apart from over stressing the carcase and wheel rim, in extreme cases the tire may blow off forcibly.
DO ensure that the machine is supported securely at all times. This is especially important when the machine is blocked up to aid wheel or fork removal.
DO take care when attempting to loosen a stubborn nut or bolt. It is generally better to pull on a wrench, rather than push, so that if you slip, you fall away from the machine rather than onto it.
DO wear eye protection when using power tools such as drill, sander, bench grinder etc.
DO use a barrier cream on your hands prior to undertaking dirty jobs - it will protect your skin from infection as well as making the dirt easier to remove afterwards; but make sure your hands aren't left slippery. Note that long-term contact with used engine oil can be a health hazard.
DO keep loose clothing (cuffs, ties etc. and long hair) well out of the way of moving mechanical parts.
DO remove rings, wristwatch etc., before working on the vehicle- especially the electrical system.
DO keep your work area tidy - it is only too easy to fall over articles left lying around.
DO exercise caution when compressing springs for removal or installation. Ensure that the tension is applied and released in a controlled manner, using suitable tools which preclude the possibility of the spring escaping violently.
DO ensure that any lifting tackle used has a safe working load rating adequate for the job.
DO get someone to check periodically that all is well, when working alone on the vehicle.
DO carry out work in a logical sequence and check that everything is correctly assembled and tightened afterwards.
DO remember that your vehicle's safety affects that of yourself and others. If in doubt on any point, get professional advice.
IF, in spite of following these precautions, you are unfortunate enough to injure yourself, seek medical attention as soon as possible.

Asbestos

Certain friction, insulating, sealing and other products - such as brake pads, clutch linings, gaskets, etc. - contain asbestos. *Extreme care must be taken to avoid inhalation of dust from such products since it is hazardous to health*. If in doubt, assume that they *do* contain asbestos.

Fire

Remember at all times that gasoline (petrol) is highly flammable. Never smoke or have any kind of naked flame around, when working on the vehicle. But the risk does not end there - a spark caused by an electrical short-circuit, by two metal surfaces contacting each other, by careless use of tools, or even by static electricity built up in your body under certain conditions, can ignite gasoline (petrol) vapor, which in a confined space is highly explosive. Never use gasoline (petrol) as a cleaning solvent. Use an approved safety solvent.

Always disconnect the battery ground (earth) terminal before working on any part of the fuel or electrical system, and never risk spilling fuel on to a hot engine or exhaust.

It is recommended that a fire extinguisher of a type suitable for fuel and electrical fires is kept handy in the garage or workplace at all times. Never try to extinguish a fuel or electrical fire with water.

Fumes

Certain fumes are highly toxic and can quickly cause unconsciousness and even death if inhaled to any extent. Gasoline (petrol) vapor comes into this category, as do the vapors from certain solvents such as trichloroethylene. Any draining or pouring of such volatile fluids should be done in a well ventilated area.

When using cleaning fluids and solvents, read the instructions carefully. Never use materials from unmarked containers - they may give off poisonous vapors.

Never run the engine of a motor vehicle in an enclosed space such as a garage. Exhaust fumes contain carbon monoxide which is extremely poisonous; if you need to run the engine, always do so in the open air or at least have the rear of the vehicle outside the workplace.

The battery

Never cause a spark, or allow a naked light near the vehicle's battery. It will normally be giving off a certain amount of hydrogen gas, which is highly explosive.

Always disconnect the battery ground (earth) terminal before working on the fuel or electrical systems (except where noted).

If possible, loosen the filler plugs or cover when charging the battery from an external source. Do not charge at an excessive rate or the battery may burst.

Take care when topping up, cleaning or carrying the battery. The acid electrolyte, even when diluted, is very corrosive and should not be allowed to contact the eyes or skin. Always wear rubber gloves and goggles or a face shield. If you ever need to prepare electrolyte yourself, always add the acid slowly to the water; never add the water to the acid.

Electricity

When using an electric power tool, inspection light etc., always ensure that the appliance is correctly connected to its plug and that, where necessary, it is properly grounded (earthed). Do not use such appliances in damp conditions and, again, beware of creating a spark or applying excessive heat in the vicinity of fuel or fuel vapor. Also ensure that the appliances meet national safety standards.

A severe electric shock can result from touching certain parts of the electrical system, such as the spark plug wires (HT leads), when the engine is running or being cranked, particularly if components are damp or the insulation is defective. Where an electronic ignition system is used, the secondary (HT) voltage is much higher and could prove fatal.

Troubleshooting

Contents

Engine doesn't start or is difficult to start

1 Starter motor doesn't rotate

Note: *Occasional starter motor stall, particularly when the engine is hot, is normal and may not mean there's a problem with the starter or the starter circuit.*

1 Engine stop switch or ignition switch Off.

2 Battery voltage low or terminals loose or corroded. The battery must be in good condition and fully charged before troubleshooting the starter. Check and clean the terminals and/or recharge the battery (Chapter 7).

3 Starter solenoid or relay defective. Check the components and circuits (Chapter 7).

4 Defective starter motor. Make sure the wiring to the starter is secure, then check the starter motor (Chapter 7).

5 Starter button not contacting. The contacts could be wet, corroded or dirty. Disassemble and clean the switch (Chapter 7).

6 Wiring open or shorted. Check all wiring connections and wiring harnesses to make sure they're clean, dry and tight. Also check for broken or frayed wires that can cause a short to ground.

7 Ignition switch defective.

8 Engine stop switch defective. Check for wet, dirty or corroded contacts. Clean or replace the switch as necessary (Chapter 7).

2 Starter motor rotates but engine doesn't turn over

1 Starter motor clutch defective. Inspect and repair or replace (Chapter 2).
2 Damaged idler or starter gears. Inspect and replace the damaged parts (Chapter 2).
3 Splined teeth in bad condition. Starter motor should be removed and the splines cleaned.

3 Starter works but engine won't turn over (seized)

Seized engine caused by one or more internally damaged components. Failure due to wear, abuse or lack of lubrication. Damage can include seized valves, valve lifters, camshaft, pistons, crankshaft, connecting rod bearings, or transmission gears or bearings. Refer to Chapter 2 for engine disassembly.

4 No fuel flow

1 No fuel in tank.
2 Fuel valve turned off.
3 Tank cap air vent obstructed. Usually caused by dirt or water. Remove it and clean the cap vent hole.
4 Fuel valve clogged. Remove the valve and clean it and the filter (Chapter 1).
5 Fuel line clogged. Pull the fuel line loose and carefully blow through it.
6 Inlet needle valve clogged. For the valve to be clogged, either a very bad batch of fuel with an unusual additive has been used, or some other foreign object has entered the tank. Many times after a machine has been stored for many months without running, the fuel turns to a varnish-like liquid and forms deposits on the inlet needle valve and jets. The carburetor should be removed and overhauled if draining the float bowl doesn't alleviate the problem.

5 Engine flooded

1 Float level incorrect. Check and adjust as described in Chapter 3.
2 Inlet needle valve worn or stuck open. A piece of dirt, rust or other debris can cause the inlet needle to seat improperly, causing excess fuel to be admitted to the float bowl. In this case, the float bowl should be cleaned and the needle and seat inspected. If the needle and seat are worn, then the leak will persist and the parts should be replaced with new ones (Chapter 3).

6 No spark or weak spark

1 Ignition switch Off.
2 Engine stop switch turned to the Off position.
3 Battery voltage low. Check and recharge the battery as necessary (Chapter 7).
4 Spark plug dirty, defective or worn out. Locate reason for fouled plug(s) using spark plug condition photos and follow the plug maintenance procedures in Chapter 1.
5 Spark plug cap or plug wires defective. Check condition. Replace either or both components if cracks or deterioration are evident (Chapter 4).
6 Spark plug cap not making good contact. Make sure each plug cap fits snugly over the plug end.
7 Ignition module or sensor defective (Chapter 4).
8 Contact breaker points defective or improperly adjusted. Refer to Chapter 1.
9 Ignition coil defective. Check the coil as described in Chapter 4.
10 Ignition or stop switch shorted. This is usually caused by water, corrosion, damage or excessive wear. The switches can be disassembled and cleaned with electrical contact cleaner. If cleaning doesn't help, replace the switches (Chapter 7).
11 Wiring shorted or broken between:
a) Ignition switch and engine stop switch
b) Ignition module and engine stop switch
c) Ignition module and ignition coil
d) Ignition coil and plug

Make sure all wiring connections are clean, dry and tight. Look for chafed and broken wires (Chapters 4 and 7).
12 Faulty condenser (contact breaker point systems). Refer to Chapter 4.

7 Compression low

1 Spark plug loose. Remove the plug and inspect the threads. Reinstall and tighten to the specified torque (Chapter 1).
2 Cylinder head bolts loose. If the cylinder head is suspected of being loose, then there's a chance the gasket or head is damaged if the problem has persisted for any length of time. The head bolts should be tightened to the proper torque in the correct sequence (Chapter 2).
3 Incorrect valve clearance. This means the valve isn't closing completely and compression pressure is leaking past it. Check the lifters (Chapter 2).
4 Cylinder and/or piston worn. Excessive wear will cause compression pressure to leak past the rings. This is usually accompanied by worn rings as well. A top end overhaul is necessary (Chapter 2).
5 Piston rings worn, weak, broken, or sticking. Broken or sticking piston rings usually indicate a lubrication or carburetion problem that causes excess carbon deposits or seizures to form on the pistons and rings. Top end overhaul is necessary (Chapter 2).
6 Piston ring-to-groove clearance excessive. This is caused by excessive wear of the piston ring lands. Piston replacement is required (Chapter 2).
7 Cylinder head gasket damaged. If the head bolts loosen up, or if excessive carbon build-up on the piston crown and combustion chamber causes extremely high compression, the head gasket may leak. Retorquing the head is not always sufficient to restore the seal, so gasket replacement is necessary (Chapter 2).
8 Cylinder head warped. This is caused by overheating or improperly tightened head bolts. Machine shop resurfacing or head replacement is necessary (Chapter 2).
9 Valve spring broken or weak. Caused by component failure or wear; the spring(s) must be replaced (Chapter 2).
10 Valve not seating properly. This is caused by a bent valve (from over-revving or improper valve adjustment), burned valve or seat (improper carburetion) or an accumulation of carbon deposits on the seat (from carburetion, lubrication problems). The valves must be cleaned and/or replaced and the seats serviced if possible (Chapter 2).

8 Stalls after starting

1 Improper choke action. Make sure the choke rod is getting a full stroke and staying in the "out" position.
2 Ignition malfunction. See Chapter 4.
3 Carburetor malfunction. See Chapter 3.
4 Fuel contaminated. The fuel can be contaminated with either dirt or water, or can change chemically if the machine is allowed to sit for several months or more. Drain the tank and float bowl (Chapter 3).
5 Intake air leak. Check for loose carburetor-to-intake manifold connection or intake manifold-to-head connections (Chapter 4).
6 Idle speed incorrect. Turn idle speed screw until the engine idles at the specified rpm (Chapters 1 and 3).

9 Rough idle

1 Ignition malfunction. See Chapter 4.
2 Idle speed incorrect. See Chapter 1.
3 Carburetor malfunction. See Chapter 3.
4 Fuel contaminated. The fuel can be contaminated with either dirt or water, or can change chemically if the machine is allowed to sit for several months or more. Drain the tank and float bowl. If the problem is severe, a carburetor overhaul may be necessary (Chapters 1 and 3).
5 Intake air leak.
6 Air cleaner clogged. Service or replace air filter element (Chapter 1).

Runs poorly at low speed

10 Spark weak

1 Battery voltage low. Check and recharge battery (Chapter 7).
2 Spark plug fouled, defective or worn out. Refer to Chapter 1 for spark plug maintenance.
3 Spark plug cap or wires defective. Refer to Chapters 1 and 4 for details on the ignition system.
4 Spark plug cap not making contact.
5 Incorrect spark plug. Wrong type, heat range or cap configuration. Check and install correct plugs listed in Chapter 1. A cold plug or one with a recessed firing electrode won't function at low speeds without fouling.
6 Ignition module defective. See Chapter 4.
7 Contact breaker points defective or incorrectly gapped. See Chapter 1.
8 Ignition coil(s) defective. See Chapter 4.
9 Condenser faulty (contact breaker point system). See Chapter 4.

11 Fuel/air mixture incorrect

1 Fuel/air mixture screw(s) out of adjustment (Chapter 3).
2 Jet or air passage clogged. Remove and overhaul the carburetor (Chapter 3).
3 Air bleed holes clogged. Remove carburetor and blow out all passages (Chapter 3).
4 Air cleaner clogged, poorly sealed or missing.
5 Air cleaner-to-carburetor joint poorly sealed. Look for cracks, holes or loose clamps and replace or repair defective parts.
6 Fuel level too high or too low. Adjust the float (Chapter 3).
7 Fuel tank air vent obstructed. Make sure the air vent passage in the filler cap is open.
8 Intake manifold(s) loose. Check for cracks, holes and loose clamps or bolts.

12 Compression low

1 Spark plug loose. Remove the plug and inspect the threads. Reinstall and tighten to the specified torque (Chapter 1).
2 Cylinder head not sufficiently tightened down. If the cylinder head bolts are loose, then there's a chance the gasket and head are damaged if the problem has persisted for any length of time. The head bolts should be tightened to the proper torque in the correct sequence (Chapter 2).
3 Improper valve clearance. This means the valve isn't closing completely and compression pressure is leaking past the valve. Check the lifters (Chapter 2).
4 Cylinder and/or piston worn. Excessive wear will cause compression pressure to leak past the rings. This is usually accompanied by worn rings as well. A top end overhaul is necessary (Chapter 2).
5 Piston rings worn, weak, broken, or sticking. Broken or sticking piston rings usually indicate a lubrication or carburetion problem that causes excess carbon deposits or seizures to form on the pistons and rings. Top end overhaul is necessary (Chapter 2).
6 Piston ring-to-groove clearance excessive. This is caused by excessive wear of the piston ring lands. Piston replacement is necessary (Chapter 2).
7 Cylinder head gasket damaged. If the head bolts loosen up, or if excessive carbon build-up on the piston crown and combustion chamber causes extremely high compression, the head gasket may leak. Retorquing the head isn't always sufficient to restore the seal, so gasket replacement is necessary (Chapter 2).
8 Cylinder head warped. This is caused by overheating or improperly tightened head bolts. Machine shop resurfacing or head replacement is necessary (Chapter 2).
9 Valve spring broken or weak. Caused by component failure or wear; the spring(s) must be replaced (Chapter 2).
10 Valve not seating properly. This is caused by a bent valve (from over-revving or improper valve adjustment), burned valve or seat (improper carburetion) or an accumulation of carbon deposits on the seat (from carburetion, lubrication problems). The valves must be cleaned and/or replaced and the seats serviced if possible (Chapter 2).

13 Poor acceleration

1 Timing not advancing. Check the mechanical advance mechanism for proper operation (Chapter 4).
2 Engine oil viscosity too high. Using a heavier oil than recommended in Chapter 1 can damage the oil pump or lubrication system and cause drag on the engine.
3 Brakes dragging. Usually caused by debris which has entered the brake piston sealing boot, or from a warped disc or bent axle. Repair as necessary (Chapter 6).

Runs poorly or no power at high speed

14 Firing incorrect

1 Timing not advancing.
2 Spark plug fouled, defective or worn out. See Chapter 1 for spark plug maintenance.
3 Spark plug cap or wire defective. See Chapters 1 and 4 for details on the ignition system.
4 Spark plug cap not making good contact. See Chapter 4.
5 Incorrect spark plug. Wrong type, heat range or cap configuration. Check and install correct plugs listed in Chapter 1. A cold plug or one with a recessed firing electrode won't function at low speeds without fouling.
6 Ignition module defective. See Chapter 4.
7 Ignition coil(s) defective. See Chapter 4.

15 Fuel/air mixture incorrect

1 Main jet clogged. Dirt, water and other contaminants can clog the main jet. Clean the fuel valve filter, the float bowl area, and the jets and carburetor orifices (Chapter 3).
2 Main jet wrong size. The standard jetting is for sea level atmospheric pressure and oxygen content.
3 Throttle shaft-to-carburetor body clearance excessive. Refer to Chapter 3.
4 Air bleed holes clogged. Remove and overhaul carburetor (Chapter 3).
5 Air cleaner clogged, poorly sealed or missing.
6 Air cleaner-to-carburetor joint poorly sealed. Look for a torn gasket or a warped mating surface.
7 Fuel level too high or too low. Adjust the float (Chapter 3).
8 Fuel tank air vent obstructed. Make sure the air vent passage in the filler cap is open.

9 Carburetor intake manifold loose. Check for cracks, holes and loose clamps or bolts.
10 Fuel valve clogged. Remove the valve and clean it and the filter (Chapter 1).
11 Fuel line clogged. Pull the fuel line loose and carefully blow through it.

16 Compression low

1 Spark plug loose. Remove the plug and inspect the threads. Reinstall and tighten to the specified torque (Chapter 1).
2 Cylinder head not sufficiently tightened down. If the cylinder head bolts are loose, then there's a chance the gasket and head are damaged if the problem has persisted for any length of time. The head bolts should be tightened to the proper torque in the correct sequence (Chapter 2).
3 Improper valve clearance. This means the valve isn't closing completely and compression pressure is leaking past the valve. Check the lifters (Chapter 2).
4 Cylinder and/or piston worn. Excessive wear will cause compression pressure to leak past the rings. This is usually accompanied by worn rings as well. A top end overhaul is necessary (Chapter 2).
5 Piston rings worn, weak, broken, or sticking. Broken or sticking piston rings usually indicate a lubrication or carburetion problem that causes excess carbon deposits or seizures to form on the pistons and rings. Top end overhaul is necessary (Chapter 2).
6 Piston ring-to-groove clearance excessive. This is caused by excessive wear of the piston ring lands. Piston replacement is necessary (Chapter 2).
7 Cylinder head gasket damaged. If the head bolts loosen up, or if excessive carbon build-up on the piston crown and combustion chamber causes extremely high compression, the head gasket may leak. Retorquing the head isn't always sufficient to restore the seal, so gasket replacement is necessary (Chapter 2).
8 Cylinder head warped. This is caused by overheating or improperly tightened head bolts. Machine shop resurfacing or head replacement is necessary (Chapter 2).
9 Valve spring broken or weak. Caused by component failure or wear; the spring(s) must be replaced (Chapter 2).
10 Valve not seating properly. This is caused by a bent valve (from over-revving or improper valve adjustment), burned valve or seat (improper carburetion) or an accumulation of carbon deposits on the seat (from carburetion, lubrication problems). The valves must be cleaned and/or replaced and the seats serviced if possible (Chapter 2).

17 Knocking or pinging

1 Carbon build-up in combustion chamber. Use of a fuel additive that will dissolve the adhesive bonding the carbon particles to the piston crown and combustion chamber is the easiest way to remove the build-up. Otherwise, the cylinder head will have to be removed and decarbonized (Chapter 2).
2 Incorrect or poor quality fuel. Old or improper grades of gasoline can cause detonation. This causes a knocking or pinging sound. Drain old gas and always use the recommended fuel grade.
3 Spark plug heat range incorrect. Uncontrolled detonation indicates the plug heat range is too hot. The plug in effect becomes a glow plug, raising cylinder temperatures. Install the proper heat range plug (Chapter 1).
4 Improper fuel/air mixture. This will cause the engine to run hot, which leads to detonation. Clogged jets or an air leak can cause this imbalance. See Chapter 3.
5 Incorrect ignition timing (too far advanced) (Chapter 1).
6 Fuel octane rating too low.
7 Faulty vacuum operated electric switch (VOES).

18 Miscellaneous causes

1 Throttle valve doesn't open completely. Adjust the cable slack (Chapter 3).
2 Clutch slipping. Caused by a cable that's improperly adjusted, or snagging or by damaged, loose or worn clutch components. Refer to Chapters 1 and 2 for adjustment and overhaul procedures.
3 Ignition timing not advancing.
4 Engine oil viscosity too high. Using a heavier oil than recommended in Chapter 1 can damage the oil pump or lubrication system and cause drag on the engine.
5 Brakes dragging. Usually caused by debris which has entered the brake piston sealing boot, or from a warped disc or bent axle. Repair as necessary.

Overheating

19 Firing incorrect

1 Spark plug fouled, defective or worn out. See Chapter 1 for spark plug maintenance.
2 Incorrect spark plug.
3 Improper ignition timing. Timing that's retarded will cause high cylinder temperatures and lead to overheating (Chapter 1).

20 Fuel/air mixture incorrect

1 Main jet clogged. Dirt, water and other contaminants can clog the main jet. Clean the fuel valve filter, the float bowl area and the jets and carburetor orifices (Chapter 3).
2 Main jet wrong size. The standard jetting is for sea level atmospheric pressure and oxygen content.
3 Air cleaner poorly sealed or missing.
4 Air cleaner-to-carburetor joint poorly sealed. Look for a broken gasket or a warped mating surface.
5 Fuel level too low. Adjust the float (Chapter 3).
6 Fuel tank air vent obstructed. Make sure the air vent passage in the filler cap is open.
7 Carburetor intake manifold loose. Check for cracks, holes and loose clamps or bolts.

21 Compression too high

1 Carbon build-up in combustion chamber. Use of a fuel additive that will dissolve the adhesive bonding the carbon particles to the piston crown and combustion chamber is the easiest way to remove the build-up. Otherwise, the cylinder heads will have to be removed and decarbonized (Chapter 2).
2 Improperly machined head surface or installation of incorrect gasket during engine assembly. Check Specifications (Chapter 2).

22 Engine load excessive

1 Clutch slipping. Caused by an out of adjustment or snagging cable or damaged, loose or worn clutch components. Refer to Chapters 1 and 2 for adjustment and overhaul procedures.
2 Engine oil viscosity too high. Using a heavier oil than recommended in Chapter 1 can damage the oil pump or lubrication system as well as cause drag on the engine.
3 Brakes dragging. Usually caused by debris which has entered the brake piston sealing boot, or from a warped disc or bent axle. Repair as necessary.

23 Lubrication inadequate

1 Engine oil level too low. Friction caused by intermittent lack of lubrication or from oil that is "overworked" can cause overheating. The oil provides a definite cooling function in the engine. Check the oil level (Chapter 1) and the oil pump (Chapter 2).
2 Poor quality engine oil or incorrect viscosity or type. Oil is rated not only according to viscosity but also according to type. Some oils aren't rated high enough for use in the engine. Check the Specifications Section and change to the correct oil (Chapter 1).

24 Miscellaneous causes

1 Engine fins packed with mud or dirt. The cooling fins can be blocked by a build-up of mud and cause a decrease in cooling. Clean the cylinder and head area.
2 Engine cooling fins on head and cylinder painted. Most paints actually seal in heat, so a cylinder/head paint should be one specifically designed for that use.
3 Valves not seating properly (being held off seat) or valve seats/faces in poor condition.
4 Carbon deposits.
5 Defective vacuum operated electric switch (VOES).
6 Modified exhaust system. Most aftermarket exhaust systems cause the engine to run leaner, which makes them run hotter. When installing an accessory exhaust system, always rejet the carburetor.

Clutch problems

25 Clutch slipping

1 No clutch lever play. Adjust clutch lever free play according to the procedure in Chapter 1.
2 Friction plates worn or warped. Overhaul the clutch assembly (Chapter 2).
3 Steel plates worn or warped (Chapter 2).
4 Clutch springs broken or weak. Old or heat-damaged (from slipping clutch) springs should be replaced with new ones (Chapter 2).
5 Clutch release not adjusted properly. See Chapter 1.
6 Clutch cable hanging up. Caused by a frayed cable or kinked housing. Replace the cable. Repair of a frayed cable isn't recommended.
7 Clutch release mechanism defective. Check the shaft, cam, actuating arm and pivot. Replace any defective parts (Chapter 2).
8 Clutch hub or drum unevenly worn. This causes improper engagement of the discs. Replace the damaged or worn parts (Chapter 2).

26 Clutch not disengaging completely

1 Clutch lever play excessive. adjust at lever or at engine (Chapter 1).
2 Clutch plates warped or damaged. This will cause clutch drag, which in turn causes the machine to creep. Overhaul the clutch (Chapter 2).
3 Clutch spring tension uneven. Usually caused by a sagged or broken spring. Check and replace the springs (Chapter 2).
4 Engine oil deteriorated. Old, thin, worn out oil will not provide proper lubrication for the discs, causing the clutch to drag. Replace the oil and filter (Chapter 1).
5 Engine oil viscosity too high. Using a heavier oil than recommended in Chapter 1 can cause the plates to stick together, putting a load on the engine. Change to the correct weight oil (Chapter 1).
6 Clutch drum seized on shaft. Lack of lubrication, severe wear or damage can cause the drum to seize on the shaft. Overhaul of the clutch, and perhaps transmission, may be necessary to repair damage (Chapter 2).
7 Clutch release mechanism defective. Worn or damaged release mechanism parts can stick and fail to apply force to the pressure plate. Overhaul the clutch cover components (Chapter 2).
8 Loose clutch hub nut. Causes drum and hub misalignment putting a load on the engine. Engagement adjustment continually varies. Overhaul the clutch (Chapter 2).

Gear shifting problems

27 Doesn't go into gear or lever doesn't return

1 Clutch not disengaging. See Section 26.
2 Shift fork(s) bent or seized. Often caused by dropping the machine or from lack of lubrication. Overhaul the transmission (Chapter 2).
3 Gear(s) stuck on shaft. Most often caused by a lack of lubrication or excessive wear in transmission bearings and bushings. Overhaul the transmission (Chapter 2).
4 Shift cam binding. Caused by lubrication failure or excessive wear. Replace the cam and bearings (Chapter 2).
5 Shift lever return spring weak or broken (Chapter 2).
6 Shift lever broken. Splines stripped out of lever or shaft, caused by allowing the lever to get loose or from dropping the machine. Replace parts as necessary (Chapter 2).
7 Shift mechanism pawl broken or worn. Full engagement and rotary movement of shift cam results. Replace shaft assembly (Chapter 2).
8 Pawl spring broken. Allows pawl to "float", causing sporadic shift operation. Replace spring (Chapter 2).

28 Jumps out of gear

1 Shifter rod not adjusted properly.
2 Shift cam worn, damaged or adjusted incorrectly.
3 Shift fork(s) worn. Overhaul the transmission (Chapter 2).
4 Gear groove(s) worn. Overhaul the transmission (Chapter 2).
5 Gear dogs or dog slots worn or damaged. The gears should be inspected and replaced. No attempt should be made to service the worn parts.
6 Damaged gears.

29 Overshifts

1 Pawl spring weak or broken (Chapter 2).
2 Shift cam follower not functioning (Chapter 2).

Abnormal engine noise

30 Knocking or pinging

1 Carbon build-up in combustion chamber. Use of a fuel additive that will dissolve the adhesive bonding the carbon particles to the piston crown and combustion chamber is the easiest way to remove the build-up. Otherwise, the cylinder head will have to be removed and decarbonized (Chapter 2).

2 Incorrect or poor quality fuel. Old or improper fuel can cause detonation. This causes the piston to rattle, thus the knocking or pinging sound. Drain the old gas and always use the recommended grade fuel (Chapter 3).
3 Spark plug heat range incorrect. Uncontrolled detonation indicates the plug heat range is too hot. The plug in effect becomes a glow plug, raising cylinder temperatures. Install the proper heat range plug (Chapter 1).
4 Improper fuel/air mixture. This will cause the engine to run hot and lead to detonation. Clogged jets or an air leak can cause this imbalance. See Chapter 3.
5 Incorrect ignition timing (Chapter 1).

31 Piston slap or rattling

1 Cylinder-to-piston clearance excessive. Caused by improper assembly. Inspect and overhaul top end parts (Chapter 2).
2 Connecting rod bent. Caused by over-revving, trying to start a badly flooded engine or from ingesting a foreign object into the combustion chamber. Replace the damaged parts (Chapter 2).
3 Piston pin or piston pin bore worn or seized from wear or lack of lubrication. Replace damaged parts (Chapter 2).
4 Piston ring(s) worn, broken or sticking. Overhaul the top end (Chapter 2).
5 Piston seizure damage. Usually from lack of lubrication or overheating. Replace the pistons and bore the cylinders, as necessary (Chapter 2).
6 Connecting rod big and/or small end clearance excessive. Caused by excessive wear or lack of lubrication. Replace worn parts.

32 Valve noise

1 Incorrect pushrod length or bent pushrod.
2 Valve spring broken or weak. Check and replace weak valve springs (Chapter 2).
3 Valve(s) sticking in guide(s) or rocker arm binding on shaft.
4 Hydraulic tappets malfunctioning.
5 Camshaft lobes or gear or valve train components worn or damaged. Lack of lubrication at high rpm is usually the cause of damage. Insufficient oil or failure to change the oil at the recommended intervals are the chief causes (Chapter 1).
6 Low oil pressure. Usually caused by oil pump not functioning properly or obstructed oil screen.

33 Other noise

1 Cylinder head gasket leaking. This will usually produce a pop or "whisper" each time the affected cylinder reaches the compression stroke. It's usually accompanied by wetness around the leak. If the leak is into an oil return channel, then the crankcase will be pressurized and the oil will quickly be contaminated. Replace the head gasket and check the head for warpage (Chapter 2).
2 Exhaust pipe leaking at cylinder head connection. Caused by improper fit of pipe(s) or loose exhaust flange. Sometimes confused with a leaking head gasket. All exhaust fasteners should be tightened evenly and carefully. Failure to do this will lead to a leak (Chapter 3).
3 Crankshaft runout excessive. Caused by a bent crank (from overrevving) or damage from an upper cylinder component failure. Can also be attributed to dropping the machine on either of the crankshaft ends.
4 Engine mount bolts loose. Tighten all bolts to the specified torque (Chapter 2).
5 Crankshaft bearings worn (Chapter 2).
6 Primary chain tensioner out of adjustment or defective. Adjust or replace as necessary (Chapters 1 and 2).

Abnormal driveline noise

34 Clutch noise

1 Clutch drum/friction plate clearance excessive (Chapter 2).
2 Loose or damaged clutch pressure plate and/or bolts(Chapter 2).

35 Transmission noise

1 Bearings worn. Also includes the possibility that the shafts are worn. Overhaul the transmission (Chapter 2).
2 Gears worn or chipped (Chapter 2).
3 Metal chips jammed in gear teeth. Probably pieces from a broken clutch, gear or shift mechanism that were picked up by the gears. This will cause early bearing failure (Chapter 2).
4 Transmission oil level too low. Causes a howl from transmission. Also affects engine power and clutch operation (Chapter 1).

Abnormal frame and suspension noise

36 Front end noise

1 Low fork oil level or improper viscosity oil. This can result in a "spurting" sound and is usually accompanied by irregular fork action (Chapter 5).
2 Spring weak or broken. Makes a clicking or scraping sound. Fork oil, when drained, will have a lot of metal particles in it (Chapter 5).
3 Steering head bearings loose or damaged. Clicks when braking. Check and adjust or replace as necessary (Chapter 5).
4 Fork clamps loose. Make sure all fork clamp pinch bolts are tight (Chapter 5).
5 Fork tube bent. Good possibility if machine has been dropped. Replace tube with a new one (Chapter 5).
6 Front axle or axle clamp nuts loose. Tighten to the specified torque (Chapter 5).

37 Shock absorber noise

1 Fluid level incorrect. Indicates a leak caused by defective seal. Shock will be covered with oil. Replace shock (Chapter 6).
2 Defective shock absorber with internal damage. This is in the body of the shock and cannot be remedied. The shock must be replaced with a new one (Chapter 6).
3 Bent or damaged shock body. Replace the shock with a new one (Chapter 6).

38 Disc brake noise

1 Squeal caused by shim not installed or positioned correctly (Chapter 6).
2 Squeal caused by dust on brake pads. Usually found in combination with glazed pads. Clean with brake system solvent (Chapter 6).
3 Contamination of brake pads. Oil, brake fluid or dirt causing brake to chatter or squeal. Clean or replace pads (Chapter 6).
4 Pads glazed. Caused by excessive heat from prolonged use or from contamination. Do not use sandpaper, emery cloth or any other abrasive to roughen the pad surfaces as abrasives will stay in the pad material and damage the disc. A very fine flat file can be used, but pad replacement is suggested as a cure (Chapter 6).
5 Disc warped. Can cause a chattering, clicking or intermittent squeal. Usually accompanied by a pulsating lever and uneven braking. Resurface or replace the disc (Chapter 6).

Oil pressure indicator light comes on

39 Engine lubrication system

1 Engine oil pump defective (Chapter 2).
2 Engine oil level low. Inspect for leak or other problem causing low oil level and add recommended lubricant (Chapters 1 and 2).
3 Engine oil viscosity too low. Very old, thin oil or an improper weight of oil used in engine. Change to correct lubricant (Chapter 1).
4 Crankshaft and/or bearings worn. Check and replace crankshaft and bearings (Chapter 2).
5 Relief valve stuck open. Repair or replace the valve (Chapter 3).

40 Electrical system

1 Oil pressure switch defective. Replace it if it's defective.
2 Oil pressure indicator light system wiring defective. Check for pinched, shorted, disconnected or damaged wiring(Chapter 7).

Excessive exhaust smoke

41 White smoke

1 Piston oil ring worn. The ring may be broken or damaged, causing oil from the crankcase to be pulled past the piston into the combustion chamber. Replace the rings with new ones (Chapter 2).
2 Cylinders worn, cracked, or scored. Caused by overheating or oil starvation. The cylinders will have to be rebored and new pistons installed.
3 Valve stem seal damaged or worn. Replace seals with new ones (Chapter 2).
4 Valve guide(s) worn. Perform a complete valve job (Chapter 2).
5 Abnormal crankcase pressurization, which forces oil past the rings. Clogged breather usually the cause (Chapter 1).

42 Black smoke

1 Air cleaner clogged. Clean or replace the element (Chapter 1).
2 Main jet too large or loose. Compare the jet size to the Specifications (Chapter 3).
3 Fuel level too high. Check and adjust the float level as necessary (Chapter 3).
4 Inlet needle held off seat. Clean float bowl and fuel line and replace needle and seat if necessary (Chapter 3).

43 Brown smoke

1 Main jet too small or clogged. Lean condition caused by wrong size main jet or by a restricted orifice. Clean float bowl and jets (Chapter 3).
2 Fuel flow insufficient. Fuel inlet needle valve stuck closed due to chemical reaction with old gas. Float level incorrect. Restricted fuel line. Clean line and float bowl and adjust float as necessary (Chapter 3).
3 Intake manifold loose (Chapter 3).
4 Air cleaner poorly sealed or not installed (Chapter 1).

Poor handling or stability

44 Handlebar hard to turn

1 Steering stem locknut too tight (Chapter 5).
2 Bearings damaged. Roughness can be felt as the bars are turned from side-to-side. Replace bearings and races (Chapter 5).
3 Races dented or worn. Denting results from wear in only one position (i.e., straight ahead) from impacting an immovable object or hole or from dropping the machine. Replace races and bearings (Chapter 5).
4 Steering stem lubrication inadequate. Causes are grease getting hard from age or being washed out by high pressure car washes. Disassemble steering head and repack bearings (Chapter 5).
5 Steering stem bent. Caused by hitting a curb or hole or from dropping the machine. Replace damaged part. Don't try to straighten stem (Chapter 5).
6 Front tire air pressure too low (Chapter 1).

45 Handlebar shakes or vibrates excessively

1 Tires worn or out-of-balance (Chapter 6).
2 Swingarm bearings worn. Replace worn bearings by referring to Chapter 5.
3 Rim(s) warped or damaged. Inspect wheels for runout (Chapter 6).
4 Wheel bearings worn. Worn front or rear wheel bearings can cause poor tracking. Worn front bearings will cause wobble (Chapter 6).
5 Handlebar clamp bolts loose (Chapter 5).
6 Steering stem or fork clamps loose. Tighten them to the specified torque (Chapter 5).
7 Engine mount bolts loose. Will cause excessive vibration with increased engine rpm (Chapter 2).

46 Handlebar pulls to one side

1 Frame bent. Definitely suspect this if the machine has been dropped. May or may not be accompanied by cracks near the bend. Replace the frame (Chapter 5).
2 Wheel out of alignment. Caused by improper location of axle spacers or from bent steering stem or frame (Chapter 5).
3 Swingarm bent or twisted. Caused by age (metal fatigue) or impact damage. Replace the arm (Chapter 5).
4 Steering stem bent. Caused by impact damage or from dropping the motorcycle. Replace the steering stem (Chapter 5).
5 Fork leg bent. Disassemble the forks and replace the damaged parts (Chapter 5).
6 Fork oil level uneven.

47 Poor shock absorbing qualities

1 Too hard:
- a) Excessive fork oil (Chapter 5).
- b) Fork oil viscosity too high. Use a lighter oil.
- c) Fork tube bent. Causes a harsh, sticking feeling (Chapter 5).
- d) Shock shaft or body bent or damaged (Chapter 5).
- e) Fork internal damage (Chapter 5).
- f) Shock internal damage.
- g) Tire pressure too high (Chapter 1).

2 Too soft:
- a) Fork or shock oil insufficient and/or leaking (Chapter 5).
- b) Fork oil level too low (Chapter 5).
- c) Fork oil viscosity too light (Chapter 5).
- d) Fork springs weak or broken (Chapter 5).

Braking problems

48 Brakes are spongy, don't hold (hydraulic disc brakes only)

1 Air in brake line. Caused by inattention to master cylinder fluid level or by leakage. Repair problem and bleed brakes (Chapter 6).

2 Pads or disc worn (Chapters 1 and 6).
3 Brake fluid leak. See paragraph 1.
4 Pads contaminated with oil, grease, brake fluid, etc. Clean or replace pads. Clean disc thoroughly with brake cleaner (Chapter 6).
5 Old, deteriorated or contaminated brake fluid. Drain system, replenish with new fluid and bleed the system (Chapter 6).
6 Master cylinder internal parts worn or damaged causing fluid to bypass (Chapter 6).
7 Master cylinder bore scratched. From ingestion of foreign material or broken spring. Repair or replace master cylinder (Chapter 6).
8 Disc warped. Replace disc (Chapter 6).

49 Brake lever or pedal pulsates

1 Disc warped. Replace disc (Chapter 6).
2 Axle bent. Replace axle (Chapter 6).
3 Brake caliper bolts loose (Chapter 6).
4 Brake caliper pins damaged or sticking, causing caliper to bind. Lube the pins and/or replace them if they're corroded or bent (Chapter 6).
5 Wheel warped or otherwise damaged (Chapter 6).
6 Wheel bearings damaged or worn (Chapter 6).

50 Brakes drag

1 Master cylinder piston seized. Caused by wear or damage to piston or cylinder bore (Chapter 6).
2 Lever slow or stuck. Check pivot and lubricate (Chapter 6).
3 Brake caliper binds. Caused by inadequate lubrication or damage to caliper pins (Chapter 6).
4 Brake caliper piston seized in bore. Caused by wear or ingestion of dirt past deteriorated seal (Chapter 6).
5 Brake pad damaged. Pad material separating from backing plate. Usually caused by faulty manufacturing process or from contact with chemicals. Replace pads (Chapter 6).
6 Pads improperly installed (Chapter 6).
7 Rear brake pedal free play insufficient.
8 Front brake lever free play insufficient.

Electrical problems

51 Weak or dead battery

1 Battery faulty. Caused by sulphated plates which are shorted through the sedimentation or low electrolyte level. Also, broken battery terminal making only occasional contact (Chapter 7).
2 Battery cables making poor contact (Chapter 7).
3 Load excessive. Caused by addition of high wattage lights or other electrical accessories.
4 Ignition switch defective. Switch either grounds internally or fails to shut off system. Replace the switch (Chapter 7).
5 Regulator/rectifier defective (Chapter 7).
6 Alternator stator coil open or shorted (Chapter 7).
7 Wiring faulty. Wiring grounded or connections loose in ignition, charging or lighting circuits (Chapter 7).

52 Battery overcharged

1 Regulator/rectifier defective. Overcharging is noticed when battery gets excessively warm or "boils" over (Chapter 7).
2 Battery defective. Replace battery with a new one (Chapter 7).
3 Battery amperage too low, wrong type or size. Install manufacturer's specified amp-hour battery to handle charging load (Chapter 7).

53 Alternator not charging

1 Faulty regulator-rectifier module (Chapter 7).
2 Faulty stator or rotor.
3 Weak or damaged battery.
4 Loose connections in wire harness.

Chapter 1 Tune-up and routine maintenance

Contents

Specifications

Recommended spark plugs

	Harley-Davidson type
1970 through 1974	No. 3-4
1975 and 1976	No. 5-6
1977 through early 1978	No. 5A6 or 5R6
Late 1978 and 1979	No. 5A6A or 5R6A*
1980 through 1983	No. 5R6A or 5RL
1984-on	No. 5R6A only

** Harley-Davidson No. 5A6 or 5R6 can be used in place of 5A6A or 5R6A if the plug is gapped at 0.038 to 0.043-inch (1.0 to 1.1 mm)*

Spark plug gap

1970 through 1978	0.028 to 0.033 in (0.7 to 0.8 mm)
1979-on	0.038 to 0.043 in (1.0 to 1.1 mm)

Contact breaker point gap

0.018 in (0.46 mm)

Ignition timing

Late 1978 through 1983	
Fully retarded	3-degrees BTDC
Fully advanced	35-degrees BTDC
Early 1984	
Range	5 to 50-degrees BTDC
Start	5-degrees BTDC
At fast idle	35-degrees BTDC
At 1800 to 2800 rpm	50-degrees BTDC
Late 1984-on	
Range	0 to 35-degrees BTDC
Start	
FX/Softail/Dyna models	5-degrees BTDC
FLT/FXR models	TDC (0-degrees)
At fast idle	35-degrees BTDC
At 1800 to 2800 rpm	35-degrees BTDC

Idle speed

Tillotson carburetor	900 to 1000 rpm
Bendix carburetor	700 to 900 rpm
Keihin carburetor	
1976 through early 1978	900 rpm
Late 1978 through 1983	800 to 900 rpm
1984 through 1989	
FLT and FXR	
1984 through 1988	900 to 950 rpm
1989	1000 to 1050 rpm
FX and Softail	1000 to 1050 rpm
1990-on	
FLT and FXR	1000 rpm
FX/Softail and Dynas	1000 to 1050 rpm

Fast idle speed

1500 rpm

Cylinder compression pressure

Minimum	90 psi (6.2 bars)
Maximum variation between cylinders	10 psi (0.7 bars)

Drivetrain and suspension

Primary chain slack	
Cold	5/8 to 7/8 in (16 to 22 mm)
Hot	3/8 to 5/8 in (9 to 16 mm)
Primary belt slack	3/8 to 1/2 in (9 to 13 mm)
Final drive chain slack	
1970 through 1983	1/2 in (13 mm)
1984-on	
FX/Softail models	
FXST only	1-1/8 to 1-1/4 in (28 to 32 mm)
All others	1/2 to 5/8 in (13 to 16 mm)
FLT/FXR models	1/2 in (13 mm)
Final drive belt slack	
Through 1983	5/8 to 3/4 in (16 to 19 mm)
1984 though 1990	
FX/Softail models	5/8 to 3/4 in (16 to 19 mm)
FLT/FXR models	5/16 to 3/8 in (8 to 9 mm)
FLST/C, FXST/C models	3/8 to 1/2 in (9 to 13 mm)
1991-on	
FLSTC/F/N, FXSTC/S models	3/8 to 1/2 in (9 to 13 mm)
FLT/FXR and Dyna models	5/16 to 3/8 in (8 to 9 mm)

Brakes

Front brake lever free play	
Drum brake	3/16 in (4.5 mm)
Disc brake	None
Brake drum maximum inside diameter	
Standard shoes	8.040 in (204.2 mm)
0.030-inch oversize shoes	8.06 to 8.10 in (204.7 to 205.7 mm)
Brake shoe thickness (minimum)	0.100 in or near rivets (2.5 mm)
Brake pad thickness (minimum)	1/16 in (1.6 mm)
Rear pedal height setting	See Chapter 6
Rear master cylinder free play	See Chapter 6

Clutch

Foot control free play	1/8 in (4 mm)
Lever free play	
1970 through early 1978 models	1/4 in (7 mm)
Late 1978 through 1983 models (except FLT/FXR)	1/16 in (1.6 mm)
FLT/FXR models	
1980 through 1983 (at ball end)	1/4 in (7 mm)
Early 1984 (at bracket)	1/16 in (1.6 mm)
Late 1984 through 1989 (at bracket)	1/8 to 3/16 in (4 to 4.5 mm)
1990-on (at bracket)	1/16 to 1/8 in (1.6 to 4 mm)
FX/Softail and Dyna models	
1985 through 1989 (at bracket)	1/8 to 3/16 in (4 to 4.5 mm)
1990-on (at bracket)	1/16 to 1/8 in (1.6 to 4 mm)
Clutch adjustment clearance	
1970 through early 1979 release lever-to-starter motor	3/8 to 5/8 in (9 to 16 mm)
Late 1979 through 1983 release lever-to-bearing extrusion on starter motor housing (except FLT/FXR)	13/16 in (21 mm)
Outer disc surface-to-spring collar inner edge	7/8 to 1-1/32 in (22 to 25 mm)

Tires – psi (Bars)

Pressures – cold with a solo rider*	**Front**	**Rear**
1970 through 1983		
FL models (except FLT)	20 (1.4)	24 (1.7)
FLT models	24 (1.7)	26 (1.8)
FX models (except FXR/FXWG)	24 (1.7)	26 (1.8)
FXWG models	30 (2.1)	26 (1.8)
FXR models	24 (1.7)	24 (1.7)
1984 and 1985		
FLT/C models	28 (1.9)	36 (2.5)
FLHT/C, FXRS and FXRT models	30 (2.1)	36 (2.5)
FXEF and FXSB models	30 (2.1)	32 (2.2)
FXWG and FXST models	30 (2.1)	28 early 1985 only (1.9) 32 all others (2.2)
1986 through 1990 (approximate)		
FXR models	30 (2.1)	36 (2.5)
FLT models	36 (2.5)	36 (2.5)
FLST/C, FXEF, FXSB	36 (2.5)	36 (2.5)
FXWG, FXST, FXST/C and FXSTS models	30 (2.1)	32 (2.2)
1991-on		
FXSTC, FXSTS, FXD Dynas	30 (2.1)	36 (2.5)
All other models same as for 1986 through 1990		

** 21 in front tire*

** **Note:** Pressure based on rider weighing 150 lbs. (68 kg). Increase pressure in rear tire 2 psi (0.14 Bar) and pressure in front tire 1 psi (0.07 Bar) for each additional 50 lbs. (23 kg) of weight.*

Tire tread depth (minimum)	
Front	1/16 in (1.6 mm)
Rear	3/32 in (2.5 mm)

Torque specifications

	Ft-lbs (unless otherwise indicated)	Nm
Oil tank drain plug	10	14
Spark plugs	18 to 22	24 to 30
Axle nuts		
1970 through 1983	50	68
1984-on (front)		
FLT models	50 to 55	68 to 75
FXR and Dyna models	50	68
FXSTS models	60 to 65	81 to 88
All others	45 to 50	61 to 68
1984-on (rear)	60 to 65	81 to 88
Brake caliper mounting bolts		
1974 through 1977	130 in-lbs	17
1978 through 1983 (except FLT/FXR models)	120 in-lbs	14
1980 through 1983		
FLT models	90 in-lbs	10
FXR models	12 to 15	16 to 20
1984-on (front)		
FXST/C, FLST/C/F/N, FLT, FXR and FXD Dyna models	25 to 30	34 to 41
FXSTS model		
Top	42 to 46	57 to 62
Bottom	25 to 30	34 to 41
1984-on (rear)		
FLT/FXR models	15 to 20	20 to 27
All others		
1984 through mid 1987	12 to 15	16 to 20
Mid-1987-on	15 to 20	20 to 27
Fork cap bolts	11	15
Transmission drain plug		
Four-speed models	9 to 15	12 to 20
Five-speed models		
FLT through 1992 and FXR	7	9
FLT 1993-on	14 to 30	19 to 40
Softails and Dynas	0.16 to 0.18 in (4.0 to 4.5 mm) above surface of housing	
Brake disc-to-hub screws through 1983		
FLT and FXR models	34 to 42	46 to 57
All others		
Spoke wheel (16 and 21-inch)	23 to 27	31 to 37
Spoke wheel (19-inch)	16 to 19	22 to 26
Cast wheel (16-inch)	23 to 27	31 to 37
Cast wheel (19-inch)	14 to 16	19 to 22
Brake disc-to-hub screws (front) – 1984-on		
FLT/FXR models		
1984 through 1990	16 to 18	22 to 24
1991-on	16 to 24	22 to 33
FX/Softail models		
FXSTS	16 to 18	22 to 24
All others	16 to 24	22 to 33
FXD Dyna models		
1991 model with spoked wheel	16 to 18	22 to 24
All other models	16 to 24	22 to 33
Brake disc-to-hub bolts (rear) – 1984-on		
FLT		
1984 through 1990	24 to 30	33 to 41
1991	30 to 35	41 to 47
FXR (1984 through 1991)	23 to 27	31 to 37
FLT/FXR rear (1992-on)	30 to 45	41 to 61
FX/Softail (through 1990)	23 to 27	31 to 37
FX/Softail (1991)		
Spoked wheel	16 to 24	22 to 33
Disc wheel	23 to 27	31 to 37
FX/Softail (1992-on) and FXD Dyna (1993-on)		
Allen head screws	23 to 27	31 to 37
Torx head screws	30 to 45	41 to 61
FXD Dyna (1991 to 1992)	23 to 27	31 to 37

Recommended lubricants and fluids

Engine oil

Type	Harley-Davidson or equivalent
Viscosity	
1970 through 1983	
Normal (20 to 90-degrees F)	SAE 20W50
Below 40-degrees F	SAE 30
Above 40-degrees F	SAE 40
Severe operating conditions	SAE 60
1984-on	SAE 20W50 or 10W40
Oil tank capacity	Approximately 4 US quarts (3.33 Imp quarts, 4.55 liters)
Primary chaincase oil (wet clutch)	Harley-Davidson Primary Chaincase Lubricant (part no. 99887-84)

Transmission oil

Type	
1970 through 1983	Harley-Davidson Super Premium or equivalent 20/50
1984-on	Harley-Davidson Transmission Lubricant (part no. 99892-84)
Capacity (approx.)	
FLT/FXR models through 1990	1 US pt, 17 Imp oz, 473 cc
All others and 1991-on FLT/FXR	20 to 24 US oz, 21 to 25 Imp oz, 591 to 710 cc

Front fork oil

Type		
1970 through 1983		
FLT, FXR, FLHT and FXRS	Harley-Davidson Type E	
All others	Harley-Davidson Type B	
1984-on (all)	Harley-Davidson Type E	
Capacity		
1970 through 1983	Wet	Dry
Through 1972 FX, FXE and FXS	5-1/2 US oz, 5.7 Imp oz, 163 cc	6-1/2 US oz, 6.8 Imp oz, 192 cc
Through early 1977 FL and FLH	6-1/2 US oz, 6.8 Imp oz, 192 cc	7 US oz, 7.3 Imp oz, 207 cc
Late 1977 and later FLT, FLH and FLHS	7-3/4 US oz, 8.1 Imp oz, 229 cc	8-1/2 US oz, 8.8 Imp oz, 251 cc
1973 and later FXR, FXB, FXSB, FXE, FXEF and FXS	5 US oz, 5.2 Imp oz, 148 cc	6 US oz, 6.2 Imp oz, 177 cc
FXWG	9-1/4 US oz, 9.6 Imp oz, 274 cc	10 US oz, 10.4 Imp oz, 296 cc
1984-on		
FXEF	5-1/2 US oz, 5.7 Imp oz, 163 cc	6-1/2 US oz, 6.8 Imp oz, 192 cc
FXSB	5-3/4 US oz, 6.0 Imp oz, 170 cc	6-3/4 US oz, 7.0 Imp oz, 200 cc
FLT	7-3/4 US oz, 8.1 Imp oz, 229 cc	8-1/2 US oz, 8.8 Imp oz, 251 cc
FXR and FXRS		
1984 through 1987	6-1/4 US oz, 6.5 Imp oz, 185 cc	7 US oz, 7.3 Imp oz, 207 cc
1988-on	9.2 US oz, 9.6 Imp oz, 272 cc	10.2 US oz, 10.6 Imp oz, 302 cc
FXRD and FXRT		
1984 through 1987	7 US oz, 7.3 Imp oz, 207 cc	7-3/4 US oz, 8.1 Imp oz, 229 cc
1988-on	10-1/2 US oz, 10.9 Imp oz, 311 cc	11-1/2 US oz, 12.0 Imp oz, 340 cc
FXLR, FXDB, FXDC and FXDL	9.2 US oz, 9.6 Imp oz, 272 cc	10.2 US oz, 10.6 Imp oz, 302 cc
FXRSE and FXRS-SP	10-1/2 US oz, 10.9 Imp oz, 311 cc	11-1/2 US oz, 12.0 Imp oz, 340 cc
FXSTC, FXDWG	10.2 US oz, 10.6 Imp oz, 302 cc	11.2 US oz, 11.7 Imp oz, 331 ccz
FLSTC/F/N	11.5 US oz, 12.0 Imp oz, 340 cc	12.5 US oz, 13.0 Imp oz, 370 cc

Brake fluid

Brake fluid	DOT 5*

Final drive chain

Final drive chain	Harley-Davidson Chain Spray or equivalent chain lube

Cables

Clutch, throttle, brake and choke	Light machine oil or cable lube
Speedometer and tachometer	Lightweight grease
General lubrication	Light machine oil

Compensating sprocket rubber dampers

Compensating sprocket rubber dampers	Harley-Davidson POLY-OIL

** Do not mix DOT 5 fluid with DOT 3 or DOT 4. If you aren't sure which type of fluid is in the system, have it flushed and filled with DOT 5 by a Harley dealer.*

1 Harley-Davidson Big Twins Routine maintenance intervals

Note: *The pre-ride inspection outlined in the owner's manual covers checks and maintenance that should be carried out on a daily or pre-ride basis. It's condensed and included here to remind you of its importance. Always perform the pre-ride inspection at every maintenance interval (in addition to the procedures listed). The intervals listed below are the shortest intervals recommended by the manufacturer for each particular operation during the model years covered in this manual. Your owners manual may have different intervals for your model.*

Daily or before riding

Check the engine oil level
Check the fuel level in the tank and look for fuel leaks
Check the operation of both brakes – look for fluid leakage (hydraulic disc brakes only) and adjust brake lever free play if necessary
Check the tires for damage, the presence of foreign objects and correct air pressure
Check the throttle for smooth operation and correct free play
Check for proper operation of the headlight, taillight, brake light, turn signals, indicator lights and horn

After the initial 500 miles

Change the engine oil and oil filter
Clean the tappet oil screen
Check the idle speed
Check the throttle and choke adjustments
Service the air cleaner
Change the transmission oil and clean the magnetic drain plug
Check the brake fluid level (disc brake models)
Grease the speedometer drive gear
Check the electrolyte level in the battery and clean the terminals
Inspect the oil lines and the brake hydraulic system for leaks
Check the adjustment of the brakes
Test the operation of the electrical components and switches
Check the adjustment of the clutch
Inspect the condition and operation of the forks and shocks
Check the tightness of the spokes (if applicable)
Inspect the brake disc and pads for wear
Clean the fuel filter screen on the fuel tank valve
Change the fork oil
Inspect the fuel lines and fittings for leaks
Lubricate and adjust the drive chain
Check the enclosed drive chain oil level (if applicable)
Check and adjust the primary chain
Adjust the chain oiler (if applicable)
Check and adjust the primary and final drive belts
Lubricate the clutch, choke and throttle cables
Lubricate the clutch and front brake hand lever pivots
Lubricate the seat suspension bushings, the seat post roller and bolt
Lubricate the seat bar bearings and the seat post
Lubricate the foot shift lever bearings
Lubricate the swingarm pivot bearings
Check the engine mounts and stabilizer links
Check the vacuum in the primary chaincase, if applicable

Every 300 miles

Lubricate the drive chain

Every 1000 miles

Service the air cleaner
Adjust the drive chain or belt
Adjust the brakes
Check the hydraulic brake fluid level
Inspect the brake discs and pads
Check the adjustment of the clutch
Check the fuel lines and fittings for leaks
Inspect the oil and hydraulic brake lines for leaks
Lubricate the clutch and front brake hand lever pivots
Lubricate the clutch, throttle and choke cables
Grease the speedometer drive gear
Check the tightness and condition of all visible fasteners
Inspect the tires for wear and proper inflation
Check the tightness of the spokes, if applicable
Adjust the primary chain or belt
Grease the rear brake pedal bearing
Grease the foot shift lever bearing
Grease the foot clutch pedal bearing (if equipped)
Lubricate the seat post roller and bolt and the seat suspension bushings

Every 2000 miles

Check and adjust the idle speed and choke setting
Check the throttle and choke
Check the engine mounts and stabilizer links
Test the operation of the electrical components and switches
Clean and adjust the chain oiler (if equipped)
Change the engine oil and filter
Check the transmission oil level
Check the battery electrolyte level
Check the condition of the spark plugs
Check the condition of the contact breaker points (if equipped)
Check and adjust the ignition timing
Clean the tappet oil screen
Check and adjust the tension of the primary chain or belt
Clean the fuel filter screen on the fuel tank valve
Grease the seat post and seat bar bearings
Grease the swingarm pivot bearings

Every 5000 miles

Replace the spark plugs
Change the transmission oil and clean the magnetic drain plug
Change the primary chaincase oil (wet clutch)
Grease the internal spiral of the throttle sleeve and the speedometer and tachometer cables
Inspect the shock absorbers and bushings
Check the steering head and swingarm bearing free play
Change the fork oil
Grease the contact breaker point cam and ignition advance unit

Every 10 000 miles

Repack the wheel bearings
Repack the swingarm bearings
Repack the steering head bearings
Lubricate the belt drive compensating sprocket rubber dampers
Adjust the springer fork rocker bearings

3.3 Pull the dipstick out of the oil tank . . .

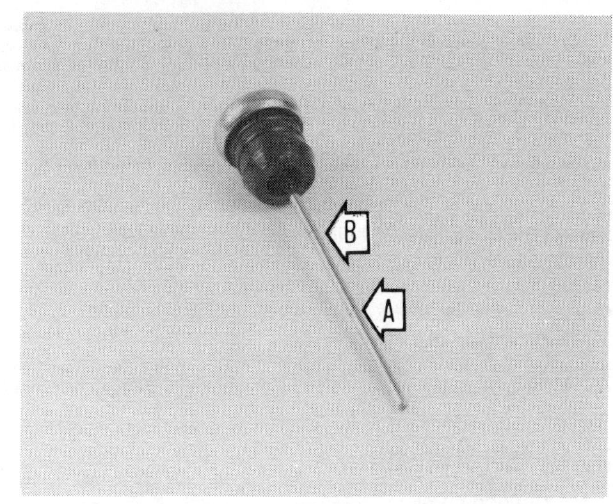

3.4 . . . and check the oil level – it should be between the lower mark (A) and the upper mark (B)

2 Introduction

This Chapter covers in detail the checks and procedures necessary for the tune-up and routine maintenance of your motorcycle. Section 1 includes the routine maintenance schedule, which is designed to keep the machine in proper running condition and prevent possible problems. The remaining Sections contain detailed procedures for carrying out the items listed on the maintenance schedule, as well as additional maintenance information designed to increase reliability.

Since routine maintenance plays such an important part in keeping the motorcycle in safe condition and operating at its optimum, this Chapter should be used as a comprehensive check list. For the rider who does all his own maintenance, these lists outline the procedures and checks that should be done on a routine basis.

Deciding where to start or plug into the routine maintenance schedule depends on several factors. If you have a motorcycle whose warranty has recently expired, and if its been maintained according to the warranty standards, you may want to pick up routine maintenance as it coincides with the next mileage or calendar interval. If you have owned the machine for some time but have never performed any maintenance on it, then you may want to start at the nearest interval and include some additional procedures to ensure that nothing important is overlooked. If a major engine overhaul has just been done, then you may want to start the maintenance routine from the beginning. If you have a used machine and have no knowledge of its history or maintenance record, it may be a good idea to combine all the checks into one large service initially and then settle into the prescribed maintenance schedule. Note that the procedures normally associated with ignition and carburetion tune-ups are included in the routine maintenance schedule. A regular tune-up will ensure good engine performance and help prevent engine damage due to improper carburetion and ignition timing.

The Sections detailing the maintenance and inspection procedures are written as step-by-step comprehensive guides to the actual performance of the work. References to additional information in applicable Chapters is also included and shouldn't be overlooked.

The first step of this or any maintenance plan is to prepare yourself before the actual work begins. Read through the appropriate Sections for all work to be done before you begin. Gather up all necessary parts and tools. If it appears that you could have a problem during a particular job, don't hesitate to ask advice from your local dealer parts or service department.

Before beginning any actual maintenance or repair, the machine should be cleaned thoroughly, especially around the oil filter, spark plugs, air cleaner, carburetor, etc. Cleaning will help ensure that dirt doesn't contaminate the engine and will allow you to detect wear and damage that could otherwise easily go unnoticed.

3 Fluid levels – check

Engine oil

Refer to illustrations 3.3 and 3.5

1 The oil level should be checked with the engine off, but at normal operating temperature.

2 On all models except the FXR, 1991 and later Dynas, and 1993 and later FLTs, the level should be checked with the motorcycle held upright. On these models, check the level with the motorcycle on its sidestand. In all cases, the motorcycle should be standing on level ground.

3 Remove the oil filler plug/dipstick, located either under the seat or on the side of the oil tank, or on later models on the oil pan (take care to avoid scalding your hands) **(see illustration)**.

4 Wipe the dipstick clean, identify the oil level marks and reinsert in the oil tank/pan. Withdraw the dipstick and read off the oil level **(see illustration)**. If the level is near or below the lower mark, add the recommended oil to bring it up to the Full mark. (On Softails where there is no Full mark, fill to the base of the filler plug.)

5 Reinstall the oil filler plug/dipstick.

Brake fluid

Refer to illustrations 3.8a, 3.8b, 3.8c and 3.10

6 In order to ensure proper operation of the hydraulic disc brakes, the fluid level in the master cylinder must be properly maintained.

7 Turn the handlebars until the top of the master cylinder is as level as

3.8a Rear brake master cylinder – pre-1980 models

3.8b On some later models, the rear brake master cylinder fluid reservoir is mounted near the brake pedal, . . .

3.8c . . . while on others it's mounted on the upper frame rail, under the left side cover

3.8d Later model reservoirs have a sightglass for checking fluid level

possible. If necessary, loosen the brake lever clamp and rotate the master cylinder assembly slightly on the handlebar to make it level.

8 On some later models the brake master cylinders are equipped with a sight glass for checking the fluid level. Earlier models require the cover to be removed to check the fluid level **(see illustrations)**.

9 Before the master cylinder cap/cover is removed, cover the gas tank to protect it from brake fluid spills (which will damage the paint) and remove all dust and dirt from the area around the cap.

10 Remove the screws and lift off the cap/cover and rubber diaphragm **(see illustration)**. **Caution:** *Do not operate the brake lever with the cap/cover removed. If the level is low, more fluid must be added.*

11 Add new, clean brake fluid of the recommended type until the level is about 1/4-inch from the top of the reservoir. Do not mix different brands of brake fluid in the reservoir – they may not be compatible.

12 Replace the rubber diaphragm and cap/cover. Tighten the screws evenly, but don't over-tighten them.

13 If the brake fluid level was low, check the entire system for leaks. In the unlikely event the reservoir was completely empty, the system should be bled as described in Section 33.

14 Wipe any spilled fluid off the reservoir body and reposition and tighten the brake lever and master cylinder assembly if it was moved.

Battery electrolyte

Refer to illustrations 3.15a and 3.15b

Caution: *Be extremely careful when handling or working around the battery. The electrolyte is very caustic and an explosive gas is given off when the battery is charging.*

15 On early models, to check the level of the electrolyte in the battery, remove the caps from the top of each cell **(see illustration)**. The electrolyte level should be up to the triangle or circle at the base of each cell. On later models, you'll have to remove the right side cover to get at the battery, which has a transparent case. The electrolyte level should be between the upper and lower level marks on the case **(see illustration)**.

16 To fill the battery, remove each cell cap and add enough distilled water to each cell to bring the level to the proper height – don't overfill it. Also, don't use tap water, except in an emergency, as it will shorten the service life of the battery. The cell holes are quite small so it may help to use a plastic squeeze bottle with a small spout to add the water.

17 Periodically, the battery should receive a thorough inspection (including a check of the specific gravity of the electrolyte). Refer to Chapter 7 for these procedures.

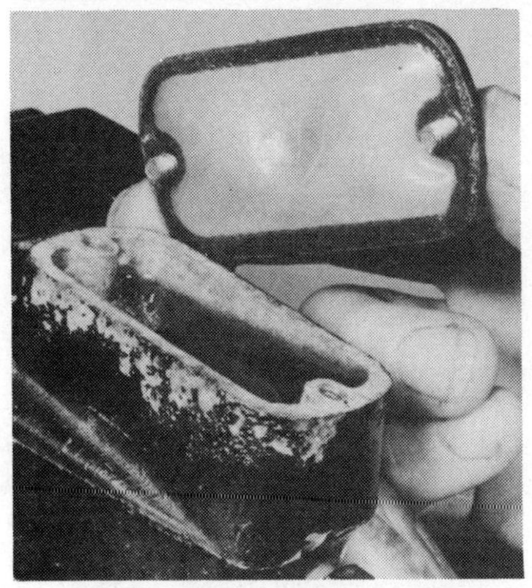

3.10 After cleaning it to avoid contamination of the brake system, remove the cap or cover from the master cylinder to check the fluid level

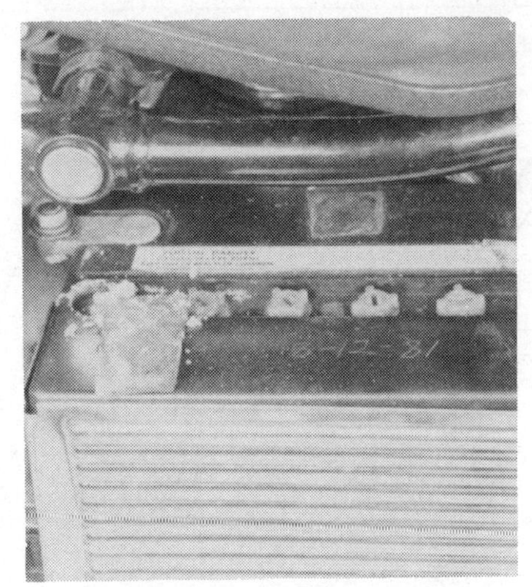

3.15a On some batteries the cell caps must be removed to check the electrolyte level, . . .

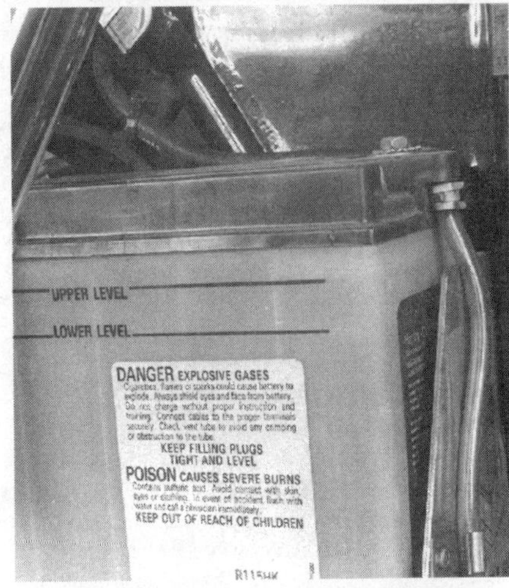

3.15b . . . while on others, after the side cover is removed, the electrolyte level can be easily checked – the case is translucent and the acceptable range is clearly marked on it (add distilled water only if the electrolyte level is low)

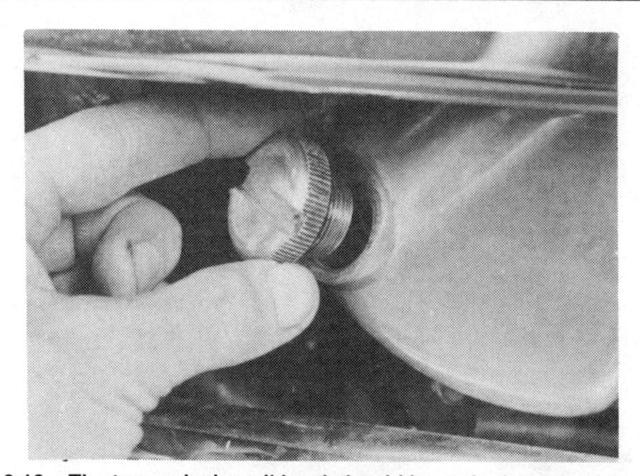
3.19 The transmission oil level should be at the bottom of the filler plug opening

4.1a The air cleaner cover is held in place with one or more Allen head screws on most models

Transmission oil

Refer to illustration 3.19

18 Hold the motorcycle in an upright position throughout the inspection procedure.

19 Remove the oil filler plug and the oil level plug. If there's only one plug, it serves as both the level and the filler plug. Some models have a dipstick attached to the oil filler plug. The oil level should be at the bottom of the plug opening or between the two marks on the dipstick **(see illustration)**.

20 If necessary, add enough oil of the recommended type to fill the transmission to the proper level – don't overfill it.

21 Install the oil level and filler plugs.

Enclosed drive chain oil

22 Hold the motorcycle in an upright position throughout the inspection procedure.

23 Remove the oil level plug at the rear of the drive chain enclosure, near the bottom. The oil level should be at the bottom of the plug opening.

24 If the oil level is low, reinstall the oil level plug.

25 Remove the saddlebag from the left side of the motorcycle and unscrew the oil fill plug from the top of the chain enclosure. Hold the motorcycle upright and remove the level plug again.

26 Add the proper grade of oil until it just begins to run out of the oil level hole. Don't add too much.

27 Install the two plugs and reinstall the saddlebag on the motorcycle.

Primary chaincase oil (wet clutch models)

Note: *Models with belt primary drive have no lubricant supply, and chain drive models with a dry clutch have an oil feed off the engine lubrication system.*

28 Check the level with the machine standing upright on level ground.

29 Remove the screws which secure the circular clutch cover in the primary chaincase. Remove the cover and its O-ring.

30 The oil should be level with the clutch cover opening on models through 1989, or level with the bottom of the clutch diaphragm spring from 1990 onwards.

31 Add oil if necessary and refit the cover and O-ring.

4 Air cleaner – servicing

Refer to illustrations 4.1a and 4.1b

1 In order to gain access to the air cleaner element, remove the plated

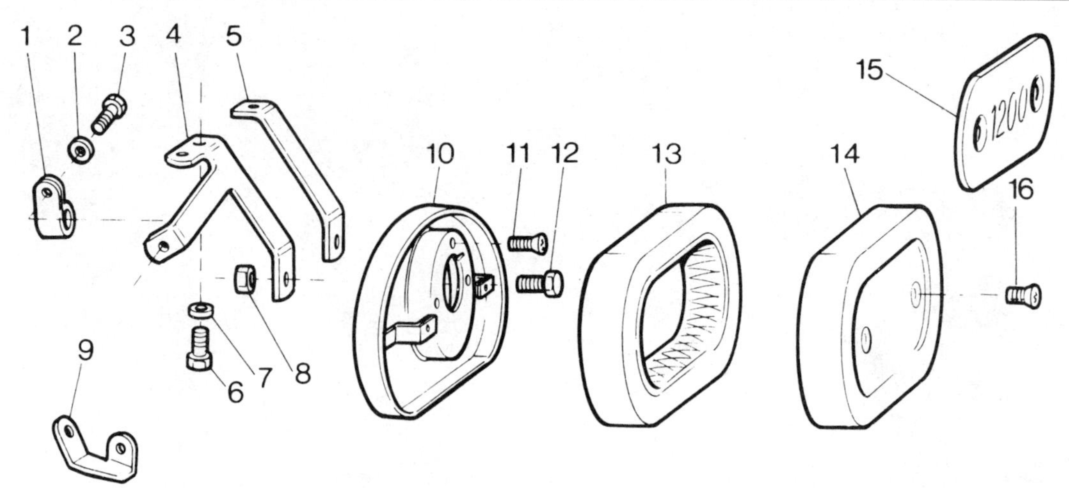

4.1b Typical air cleaner assembly components – exploded view

1 Choke cable clip – FL/FLH
2 Lock washer
3 Screw
4 Support bracket – FL/FLH
5 Support bracket – FX
6 Bolt
7 Lock washer
8 Nut
9 Bracket
10 Air cleaner box backplate
11 Screw
12 Bolt
13 Air filter element
14 Cover
15 Trim
16 Screw

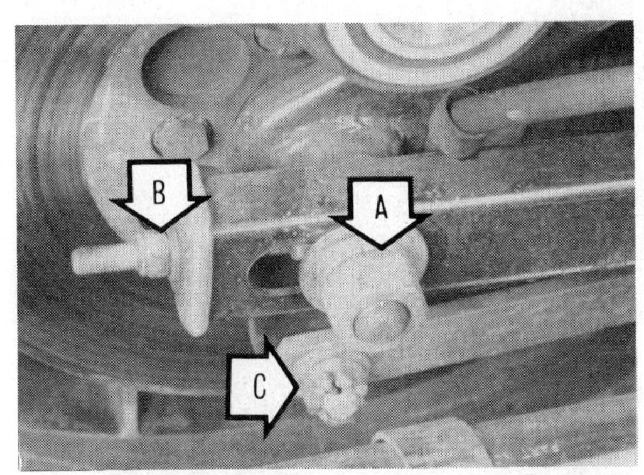

5.6 Loosen the axle nut (A) and the brake caliper anchor nut (C) before adjusting the chain tensioner (B) (typical of most models)

cover attached to the air cleaner assembly by screws or Allen head bolt(s). This will expose the filter element and screen **(see illustrations)**.

2 The air cleaner element used on 1970 and 1971 models is made of metal mesh. It should be removed, washed in a non-flammable solvent and saturated with clean engine oil after its been allowed to dry. This type of service should be performed at least every 1000 miles and more frequently if the machine is used in very dusty conditions.

3 Later models have a foam-type filter. If a film of dirt has built up on the outer surface or if light spots appear on the surface, the filter should be cleaned and re-oiled. If the filter element is cracked, torn or distorted so it doesn't fit the screen, install a new one.

4 On 1972 through 1983 models, the plastic foam air filter element should be removed from the protective screen and washed in a non-flammable solvent, then allowed to dry. Next, immerse it in clean engine oil. Allow it to soak until the element is uniform in color. If necessary, work the oil into the filter with your hands. Wring out the filter to remove the excess oil, then replace the element on the screen so the three grooves face the screen. Install the cover with the retaining screws or bolts.

5 On 1984 through 1990 models, remove the foam element from the screen and wash it with hot water and soap. After it's dry, apply 1-1/2 tablespoons of engine oil to it with an atomizer or work it into the filter by hand. Squeeze out any excess oil, then reinstall the filter, cover and bolt.

6 1991 and later models use a paper/wire mesh filter. To clean, wash in lukewarm water and a mild detergent. Let it dry naturally, or blow dry from the inside of the element out. Refit the element when completely dry. Take care when installing the filter on 1993 and later models that the crankcase breather connectors fit into the holes in the back of the element and over the bleed bolt heads.

7 Never run the machine without the air cleaner attached or without the air filter element in place. The carburetion is set up to take into account the presence of the air cleaner and will be affected if the settings aren't changed to compensate.

8 A clogged or split air cleaner element will also have an adverse effect on carburetion and engine performance. It's better to give the air cleaner more frequent attention than necessary rather than to neglect it altogether. It's an item that shouldn't be ignored during the normal service routine.

5 Drive chain – check, adjustment and lubrication

Check

1 A neglected drive chain won't last long and can quickly damage the countershaft and rear wheel sprockets. Routine chain adjustment and lubrication isn't difficult and will ensure maximum chain and sprocket life.

2 To check the chain, place the motorcycle upright with a rider sitting on it and shift the transmission into Neutral. Make sure the ignition switch is Off.

3 Check for the specified free play (slack) at the lower chain run, midway between the sprockets. On models with an enclosed chain, you'll

5.10 Apply chain lubricant to the joints between the side plates and the rollers – not in the center of the rollers (raise the rear wheel off the ground, hold the plastic nozzle near the edge of the chain and turn the wheel by hand as the lubricant sprays out – repeat the procedure on the inside edge of the chain)

have to remove the lower rubber boot from the chain housing to check the chain tension. The boot can be detached by removing the two screws and the mounting plate. Chains usually don't wear evenly, so rotate the rear wheel and check the free play in a number of places. As wear occurs, the chain will actually get longer, which means that adjustment usually involves removing some slack from the chain. In some cases where lubrication has been neglected, corrosion and galling may cause the links to bind and kink, which effectively shortens the chain"s length. If the chain is tight between the sprockets, rusty or kinked, it's time to replace it with a new one.

4 After checking the slack, grasp the chain where it wraps round the rear sprocket (this won't be possible on enclosed chains) and try to pull it away from the sprocket. If more than 1/4-inch of play is evident, the chain is excessively worn and should be replaced with a new one.

Adjustment

Refer to illustration 5.6

5 Rotate the rear wheel until the chain is positioned where the least amount of slack is present.

6 Loosen the axle nut. On pre-1973 models, loosen the brake sleeve nut and brake anchor nut also. On 1973 through 1978 models, loosen the brake anchor castle nut after removing the cotter pin. Late 1978 through 1983 models must have the brake caliper anchor nut loosened to adjust the chain tension **(see illustration)**.

7 Turn the axle adjusting nuts on both sides of the rear wheel until the proper chain tension is attained. Be sure to turn both adjusting nuts the same amount to keep the rear wheel in alignment. If the adjusting nuts reach the end of their travel, the chain is probably excessively worn and should be replaced with a new one. An accurate method of checking the alignment of the rear wheel is to measure the center-to-center distance between the swingarm pivot bolt and the rear axle on both sides of the machine. When the distances are equal, the rear wheel (and thus the chain and sprockets) should be properly aligned.

8 Tighten the axle nut and the anchor bolt (the brake caliper anchor nut on rear disc brake models) and install new cotter pins where necessary. Recheck the tension of the drive chain.

Lubrication

Refer to illustrations 5.10, 5.11 and 5.15

9 Many models produced through 1982 are equipped with automatic chain oilers; refer to Section 15 for maintenance and adjustment procedures. Some models have an enclosed chain. On these models the housing is full of oil, so the chain is constantly lubricated. The remaining models must have the chain manually lubricated.

10 The best time to lubricate the chain is after the motorcycle has been ridden. When the chain is warm, the lubricant will penetrate the joints between the side plates, pins, bushings and rollers to provide lubrication of

5.11 The spring clip can be dislodged with a screwdriver or needle-nose pliers and removed from the master link pins

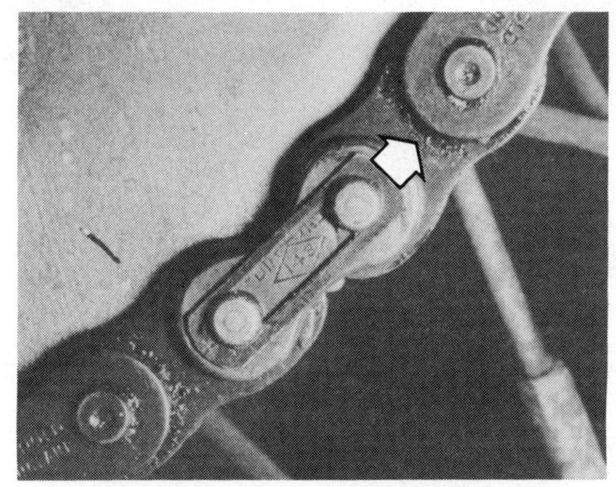

5.15 The closed end of the spring clip MUST face the direction of chain travel (arrow)

the internal load bearing areas. Use a good quality chain lubricant and apply it to the area where the side plates overlap – not the middle of the rollers. After applying the lubricant, let it soak in for a few minutes before wiping off any excess **(see illustration)**.

11 If the chain is extremely dirty, it should be removed and cleaned before it's lubricated. Remove the master link spring clip with pliers **(see illustration)**. Be careful not to bend or twist it. Slide out the master link and remove the chain from the sprockets. Clean the chain and master link thoroughly with solvent. Use a small brush to remove caked-on dirt. Wipe off the solvent, hang up the chain and allow it to dry.

12 Inspect the chain for wear and damage. Look for cracked rollers and side plates and check for excessive looseness between the links. To check for overall wear, lay the chain out on a clean, flat surface in a straight line. Push the ends together to take up all the slack between the links, then measure the overall length. Pull the chain ends apart as far as possible and measure the overall length again. Subtract the two measurements to determine the difference in the compressed and stretched lengths. If the difference, which is an indication of wear, is equal to or greater than 3-percent of the chain's nominal length, it's excessively worn and should be replaced with a new one.

13 Check the master link, especially the clip, for damage. A new master link should be used whenever the chain is reassembled.

14 Check the sprockets for wear also. If the teeth have a hooked appearance, or are excessively worn or damaged, replace the sprockets with new ones. Never put a new chain on worn sprockets or a worn chain on new sprockets. Both chain and sprockets must be in good condition or the new parts will wear rapidly. Refer to Chapter 6 for sprocket removal and installation procedures.

15 Reposition the chain on the sprockets and insert the master link. This should be done with both ends of the chain adjacent to each other on the back side of the rear wheel sprocket. **Note:** *Make sure the closed end of the master link clip points in the direction of chain travel* **(see illustration)**.

16 Lubricate and adjust the chain as previously described.

6 Brake shoes/pads – wear check

1 The brake shoes and pads should be checked at the recommended intervals and replaced when worn beyond the specified limits.

Front drum brake

2 The front brake assembly, complete with the backing plate, can be withdrawn from the front hub after the axle has been pulled out and the wheel removed from the forks. Refer to Chapter 6 for the recommended wheel removal procedure.

3 Examine the brake shoe linings. If they're thin or worn unevenly, they should be replaced with new ones.

Front disc brake

Refer to illustrations 6.4 and 6.5

4 The front brake pads, on some models, can be examined for wear by looking through the opening at the rear of the caliper **(see illustration)**. If in doubt about the condition of the brake pads, remove the caliper(s) and

6.4 The brake pad lining thickness (arrows) can be checked on most models without removing the caliper

6.5 If the pads are allowed to wear to this extent, you risk damage to the disc(s)

7.5 On models through early 1984, adjust the clutch cable at the bracket on the engine

measure the thickness of the pad lining material. Refer to Chapter 6 for caliper removal procedures.

5 Check the pads for wear, damage and looseness. Make sure the metal backing pad is flat and not distorted. If one pad has worn until the lining is less than 1/16-inch (1.6 mm) thick, replace both pads with new ones **(see illustration)**.

Rear drum brake

6 Refer to Chapter 6 for the correct procedure to remove the rear wheel in order to examine the condition of the rear drum brake linings. If the linings are worn unevenly or worn to the specified service limit, they should be replaced with new ones.

Rear disc brake

7 On some models the caliper must be disassembled to examine the condition of the rear brake pads, while other models merely require the caliper to be removed from the mounting bracket to gain access to the brake pads. Refer to Chapter 6 for the procedure required for your motorcycle. Remove the brake pads and check their condition. The information in Step 5 also applies to rear brake pads.

8 Check the condition of the caliper mounting pins and boots. Clean the boots and apply silicone grease to the pin sliding surfaces.

9 Check and lubricate the rear brake linkage.

7 Clutch – adjustment

Foot clutch control pedal

1 Place the foot pedal in the "heel down" (disengaged) position and check to make sure the clutch lever strikes the transmission case cover. The length of the foot pedal rod must be adjusted until it just covers the foot pedal bearing cover so the bearing cover doesn't bend the rod down.

2 Remove the clutch cover from the primary chaincase cover.

3 With the foot pedal in the "toe down" (engaged) position, loosen the clutch adjustment locknut. Use a screwdriver to turn the pushrod adjusting screw. The pushrod should be adjusted so there's about 1/8-inch of free play at the end of the clutch lever before the clutch disengages. Turn the screw counterclockwise for more lever movement, or to the right (clockwise) for less movement.

4 Hold the adjusting screw stationary and tighten the locknut.

Dry clutch models (through early 1984)

Clutch lever

Refer to illustration 7.5

5 If the free play at the clutch lever on the handlebar isn't as specified, it

7.9 Remove the screws and detach the clutch cover from the primary chaincase

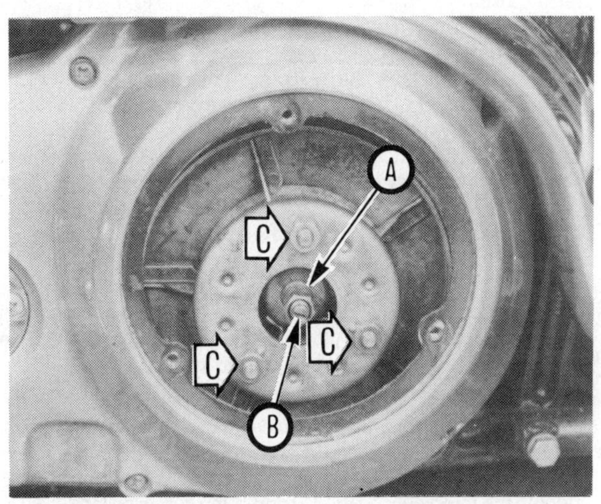

7.10 Clutch adjustment details – through early 1984 models

A	*Locknut*	*C*	*Nuts (for tightening clutch plates/discs)*
B	*Adjusting screw*		

must be adjusted. Loosen the locknut on the adjusting sleeve (you'll have to keep the swivel nut from turning on 1982, 1983 and early 1984 FLT and FXR models). Turn the adjusting sleeve either in or out of the mounting bracket until the desired free play is attained **(see illustration)**.

6 Tighten the locknut and recheck the free play adjustment.

Clutch

Refer to illustrations 7.9, 7.10 and 7.11

7 If the sleeve can't be adjusted enough to achieve the specified free play, the cable may be stretched beyond adjustment, requiring a new cable, or the clutch itself may need adjustment.

8 Loosen the locknut on the cable adjusting sleeve and screw the adjusting sleeve into the bracket as far as possible. On 1982, 1983 and early 1984 FLT and FXR models, the ball end of the cable must be removed from the release arm.

9 Remove the screws securing the clutch cover to the primary chaincase and detach the cover **(see illustration)**.

10 Loosen the locknut on the adjusting screw in the center of the clutch **(see illustration)**.

11 On 1970 through 1978 models, turn the adjusting screw until the clutch release lever is the specified distance from the starter motor hous-

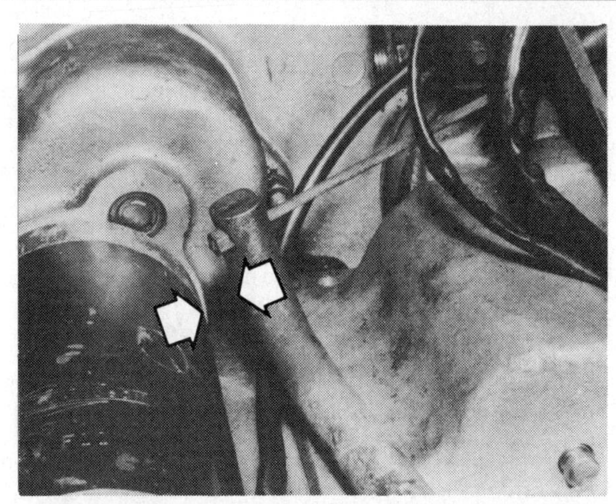

7.11 Measure the distance between the clutch release lever and the starter motor (1970 through 1978 models)

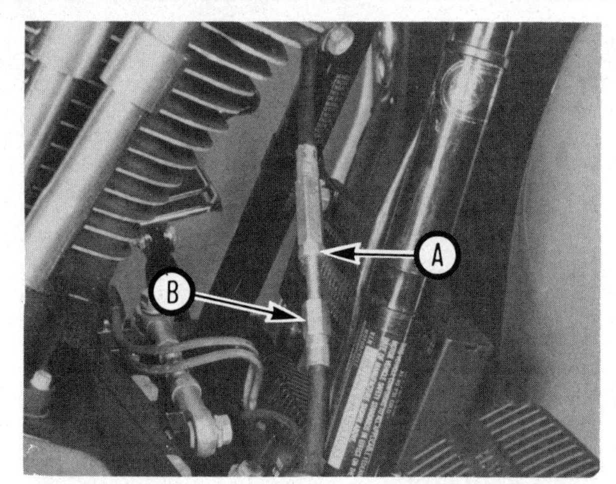

7.21 On 1987 and later models, the clutch cable adjuster is part of the cable housing

A Locknut B Adjuster

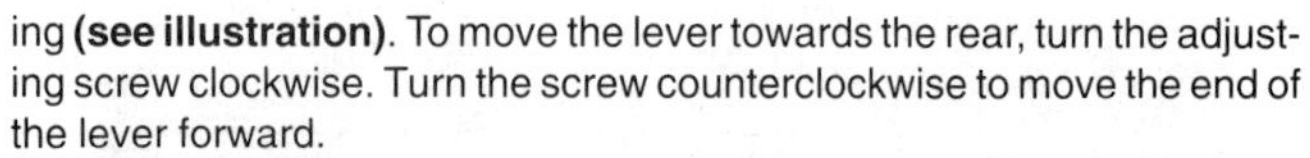

ing **(see illustration)**. To move the lever towards the rear, turn the adjusting screw clockwise. Turn the screw counterclockwise to move the end of the lever forward.

12 To adjust the clutch on late 1978 and later models (except FLT and FXR), turn the adjusting screw in until it makes contact with the pushrod, then back the screw out 1/8-turn.

13 Later FLT and FXR models (1980 through late 1984) are adjusted in a similar manner. Turn the adjusting screw clockwise until there's no free play in the release arm, then back the screw out 1/4-turn.

14 Tighten the locknut while holding the adjusting screw to keep it from moving.

15 Attach the clutch cable, if it was removed. Adjust the cable or rod until the recommended free play is achieved at the clutch lever or foot pedal.

16 Install the clutch cover on the chaincase with a new gasket coated with gasket sealer. The cover must be installed airtight.

Clutch discs

17 If the clutch still slips after the cable and clutch control are adjusted, the clutch discs must be adjusted.

18 Remove the clutch cover from the chaincase **(see illustration 7.9).**

19 Tighten the three nuts on the outside of the spring collar 1/2-turn at a time until the clutch holds **(see illustration 7.10).** With the transmission in Neutral, crank the engine to see if the clutch is holding. DO NOT tighten the nuts any more than necessary to make the clutch hold. Measure the distance between the outer disc and the spring collar. If it's less than 7/8-inch, the clutch plates must be replaced with new ones.

20 Install the clutch cover (Step 16).

Wet clutch models (late 1984-on)

Refer to illustrations 7.21, 7.23 and 7.25

21 On 1986 and earlier models, disconnect the clutch cable from the transmission release lever. On 1987 and later models, loosen the locknut at the cable adjuster and turn the adjuster all the way in to provide slack in the cable **(see illustration).**

22 Remove the clutch cover from the primary chaincase **(see illustration 7.9).**

23 Loosen the locknut on the adjusting screw in the center of the clutch **(see illustration).**

24 Turn the adjusting screw out (counterclockwise) to provide clutch pushrod free play.

25 Place a straightedge across the face of the diaphragm spring, against the adjuster plate **(see illustration)**. Using a feeler gauge, measure the gap between the outer edge of the spring and the straightedge. If it's great-

7.23 Clutch adjustment details – late 1984 and later

A Locknut B Adjusting screw

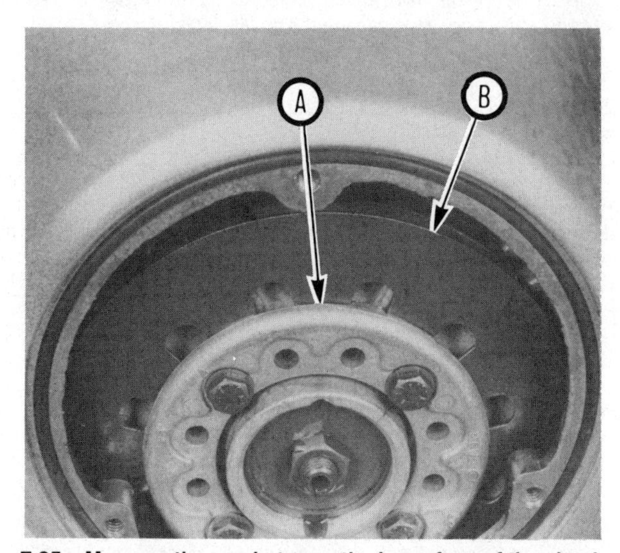

7.25 Measure the gap between the inner face of the clutch adjuster plate (A) and the diaphragm spring edge (B) with a straightedge and feeler gauge(s) – the spring should be flat or very slightly concave, not bowed out (late 1984-on)

er than 0.010-inch (0.25 mm) (which means the spring is convex, or bowed out at the center), the spring compression must be adjusted. This is done by repositioning the adjuster plate. If the spring is bowed out more than 0.010-inch (0.25 mm) in the center, the plate is moved to the next hole position to compress the spring more. If the spring is dished more than 0.010-inch (0.25 mm) at the center (concave), the plate is moved to relieve the compression. The recommended spring position is flat to 0.010-inch (0.25 mm) concave.

26 To make the adjustment, loosen the adjuster plate bolts in 1/4-turn increments until the spring tension is relieved, then remove them. Reposition the plate to compress or release the spring, as required, then install the bolts. **Note:** *If the spring can't be properly adjusted, the clutch plates must be removed and measured (Chapter 2). If the thickness is within the specified limits, an additional steel plate can be added to the clutch assembly. If the spring is not properly adjusted, the clutch will fail prematurely.*

27 Tighten the adjuster plate bolts to 8 ft-lbs (11 Nm) in a criss-cross pattern, then recheck the clearance as described in Step 25.

28 If it's correct, remove the bolts once again (Step 26) and apply Loctite 222 to the threads, then reinstall them for the final time (Step 27).

29 Turn the adjusting screw in (clockwise) until the pushrod free play is removed.

30 Back out the adjusting screw 3/4-turn and tighten the locknut. Hold the screw to keep it from moving as the nut is tightened.

31 On 1986 and earlier models, reconnect the cable at the transmission and adjust it to produce the specified free play at the hand lever. **Note:** *On 1985 and 1986 FXWG and 1986 FXST models, loosen the locknut on the cable adjuster and turn the adjuster to position the release lever 13/16-inch (21 mm) from the transmission cover tower. Apply light pressure to the lever to eliminate pushrod freeplay.*

32 On 1987 and later models, squeeze the handlebar lever as far as possible three times, to set the ball and ramp release mechanism, then turn the adjuster in the cable to provide the specified free play at the hand lever. Tighten the adjuster locknut and make sure the rubber boot is repositioned.

8 Fuel system – check

1 The condition of the fuel system components, hoses and connections should be checked periodically to reduce the likelihood of a fuel leak developing. If the smell of gasoline is noticed while riding or after the motorcycle has been parked, the system should be checked immediately. **Warning:** *If a fuel odor is detected, be sure to work in a well-ventilated area and don't allow open flames (cigarettes, appliance pilot lights, etc.) or unshielded light bulbs in or near the work area.*

2 Inspect the area around the fuel tank, fuel valve and underneath the carburetor for evidence of leaks and damage. Carefully inspect all fuel lines to make sure they're tightly connected to the fittings and not cracked or otherwise deteriorated. If so, they should be replaced immediately with new ones. Check all hose clamps to make sure they're tight.

3 If there's leakage from the fuel valve, make sure it's securely attached to the fuel tank. If a leak persists between the valve and tank, drain the tank (referring to Section 23, if necessary), remove the valve and apply thread sealing tape to the threads of the valve. If the leak is coming from the body of the valve, it'll have to be replaced with a new one.

4 If leakage is occurring at the carburetor, it indicates defective carburetor gaskets. The carburetor should be removed and disassembled as described in Chapter 3 to locate the problem.

5 Check all evaporative emission system hoses and components for damage and deterioration (later California models only).

9 Brake system – general check

1 A routine general check of the brakes will ensure that any problems are discovered and remedied before the rider's safety is jeopardized.

2 On early models, check the brake shoes for excessive wear and the lever, brake cable or rod and pedal for loose connections, excessive play, distortion and damage. Replace any damaged parts with new ones. Refer to Section 6 for the brake shoe wear check.

3 On disc brake systems, carefully examine the master cylinder, the hoses and the calipers for evidence of brake fluid leakage. Pay particular attention to the hoses. If they're cracked, abraded, or otherwise damaged, replace them with new ones. If leaks are evident at the master cylinder or caliper(s) they should be rebuilt by referring to the appropriate Sections in Chapter 6.

4 Check the disc brake lever/pedal for proper operation. It should feel firm and return to its original position when released. If it feels spongy, or if lever travel is excessive, the system may have air trapped in it. Refer to Section 33 and bleed the brakes.

5 Check the brake pads for excessive wear by referring to Section 6.

6 Examine the brake discs for cracks and evidence of scoring. Measure the thickness of each disc and compare it to the Specifications in Chapter 6. Any disc worn beyond the allowable limit must be replaced with a new one.

7 If the brake lever/pedal pulsates when the brakes are applied during operation of the machine, the disc(s) may be warped. Attach a dial indicator setup to the fork slider or swingarm and check the disc runout. If the runout is greater than specified in Chapter 6, replace the disc with a new one. If a dial indicator isn't available, a dealer service department or motorcycle repair shop can make this check for you.

8 Make sure both brake light switches operate properly.

10 Lubrication – general

Refer to illustrations 10.6, 10.7 and 10.9

1 Since the controls, cables and various other components of a motorcycle are exposed to the elements, they should be lubricated periodically to ensure proper operation.

2 The clutch and brake lever pivots should be lubricated with light oil. Don't apply too much oil to the pivots, as the oil will attract dirt, which could cause the controls to bind.

3 The throttle, clutch, front brake and choke cables should be treated with a commercially available cable lubricant, which is specially formulated for use on motorcycle control cables. Small adapters for pressure lubricating the cables with spray can lubricants are available and work very well. **Caution:** *DO NOT lubricate the enrichener cable for the CV carburetor used on 1990 and later models.*

4 Speedometer and tachometer cables should be removed from their housings and lubricated with a very light grease or cable lubricant.

5 If the throttle operates roughly, a light coat of grease should be applied to the inside of it, as described in Section 25.

6 The pivot points of the rear brake pedal, foot clutch control pedal, gear shift lever, footpegs and sidestand should all be lubricated with multi-purpose grease **(see illustration)**.

10.6 Use a grease gun to inject grease into the brake pedal fitting, . . .

1

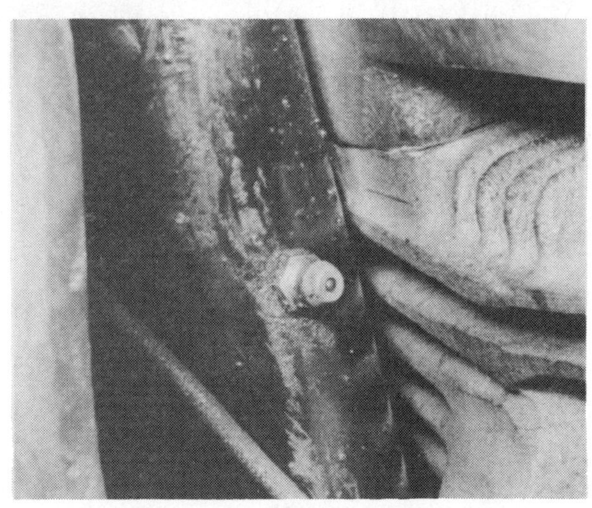

10.7 . . . the seat post fitting (not all models) . . .

10.9 . . . and the shift lever fitting

7 If applicable, lubricate the seat post, seat bar bearings, seat suspension bushings and the seat post roller and bolt **(see illustration)**.
8 Some models have a grease fitting installed in the swingarm. Use a hand-operated grease gun to fill the area between the bearings with grease at the recommended intervals.
9 On models equipped with a grease fitting on the shift lever or its linkage pivot points **(see illustration)**, grease the fittings at the recommended intervals.

11 Clutch and front brake cable – lubrication

1 Lubrication of the clutch cable and the front brake cable (on earlier motorcycles without disc brakes) is a simple operation which will help prevent premature wear of the cables.
2 Both cables can be serviced in the same way. Loosen the locknut on the cable adjuster, then loosen the cable adjuster as far as possible by turning it in a clockwise direction (refer to Section 7 if necessary).
3 Pull the cable housing out of the lever mount and position the cable so the end can be removed from the lever.
4 Spray cable lubricant into the cable housing until it comes out the other end. Special adapters are available for pressure lubricating the cables and can be purchased at most motorcycle shops.
5 When the cables are lubricated, attach the end to the lever and pull the cable housing into position in the mount.
6 Adjust the cables for the specified amount of free play, then secure the adjustment by tightening the locknut.

12 Fasteners – check

1 Since vibration of the machine tends to loosen fasteners, all nuts, bolts, screws, etc. should be periodically checked for tightness. **Note:** *DO NOT tighten the cylinder head bolts.*
2 If a torque wrench is available, use it along with the torque Specifications at the beginning of the appropriate Chapter.
3 Be sure to check the tightness of all the engine mounting bolts and the engine mount stabilizer(s) at the recommended intervals.

13 Tires/wheels/spokes – general check

1 Routine tire and wheel checks should be made with the realization that your safety depends to a great extent on their condition.
2 Check the tires carefully for cuts, tears, embedded nails or other sharp objects and excessive wear. Operation of the motorcycle with excessively worn tires is extremely hazardous, as traction and handling are directly affected. Check the tread depth at the center of the tire and replace worn tires with new ones when the tread is worn excessively.
3 Repair or replace punctured tires and tubes as soon as damage is noted. Don't try to patch a torn tire, as wheel balance and tire reliability may be impaired.
4 Check the tire pressures when the tires are cold and keep them properly inflated. Proper air pressure will increase tire life and provide maximum stability and ride comfort. Keep in mind that low tire pressures may cause the tire to slip on the rim or come off, while high tire pressures will cause abnormal tread wear and unsafe handling.
5 The cast alloy wheels used on some models are virtually maintenance free, but they should be kept clean and checked periodically for cracks, dented rims and other damage. Never attempt to repair damaged cast wheels; they must be replaced with new ones.
6 On machines equipped with wire spoked wheels, periodic checks are extremely important. Inspect the rims for dents and cracks. Check all spokes for damage (such as cracks and distortion) and make sure they're tight. To check spoke tightness, strike each one lightly with a screwdriver or other metal tool and listen to the sound that's produced. A crisp, ringing sound indicates a properly tensioned spoke. If a dull thud is produced, the spoke is loose.
7 A spoke wrench of the proper size should be used to tighten the loose spokes. Also, lubricate the nipples (at the point where the spokes thread into them and at the rim) with light oil before tightening the spokes. Don't over-tighten them, as wheel concentricity, roundness and side play will be affected.
8 Refer to Chapter 6 for the procedures to follow for checking and truing wheels.

14 Idle speed – adjustment

Refer to illustration 14.3

1 The idle speed should be checked and adjusted when it's obviously too high or too low. Before adjusting the idle speed, be sure the ignition timing is set correctly and the spark plug gaps are correct.
2 The engine should be at normal operating temperature, which is usually reached after 10 or 15 minutes of stop and go riding. Place the motorcycle on the kickstand and make sure the transmission is in Neutral.
3 Turn the throttle stop screw until the specified speed is obtained **(see illustration)**.
4 If a smooth, steady idle cannot be obtained, the fuel/air mixture may be incorrect. Refer to Chapter 3 for fuel/air mixture adjustment procedures.

14.3 The throttle stop screw (arrow) is turned to change the idle speed (a stubby screwdriver will be required to reach it)

15.1 Automatic chain oiler adjustment screw (arrow)

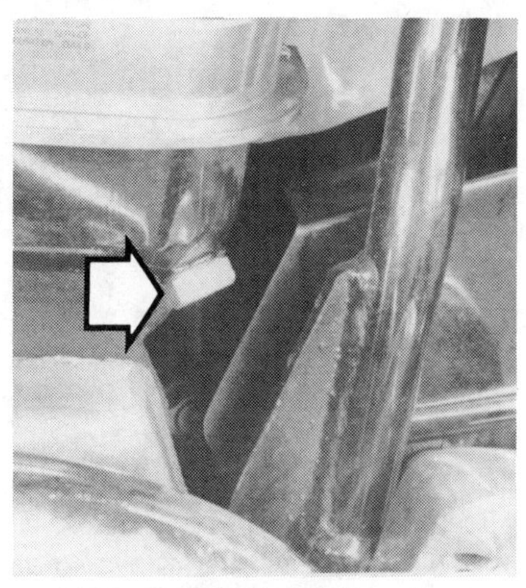
16.2 Typical oil tank drain plug location (arrow)

15 Automatic drive chain oiler – maintenance and adjustment

Refer to illustration 15.1

1 Many early models are equipped with an automatic rear drive chain oiler **(see illustration)**. On some, but not all models, an extension of this system lubricates the primary drive chain where a dry clutch is fitted.

2 Turn the oiler adjusting screw in until it bottoms on the seat. **Note:** *Keep track of the number of turns required to bottom the adjuster.*

3 Completely unscrew the adjuster and blow the orifice out with compressed air.

4 Install the adjusting screw and turn it in until it bottoms, then back it out the required number of turns to its original position. The normal setting is 1/4-turn open.

5 The oiler should release two or three drops of oil per minute. Turn the adjusting screw in if less oil is desired; turn it out if more oil is needed.

16 Engine oil, filter and primary chaincase oil – change

Refer to illustrations 16.2, 16.5a, 16.5b, 16.6, 16.7a and 16.7b

1 Consistent routine oil and filter changes are the single most important maintenance procedure you can perform on a motorcycle. The oil not only lubricates the internal parts of the engine, but it also acts as a coolant, a cleaner, a sealant and a protectant. Because of these demands, the oil takes a terrific amount of abuse and should be replaced often with new oil of the recommended grade and type. Saving a little money on the difference in cost between good oil and cheap oil won't pay off if the engine is damaged.

2 Before changing the oil, warm up the engine so the oil will drain easily. Be careful when draining the oil – the exhaust pipes, the engine and the oil itself can cause severe burns. The drain plug is at the bottom of the oil tank **(see illustration)**.

3 Place a container below the oil tank to drain the used oil into. If required, a chute made out of sheet metal or cardboard can be used to guide the oil into the receptacle. This will prevent oil from spilling on the motorcycle. On 1993 and later FLT and 1991 and later Dyna models, the engine oil is held in an oil pan attached to the bottom of the transmission. Position a drain tray under the drain plug (forward most plug on the pan). It isn't necessary to drain the crankcase.

4 Remove the oil filler cap and dipstick, followed by the oil tank/pan drain plug.

5 On late 1978 and later models, the tappet oil screen must be removed and cleaned at every oil change. The screen is located near the rear cylinder tappet block on the cam case **(see illustrations)**.

6 Remove the primary chaincase drain plug on late 1978 and later models (except FXS/FXSB) and allow the oil to drain. The plug is located on the

16.5a Remove the tappet oil screen plug (arrow) with a large screwdriver, . . .

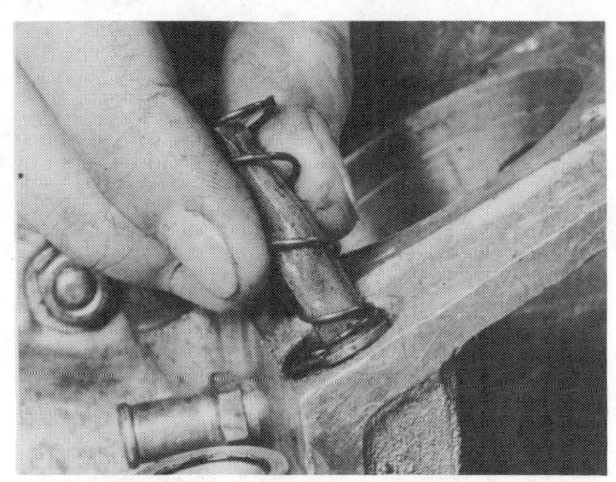
16.5b . . . then remove the screen and clean it with solvent (and compressed air, if available) – make sure the spring is in place before reinstalling the plug

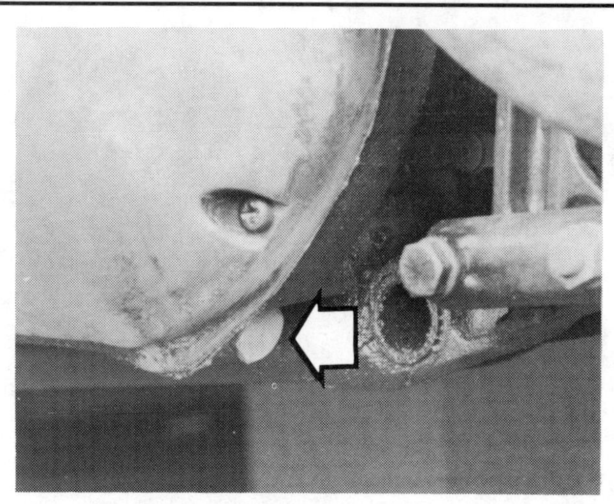

16.6 Remove the chaincase drain plug (arrow) to drain the oil out of the case

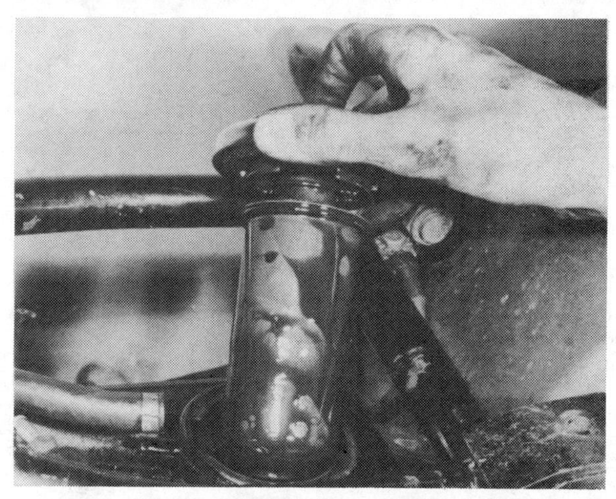

16.7a Remove the oil filter assembly from the oil tank (not all models)

bottom of the inner chaincase or chaincase cover, below the clutch cover **(see illustration)**. Clean the magnetic drain plug, reinstall it and tighten securely.

7 On models equipped with a filter in the oil tank, lift the filter out of the tank after the oil is drained **(see illustrations)**. The oil filter is housed in a cartridge and can be removed after detaching the filter clip and sealing washer from the upper end of the cartridge tube. The filter element should be replaced with a new one every time the oil is changed.

8 When replacing the element, make sure the O-ring is positioned correctly on the cartridge flange. The correct order of assembly within the cartridge is: Seal, spring, lower retainer, filter element, cap sealing washer and filter clip.

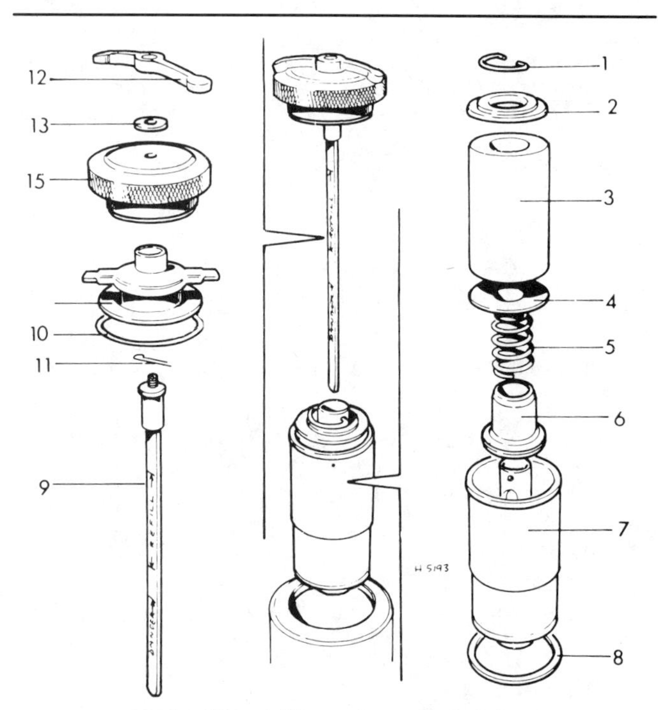

16.7b Oil tank filter unit – exploded view

1	*Filter clip*	*8*	*O-ring*
2	*Cap sealing washer*	*9*	*Dipstick*
3	*Filter element*	*10*	*Gasket*
4	*Filter lower retainer*	*11*	*Cotter pin*
		12	*Screw*
5	*Spring*	*13*	*Washer*
6	*Seal*	*14*	*Nut*
7	*Cartridge tube*	*15*	*Cap*

9 On late 1978 and later models so equipped, remove the plug from the air filter drain hose and allow the hose to drain.

10 Many models produced from 1980 on are equipped with a spin-on disposable oil filter. Locate the filter and unscrew it from the engine. Use a filter wrench that fits over the end of the filter and is turned with a 3/8-inch drive ratchet to remove the filter. If a filter wrench isn't available and the filter can't be removed by hand, one last-ditch method of removing the filter is to pierce it with a long screwdriver and twist it off. The damage to the filter doesn't matter, since it'll be replaced with a new one anyway. If additional maintenance is planned for this time period, check or service another component while the oil is allowed to drain completely.

11 Prior to installing the new spin-on filter, coat the rubber seal with clean engine oil, then screw it into place. Tighten the filter by hand an additional 1/4 to 1/2-turn (3/4 to full turn on Dyna models) after the seal first makes contact with the mounting surface.

12 The oil tank should be flushed at least every other oil change. To flush it, install the drain plug and pour approximately one quart of kerosene into the tank. Agitate the kerosene by rocking the motorcycle from side-to-side. Remove the drain plug and allow the kerosene and displaced sludge to drain out. Cleaning on those models which have an oil pan under the transmission may also be carried out by detaching the pan from the transmission (see Chapter 3).

13 Be sure the seal is in good condition before final installation of the drain plug.

14 Pour four quarts of the recommended grade of oil into the oil tank. Insert the oil filter canister, on models so equipped, and replace the oil filler cap and dipstick.

15 Where applicable, refill the primary chaincase with the specified quantity and type of oil, and check the level.

16 Start the engine and allow it to run for about a minute. Check the oil filter and drain plug(s) for leaks. Check the oil level on the dipstick and, if necessary, add enough oil to bring the level to the recommended height.

17 Spark plugs – check and replacement

Refer to illustrations 17.6a and 17.6b

1 Make sure your spark plug wrench or socket is the correct size before attempting to remove the plugs.

2 Disconnect the spark plug caps. Clean any dirt from around the base of the plugs with compressed air, a damp cloth or a brush, then remove the plugs.

3 Inspect the electrodes for wear. Both the center and side electrodes should have square edges and the side electrode should be of uniform thickness. Look for excessive deposits and a cracked or chipped insulator around the center electrode. Compare the spark plugs to the color spark plug photos in this Chapter. Check the threads, the washer and the porcelain insulator body for cracks and other damage.

Spark plug maintenance: Checking plug gap with feeler gauges

Altering the plug gap. Note use of correct tool

Spark plug conditions: A brown, tan or grey firing end is indicative of correct engine running conditions and the selection of the appropriate heat rating plug

White deposits have accumulated from excessive amounts of oil in the combustion chamber or through the use of low quality oil. Remove deposits or a hot spot may form

Black sooty deposits indicate an over-rich fuel/air mixture, or a malfunctioning ignition system. If no improvement is obtained, try one grade hotter plug

Wet, oily carbon deposits form an electrical leakage path along the insulator nose, resulting in a misfire. The cause may be a badly worn engine or a malfunctioning ignition system

A blistered white insulator or melted electrode indicates over-advanced ignition timing or a malfunctioning cooling system. If correction does not prove effective, try a colder grade plug

A worn spark plug not only wastes fuel but also overloads the whole ignition system because the increased gap requires higher voltage to initiate the spark. This condition can also affect air pollution

17.6a Spark plug manufacturers recommend using a wire-type gauge when checking the gap – if the wire doesn't slide between the electrodes with a slight drag, adjustment is required

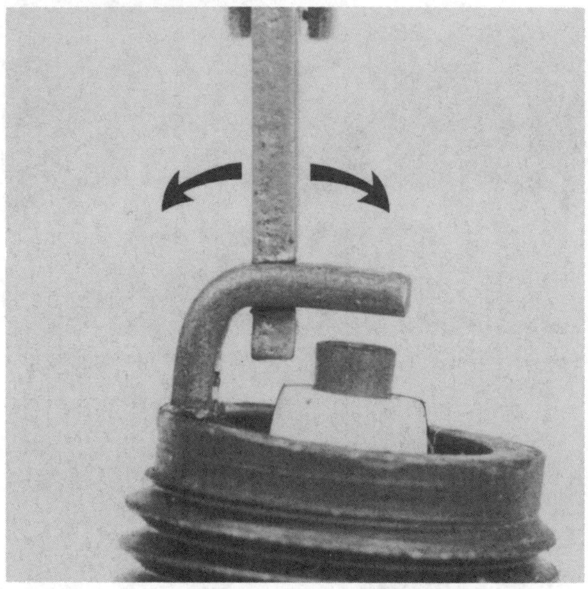

17.6b To change the gap, bend the *side* electrode only, as indicated by the arrows, and be very careful not to crack or chip the porcelain insulator surrounding the center electrode

4 If the electrodes aren't excessively worn, and if the deposits can be easily removed with a wire brush, the plugs can be regapped and reused (if no cracks or chips are visible in the insulator). If in doubt concerning the condition of the plugs, replace them with new ones, as the expense is minimal.

5 Cleaning spark plugs by sandblasting isn't recommended, since grit from the sandblasting process may remain in the plug and be dislodged after it's installed in the engine, which obviously can cause damage and increased wear.

6 Before installing new plugs, make sure they're the correct type and heat range. Check the gap between the electrodes – they're not pre-set. For best results, use a wire-type gauge rather than a flat gauge to check the gap **(see illustration)**. If the gap must be adjusted, bend the side electrode only and be very careful not to chip or crack the insulator nose **(see illustration)**.

7 Thread the plug into the head by hand. Tighten the plugs finger-tight (until the washers bottom on the cylinder head) then use a wrench to tighten them an additional 1/4-turn. Do not over tighten them.

8 Reconnect the spark plug caps.

18 Contact breaker points – check and replacement

Refer to illustrations 18.2, 18.3, 18.4, 18.7 and 18.8

1 If the contact breaker points are badly burned, pitted or worn, they should be replaced with a new set. This also applies if the fiber heel that rides on the breaker cam is badly worn.

2 To remove the points, detach the point cover **(see illustration)**, then remove the two screws that secure the base plate to the camshaft cover on the engine.

3 You'll also have to disconnect the primary wire **(see illustration)**.

4 Prior to removal, mark the base plate in relation to the distributor or camshaft cover with a scribe or permanent felt-tip pen so the assembly can be installed in the same position. This will eliminate the need to retime the ignition after reassembly. Detach the base plate – the points will come out with the plate **(see illustration)**.

5 Pull the condenser terminal off the terminal post, unhooking the moving contact point return spring at the same time. Lift off the moving contact

18.2 The breaker point cover is retained by two screws (arrows)

18.3 Detach the primary wire (1) and remove the screws (2) securing the base plate

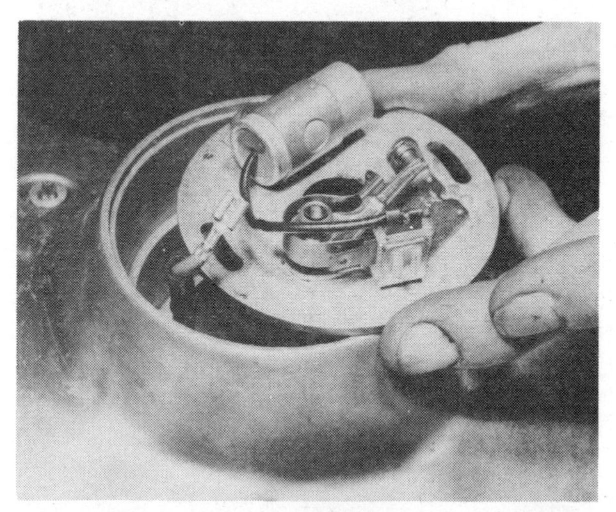

18.4 Remove the base plate with the points attached

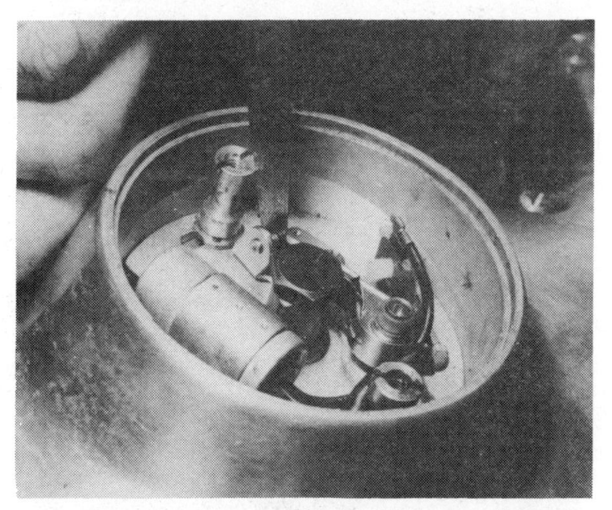

18.7 Check the point gap with a feeler gauge – if the gap is correct, the feeler gauge will just slide between the two contacts with a slight amount of drag

18.8 Lock screw (1) and breaker point adjusting slot (2) locations

19.1 Install the special clear plastic plug in the crankcase inspection hole before running the engine to check the timing

point and release the fixed contact by removing the single screw through the mounting base. **Note** *the arrangement of the insulators and other washers to prevent them from being replaced in the wrong order.*

6 Install the points in the reverse order of removal. Make sure the insulators are installed in the correct positions. It's a good idea to place a small amount of distributor cam lube on the pivot pin prior to installation of the moving contact point.

7 Check and adjust the point gap with a feeler gauge when the points are completely opened by one of the cam lobes **(see illustration)**.

8 Loosen the lock screw on the base plate of the fixed contact point and move the point by inserting a screwdriver into the adjusting slot and turning it **(see illustration)**. Adjust the points until the specified gap is obtained, then retighten the lock screw and recheck the gap.

9 Turn the engine over until the points are completely opened by the other cam lobe and check the gap. The gap should be exactly the same for both cam lobes. If it isn't, the cam is defective and must be replaced with a new one.

19 Ignition timing – check and adjustment

Refer to illustrations 19.1, 19.2a, 19.2b, 19.3a, 19.3b and 19.4

1 A timing light is the best and most accurate way to check the ignition timing, since it's done with the engine running. The timing light leads should be attached to the appropriate spark plug wire (start with the front one) and to the battery terminals. It's a good idea to replace the plug in the left crankcase with the special clear plastic factory Timing Mark View Plug (part no. HD-96295-65C) **(see illustration)**, otherwise oil spray will be a problem when the engine is running. Make sure the plug doesn't touch the flywheel.

2 Run the engine at approximately 2000 rpm (1300 to 1500 rpm on 1984 and later models) and observe the timing marks through the crankcase opening **(see illustrations)**. The front cylinder timing mark should appear stationary in the opening. As the engine is revved up, the advance mark should move into view.

19.2a On 1970 through early 1978 models, this is what the timing marks look like – ideally, they should be centered in the opening

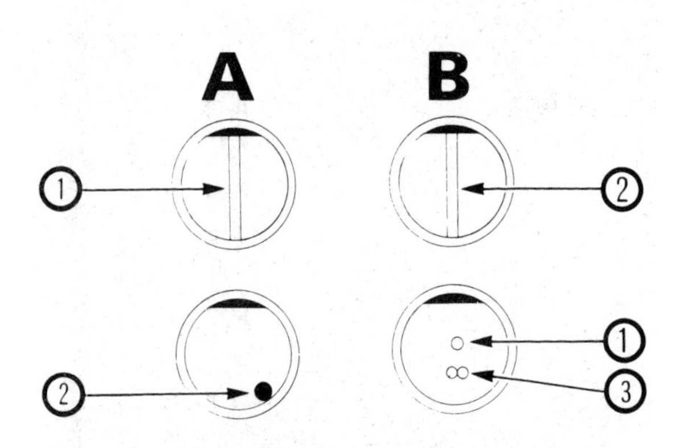

19.2b Timing marks for late 1978 through early 1980 models (A) and late 1980 on (B)

1 *Front cylinder mark*
2 *TDC for front cylinder*
3 *Rear cylinder advance mark*

19.3a On later models with electronic ignition, drill out the rivet heads and remove the outer cover, . . .

3 On early models, remove the point cover as described in Section 18 to get at the breaker plate. On later models, drill out the rivet heads **(see illustration)** and detach the outer cover, then remove the screws and take off the inner cover and gasket to get at the sensor plate **(see illustration)**. Mark the sensor plate and cover with a felt-tip pen or a scribe to ensure the sensor plate (ignition timing) can be returned to its original position if desired.

4 Adjustments can be made by loosening the contact breaker point base plate or electronic ignition sensor plate screws and rotating the plate very carefully, in small increments, with a screwdriver inserted in the slot provided **(see illustration)**. The ignition timing will change as the plate is moved. **Note:** *When checking the ignition advance on later models, be sure to check the Vacuum Operated Electric Switch (VOES) also. With the engine idling, unplug the VOES hose from the carburetor and plug the carburetor fitting. The timing should retard – the engine speed should decrease. When the hose is reattached to the carburetor, the engine speed should increase (the timing should advance). If it doesn't, check the VOES wire connection at the ignition module and the VOES ground wire connection. If they appear to be okay, the VOES may be defective and should be replaced with a new one.*

20 Primary chain – adjustment

Refer to illustrations 20.4 and 20.7

1 Adjustment of the primary chain requires the removal of the primary chain cover on 1970 through early 1978 models. On these models, place a shallow drain pan under the primary chaincase and remove the eight screws securing the case.

2 On early models with footboards, the rear pivot bolt on the footboard must be removed to allow the footboard to move down for removal of the chaincase.

3 On some FX and FXE models, the gear shift lever must be removed from the pivot shaft. The shift lever is secured to the shaft by a pinch bolt and nut. When the pinch bolt and nut are removed, the lever can be separated from the splined shaft.

4 Models produced from late 1978-on have a removable inspection cover attached to the chaincase. Remove the screws and detach the cover **(see illustration)**.

5 Check the tension of the primary chain midway between the two sprockets. The chain probably will not wear evenly, so one section of chain may be tighter then another. Because of this the chain tension should be

19.3b . . . then remove the screws (arrows) and detach the inner cover and gasket to get at the sensor plate

19.4 Mark the sensor plate (arrow), then loosen the screws (A) and insert a screwdriver into the slot (B) to change the position of the plate – move the plate a little at a time and recheck the timing

20.4 On later models, remove the cover from the outer primary chaincase to check/adjust the chain tensioner

20.7 Loosen the bolt (arrow) and move the tensioning shoe up-or-down to adjust the primary chain

adjusted at its tightest spot. On engines with an electric starter, disconnect the wires to the spark plugs and ground them against the engine. Crank the engine over to make the primary chain move so the tension can be checked at various points on the chain.

6 Adjustment of the chain is controlled by a tensioning shoe which is retained on a serrated back plate by a large center bolt.

7 With the chain at its tightest point, loosen the center bolt of the chain adjuster and move the tensioner shoe up-or-down **(see illustration)** until the desired tension is attained.

8 Tighten the center bolt and recheck the tension along the entire length of the chain.

9 Inspect the chain for broken, cracked or badly worn links. If the chain cannot be adjusted to the specified tension or if it exhibits any type of defect, it should be replaced with a new one as described in Chapter 2.

10 Carefully clean any gasket material and dirt off the primary chaincase and the inspection cover (on later models) or the chaincase cover (on early models). Attach new gaskets and install the primary case cover or inspection cover.

11 Tighten the screws securely. On early models, install the footboard and shift lever if they were removed.

12 Replenish any lost chaincase lubricant (wet clutch models) or engine oil (dry clutch models) if the chaincase cover was removed.

21 Primary belt – adjustment

1 On models equipped with a primary drive belt, the tension should be checked every 10000 miles. Only models produced from late 1981 through 1983 have belts that are adjustable. Earlier models have non-adjustable belts that must be replaced if they're stretched excessively.

2 Remove the screws securing the primary case cover to the engine and detach the cover with the gasket. On FXB/FXSB models, note the location of the one short screw.

3 Loosen the four bolts securing the primary case to the crankcase after cutting the safety wire. Pay attention to the way the safety wire is routed to ensure correct reassembly.

4 There are four nuts that secure the transmission to the transmission mounting plate that must be loosened, followed by the bolt securing the transmission to the frame tab.

5 Loosen the two starter motor nuts.

6 On most models there's a hole in the housing through which a screwdriver can be inserted to pry against the housing, increasing tension on the belt. Check the tension on the belt and increase or decrease it as necessary until the free play is within the specified limit.

7 On models without the hole drilled in the case, the tension must be adjusted by ***carefully*** prying between the alternator rotor and the primary case.

8 With the screwdriver held in place, apply tension to the belt. Tighten the two inside bolts to the specified torque, followed by the two outside bolts.

9 Install new safety wire through the inside bolts. **Caution:** *Extensive damage within the primary case could occur if the motorcycle is operated without safety wire installed.*

10 Tighten the four transmission-to-transmission mounting plate nuts to the specified torque. Tighten the transmission-to-frame tab bolt to the specified torque.

11 Tighten the starter motor nuts.

12 Clean any gasket material and dirt off the mating surfaces of the primary chaincase cover and the crankcase. Install the primary chaincase cover with a new gasket. On FXB/FXSB models, be sure to install the one shorter screw in the proper location.

13 After the primary belt is adjusted, it may be necessary to adjust the clutch cable, the rear brake pedal free play, the shifter linkage and the secondary drive belt as described in the appropriate Sections. All of these adjustments should be checked before the motorcycle is operated.

22 Final drive belt – check and adjustment

Refer to illustration 22.1

1 Deflect the belt with 10-pounds of force midway between the two sprockets **(see illustration)**. It's best to check the deflection on the bottom

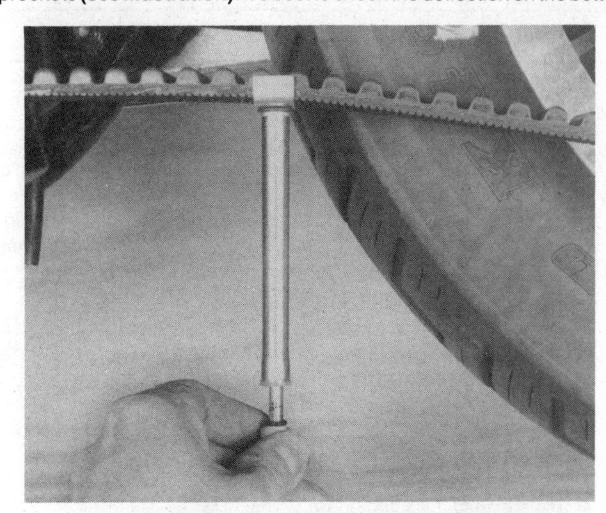

22.1 The final drive belt tension should be checked at the center of the lower run – the special gauge that accurately indicates the correct amount of force to use is shown here, although it's not absolutely necessary

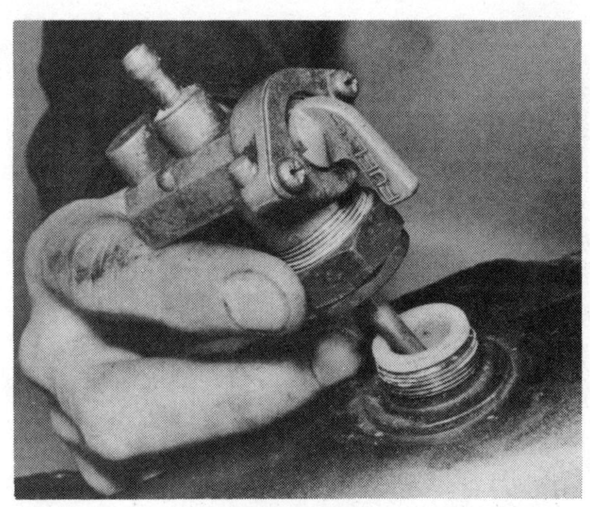

23.2a Loosen the large nut and remove the fuel valve from the tank . . .

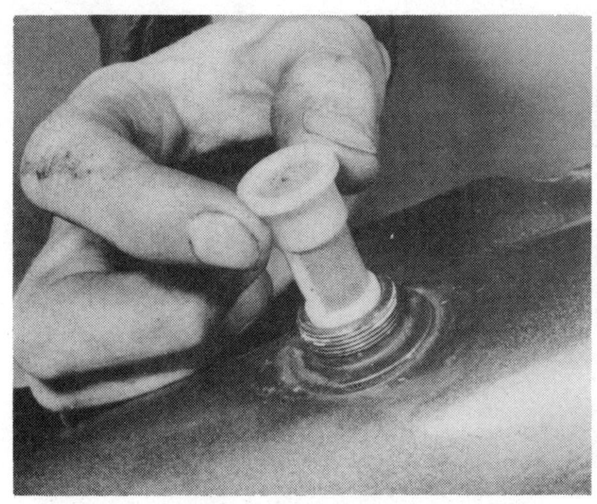

23.2b . . . to gain access to the screen-type filter for cleaning

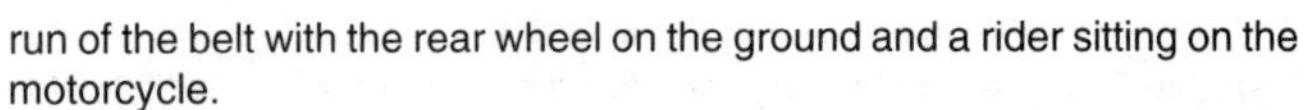

run of the belt with the rear wheel on the ground and a rider sitting on the motorcycle.

2 If the deflection is not as specified, the belt must be adjusted. Loosen the axle nut (remove the cotter pin first if one is installed) and the brake anchor nut.

3 Tighten the axle adjusting nuts to apply more tension on the belt or loosen the adjusting nuts to release tension on the belt. Turn each adjusting nut an equal number of turns to keep the rear wheel aligned. To move the wheel forward, releasing tension on the belt, it may be necessary to tap on the ends of the adjuster studs once the nuts are loosened.

4 Hold a straightedge against the outside edge of the rear wheel sprocket, near the bottom of the sprocket. With the straightedge running parallel to the drivebelt, make sure the belt is an equal distance from the straightedge for the entire length of the straightedge. Turn the axle adjusting nuts as required to align the rear wheel.

5 Check the tension of the belt again.

6 Tighten the brake anchor nut finger-tight and secure it with a new cotter pin.

7 Tighten the axle nut to the specified torque and install a new cotter pin in the axle, if so equipped.

23 Fuel filter – cleaning and replacement

Refer to illustrations 23.2a and 23.2b

Warning: *Gasoline is extremely flammable and highly explosive under certain conditions – safety precautions must be followed when working on any part of the fuel system! Don't smoke or allow open flames or unshielded light bulbs in or near the work area. Don't do this procedure in a garage with a natural gas appliance (such as a water heater or clothes dryer). Also, before starting work, disconnect the negative battery cable from the battery.*

1 Make sure the fuel control valve is in the Off position. Remove the air cleaner assembly, detach the hose from the carburetor fitting and drain the contents of the fuel tank into a gasoline container.

2 Unscrew the gland nut on the valve from the bottom of the fuel tank **(see illustration)**. The screen-type filter is attached to the fuel control valve **(see illustration)**.

3 Thoroughly clean the filter. It can be removed from the valve to be cleaned or replaced if it's damaged.

4 If the valve itself leaks, it's not practical to attempt to repair it. The complete unit should be replaced with a new one.

5 Apply thread sealant to the threads before installing the valve on the fuel tank.

24 Transmission oil – change

1 Remove the drain plug from the bottom of the transmission and allow the oil to drain completely into a shallow pan. On later models (with the engine oil tank mounted under the transmission), note that the transmission oil drain plug threads into the oil tank casting; it is the rearmost of the two drain plugs.

2 Clean the magnetic drain plug of any metallic particles; if significant amounts are found, the transmission should be dismantled for detailed inspection.

3 When all of the oil is drained, reinstall the drain plug. Be careful not to over-tighten it.

4 Remove the oil filler plug and the oil level plug, if equipped, from the right rear side of the transmission.

5 Hold the motorcycle in an upright position while filling the transmission. Pour in the specified grade of oil until the desired level is reached. On models with an oil level hole, allow the oil to flow out of the hole until the level is at the bottom of the hole.

6 Install the oil filler plug and the oil level plug, if equipped.

25 Throttle – check and lubrication

Refer to illustration 25.2

1 Make sure the throttle grip rotates easily from fully closed to fully open with the front wheel turned at various angles. The grip should return automatically from fully open to fully closed when released. If the throttle sticks, check the throttle cable for cracks or kinks in the housing. Make sure the inner cable is clean and well-lubricated and the adjustment screw located on the bottom of the lower throttle clamp isn't too tight.

2 On models with a drum-type throttle, the throttle cable(s) can be lubricated by first removing the screws that attach the top and bottom sections of the throttle assembly. Position the top section out of the way to gain access to the ends of the throttle cable(s) **(see illustration)**. The cable(s) can be lubricated with a spray lubricant specially made for motorcycle cables. As you apply the lubricant, twist the throttle back-and-forth to help move the lubricant through the cable(s). Reinstall the throttle assembly attaching screws.

3 On earlier models with spiral-type throttles, remove the handlebar end screw, screw spring and grip sleeve. Remove the roller pin and the rollers from the throttle cable plunger. Disconnect the throttle cable from the carburetor, then pull the plunger from the end of the handlebar with the cable still attached to it. Spray cable lubricant into the cable housing until it comes out of the carburetor end.

4 Assemble the throttle in the reverse order of disassembly.

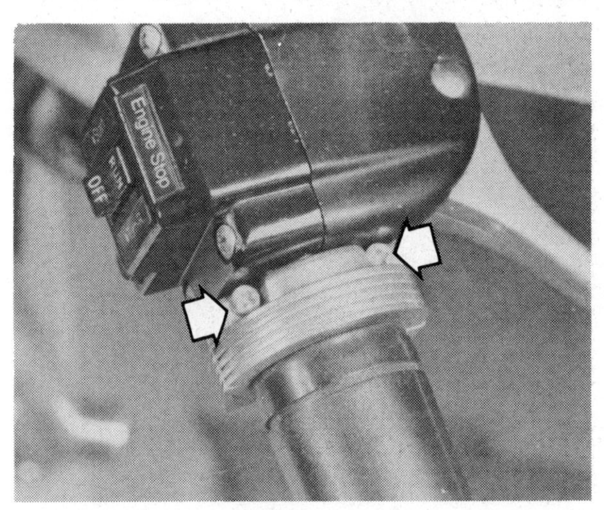

25.2 Remove the screws (arrows) securing the two halves of the throttle together

28.2 Unscrew the cap from the top of the fork leg

5 Another possible cause of a sticking throttle grip is lack of lubrication between the throttle and the handlebar. To lubricate the throttle grip, first remove it from the handlebar. Apply a thin coat of multi-purpose grease to the handlebar area that's covered by the throttle, then reinstall the throttle.

26 Suspension – inspection

1 The suspension components must be maintained in top operating condition to ensure rider safety. Loose, worn or damaged suspension parts decrease the vehicle's stability and control.
2 While standing alongside the motorcycle, lock the front brake and push on the handlebars to compress the forks several times. See if they move up-and-down smoothly without binding. If binding is felt, the forks should be disassembled and inspected as described in Chapter 5.
3 Carefully inspect the area around the fork seals for any signs of fork oil leakage. If leakage is evident, the seals must be replaced as described in Chapter 5.
4 Check the tightness of all suspension nuts and bolts to be sure none have worked loose.
5 Inspect the shocks for fluid leakage and tightness of the mounting nuts. If leakage is found, the shocks should be replaced.
6 Carefully raise the motorcycle and support it securely with the rear wheel off the ground. Make sure it can't fall off the support to either side. Grab the swingarm on each side, just ahead of the axle. Rock the swingarm from side-to-side. There should be no discernible movement at the rear. If there's a little movement or a slight clicking can be heard, make sure the pivot bolts and locknut are correctly torqued. If the pivot bolts are tight, but movement is still noticeable, then the swingarm will have to be removed and the bearings replaced, repacked and adjusted as described in Chapter 5.
7 Check the tightness of all rear suspension nuts and bolts.
8 Check all air suspension lines and fittings for damage and distortion. Make sure all fittings are tight.

27 Steering head bearings – check

1 The steering head is equipped with tapered roller-type bearings, which seldom require servicing. In extreme cases, worn or loose steering head bearings can cause steering wobble that's potentially dangerous.
2 To check the bearings, lift the motorcycle and support the machine so the front wheel is in the air.
3 Point the wheel straight ahead and slowly move the handlebars from side-to-side. Dents or roughness in the bearing races will be felt and the bars won't move smoothly.
4 Next, grasp the fork legs and try to move the wheel forward and backward. Any looseness in the steering head bearings will be felt. Refer to Chapter 5 for bearing maintenance and repair procedures.

28 Fork oil – change

Refer to illustrations 28.2, 28.3, 28.6 and 28.7

1 Support the motorcycle securely with the front wheel off the ground.
2 Remove the cap from the top of one of the fork legs **(see illustration)**.
3 Remove the drain plug from the bottom of the fork leg (near the axle), that the cap was removed from **(see illustration)**.
4 Allow the oil to drain for a few minutes, then pump the forks up-and-down to force the remainder of the oil out.
5 Install the drain plug in the bottom of the fork leg.

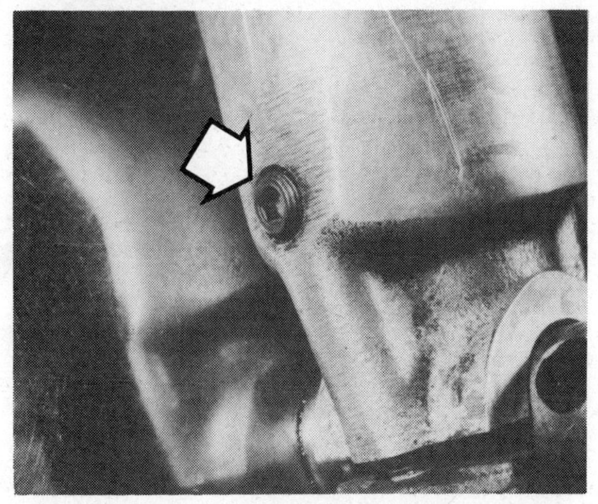
28.3 Remove the drain plug from the bottom of the fork leg (this one requires an Allen wrench, others can be removed with a Phillips screwdriver)

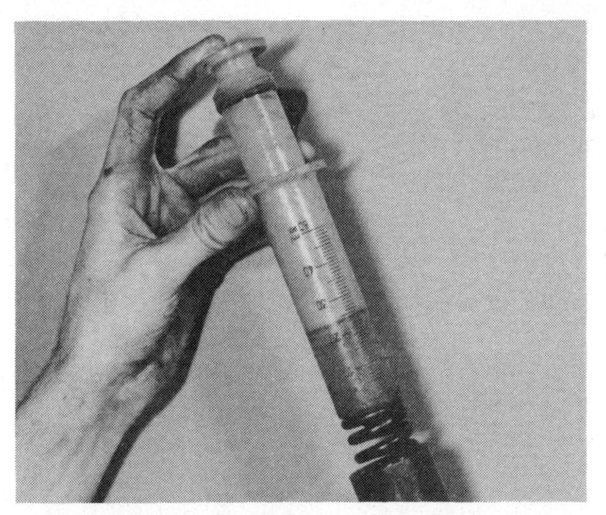

28.6 Fill the fork legs with the specified amount of fork oil (a baby bottle or measuring cup will make it easier to accurately determine the number of fluid ounces added)

28.7 Replace the fork cap seal (arrow) if it's damaged

6 Fill the fork with the specified amount of the recommended fork oil **(see illustration)**. If the fork has been disassembled, it will require additional fork oil.

7 Check the condition of the seal around the fork cap. If necessary replace the seal with a new one **(see illustration)**. Attach the fork cap to the top of the fork leg.

8 Repeat Steps 2 through 7 for the other fork leg.

29 Wheel bearings – repack

Note: *The following procedures are based on the assumption the wheel has been removed from the motorcycle. Because of detail changes made from model year-to-model year, the procedures and illustrations included here may not match the motorcycle you're working on exactly. To avoid possible problems, lay the parts out in the correct order and orientation as they're removed from the hub or make a simple sketch of the parts detailing how they fit in the hub.*

Pre-1973 models (16-inch wheel)

Refer to illustration 29.6

1 Pre-1973 front and rear wheels are equipped with permanently sealed and lubricated bearings. The bearings on these models don't require attention at any set intervals.

2 To remove the bearings, the brake drum or brake disc flange must first be removed from the wheel hub.

3 Slide the bearing spacer out of the hub.

4 Using a piece of pipe as a drift, press the bearing components out of the brake drum from the hub side.

5 Remove the bearing locknut retainer from the hub on late 1970 through 1972 models.

6 Unscrew the ball bearing locknut from the hub **(see illustration)**. The locknut has a slotted head and should be removed with a special tool. It also has left-hand threads, requiring the nut to be turned ***clockwise*** to loosen it.

7 Carefully Pry the seal out of the hub with screwdrivers and lift out the spacer.

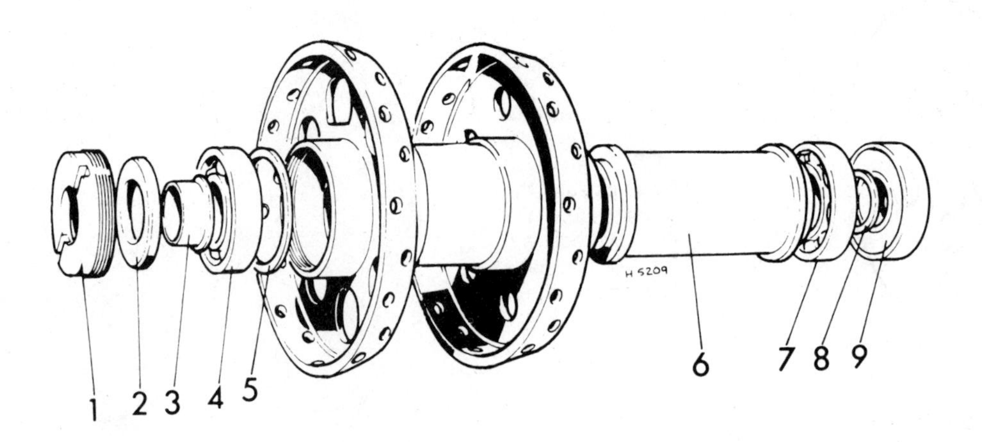

29.6 Typical 16-inch wheel hub components – exploded view (1970 through 1972 models)

1 Locknut
2 Seal
3 Outer spacer
4 Ball bearing
5 Washer
6 Spacer
7 Ball bearing
8 Spacer washer
9 Sealed ball bearing

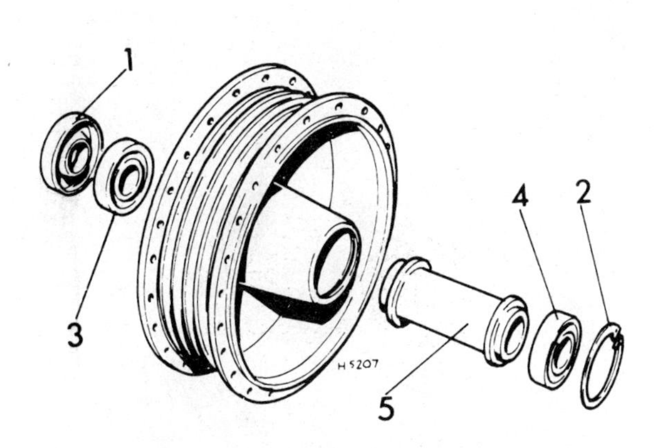

29.13 Typical front wheel hub components – exploded view (1970 through 1972 FX models shown)

1 *Seal*
2 *Snap-ring*
3 *Ball bearing*
4 *Ball bearing (brake side)*
5 *Bearing spacer*

8 The remaining bearing can now be pressed out of the drum side of the hub.
9 Inspect the bearings for wear and damage. Turn each bearing by hand to see if there's any roughness or excessive looseness between the inner and outer races. Check the lip of the seal for wear and damage.
10 Replace any defective parts with new ones.
11 Assemble the bearings and related components in the hub in the reverse order of disassembly. Fill the space on both sides of the bearing on the wheel side and the inside of the bearing on the drum side with multi-purpose grease.
12 Tighten the bearing locknut securely by striking the handle of the special tool with a hammer. When the nut is tight, stake the threads with a center punch to prevent it from loosening. On later models with a locknut retainer, simply drive the retainer into the slotted head with a chisel to hold the locknut in position.

Pre-1973 models (19-inch wheel)

Refer to illustration 29.13

13 Carefully pry the seal out of the hub with a screwdriver **(see illustration)**.
14 Remove the snap-ring from the other side of the hub using snap-ring pliers.
15 Tap the ball bearing, on the grease seal side of the hub, in until it's against the seat in the hub. This will cause the bearing on the other side of the hub to be driven out enough so the spacer between the two bearings can be moved away from the bearing on the snap-ring side.
16 Insert a drift punch through the hub from the grease seal side and drive out the bearing on the snap-ring side of the hub.
17 Remove the spacer from the center of the hub and drive the remaining bearing out of the hub.
18 Clean the bearings and related components and inspect them for wear and damage. If any of the bearings are pitted, chipped or scored, they should be replaced with new ones. Pack the bearings with fresh multi-purpose grease.
19 Position the bearing on the snap-ring side in the hub with the shielded side facing out. Press the bearing in until it's against the shoulder in the hub.
20 Secure the bearing with the snap-ring. Be sure the flat side of the snap-ring is facing the bearing.
21 Insert the spacer into the center of the hub and press the other bearing into the hub. When the bearing is seated against the shoulder of the hub, tap a new grease seal into position. The lip of the seal should be lubricated with grease or oil before installation.

1973 through 1978 models (16-inch wheel)

Refer to illustrations 29.22a, 29.22b, 29.23, 29.24, 29.30 and 29.31

22 Remove the snap-rings from both sides of the hub with snap-ring pliers **(see illustrations)**.
23 Lift the washers out of the hub **(see illustration)**.

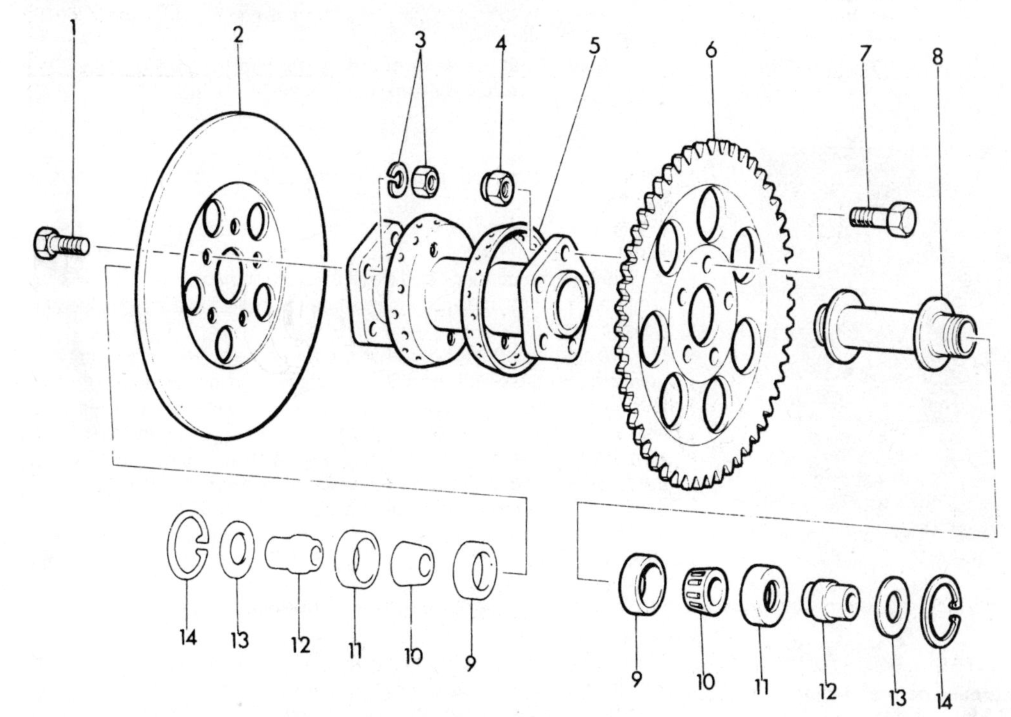

29.22a Typical rear wheel hub components – exploded view (1973 through 1983 models except FLT and FXR)

1 *Bolt*
2 *Brake disc*
3 *Nut/washer*
4 *Nut*
5 *Hub*
6 *Sprocket*
7 *Bolt*
8 *Bearing spacer*
9 *Bearing outer race*
10 *Bearing inner race*
11 *Seal*
12 *Seal collar*
13 *Washer*
14 *Snap-ring*

1

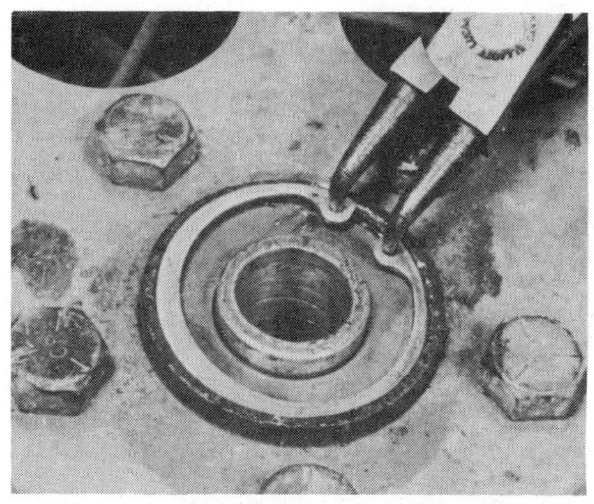

29.22b Remove the snap-ring securing the bearing in the hub, . . .

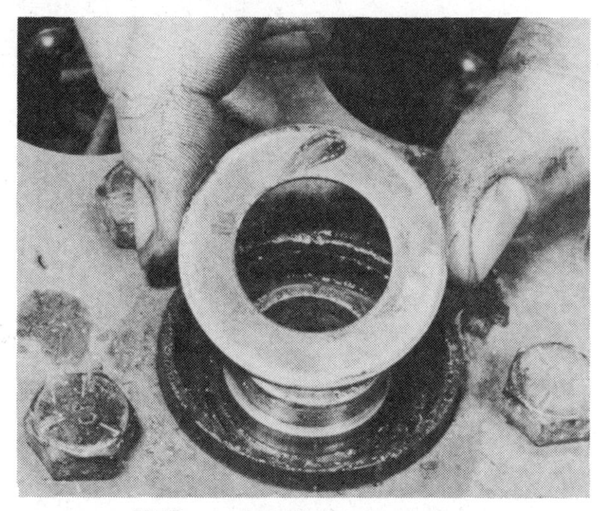

29.23 . . . then lift out the washer

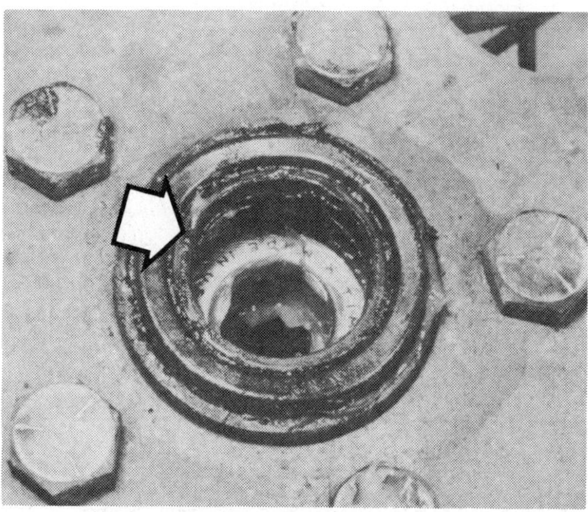

29.24 The oil seal (arrow) can now be pried out of the hub

29.30 Coat the ends of the center bearing spacer with grease before installing it in the hub

29.31 Pack the bearings with grease before placing them in the hub

24 Using a screwdriver, carefully pry the seals out of the hub **(see illustration)**.

25 Remove the spacers – note how they're installed in the hub. This will help during reassembly.

26 Remove the bearings and clean all of the components. Don't mix the parts up. Bearings and spacers must be reinstalled in the same location they're removed from.

27 The bearing outer races (bearing cups) in the hub don't have to be removed unless the bearings must be replaced with new ones. A standard bearing puller must be used to remove the races. If this tool isn't available, most motorcycle repair shops can remove the races and install the new ones. It's also possible to drive the races out of the hub with a drift punch and hammer. Working from the back side of the race, carefully tap it out of position. Be sure to move the punch around the race to drive it out straight.

28 Inspect the bearings for defects. Look for chips, pits and flat spots. Check the bearing races also. If any damage is visible, the bearing and race must be replaced as a matched set.

29 Pack the bearings with multi-purpose grease and coat the seal lips with oil or grease.

30 If new races are being installed, place them in position, then press them in until they're seated in the hub. A large socket of the correct size and a hammer can be used to drive the race into position. Apply grease to the bearing spacer and slide it into the hub **(see illustration)**.

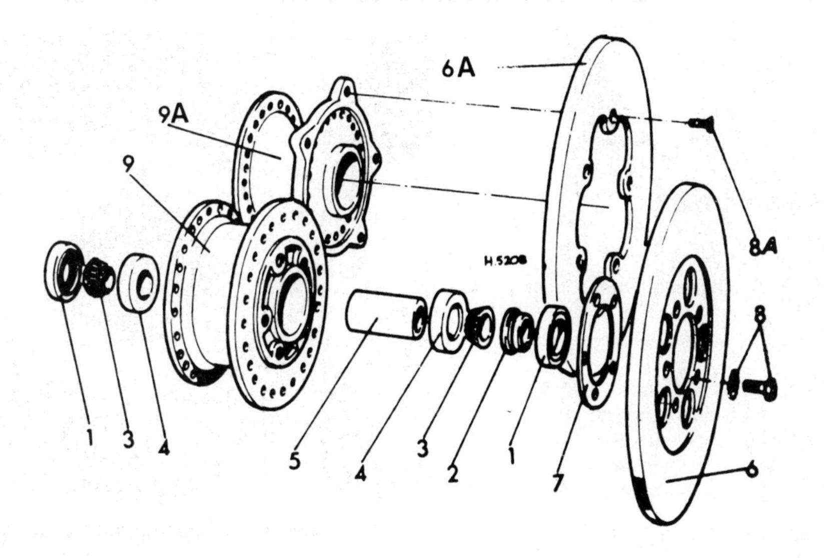

29.34 Front wheel hub components – exploded view (1973 through 1983 models except FXR, FXWG and FLT)

1 *Seal*
2 *Spacer*
3 *Bearing*
4 *Bearing race*
5 *Bearing spacer*
6 *Brake disc (1973)*
6A *Brake disc (1974-on)*
7 *Brake disc spacer (1973)*
8 *Bolt and lock washer (1973)*
8A *Allen-head screw (1974-on)*
9 *Hub (1973)*
9A *Hub (1974-on)*

31 Insert the bearing, packed with grease, into the hub **(see illustration)**.
32 Install the spacers in the hub (make sure they're facing the correct direction).
33 Press the seal into the hub until it's 1/4 to 3/16-inch below the outside edge of the hub. Set a washer in each side of the hub and secure them with the snap-rings.

1979 through 1983 models (19-inch wheel) except FLT, FXR and FXWG models

Refer to illustration 29.34

34 Carefully pry the seals out of the hub with a screwdriver **(see illustration)**.
35 Remove the spacer from the hub. Pay attention to the side of the hub the spacer is removed from and the direction it's facing when installed.
36 Lift the bearings out of the hub. Don't allow the bearings to get mixed up. They must be installed in the same location they were removed from. Remove the spacer from the center of the hub.
37 Refer to Steps 27 through 31 for bearing inspection and replacement procedures. Be sure to insert the spacer into the center of the hub.
38 Install the spacer in the hub, followed by the seals. Lubricate the seal lips before installation. The seals should be pressed flush with the outer surface of the hub.

1978 through 1983 models (21-inch wheel and all rear hubs) except FLT and FXR models

39 Remove the snap-rings from both sides of the hub with snap-ring pliers.
40 Lift out the washers and spacers. **Note** *which way the spacers are installed in the hub.*
41 Pry the seals out of the hub and lift out the bearings and the inner spacer.
42 Refer to Steps 27 through 31 for bearing inspection and replacement procedures. Be sure to install the spacer in the center of the hub.
43 Lubricate the lips of the seals and tap them into position, 3/64 to 7/32-inch below the edge of the hub.
44 Install the spacers and washers and secure them with the snap-rings.

1980 and 1981 FLT (front hub)

45 Remove the screws securing the discs to the front hub.
46 Detach the disc and bearing assembly from the right side of the hub, followed by the spacer from the center of the hub.
47 Pry the seal out of the left side of the hub, then slide the bearing inner race out of the hub.
48 Special tool no. HD-95760-69 must be used to pull the bearing from the hub.
49 Clean the bearings and associated components and inspect them for wear and damage. If any component is defective, it must be replaced with a new one. If a new bearing must be pressed into the hub, be sure to press only on the side of the bearing the numbers or letters are stamped on. Special tool no. HD-94440-81 must be used to install both the bearing and the seal.
50 Insert the inner race and the center spacer. Use moderate pressure to install the spacer. This will drive the inner race slightly out of the hub and seal. Be sure to install the spacer with the flat edge against the inner bearing race.
51 Place the right side bearing assembly on the hub, aligning the bolt holes. Set the right disc on top of the bearing assembly and secure it to the hub with the longer set of mounting screws. Coat the threads of the mounting screws with thread locking compound before installation and tighten them to the specified torque (see Chapter 6).

1982 and later FLT (front hub)

52 On 1982 models, remove the snap-rings from both sides of the hub with snap-ring pliers, then lift the washers out of each side.
53 Pay attention to the order the following parts are installed in the hub while you remove them. Keep the components from the left side separate from the components from the right side of the hub.
54 Remove the spacers from both sides and pry the seals out of the hub. Lift the bearings out of position.
55 Remove the spacer washer and spacer from the left side of the hub and the large spacer and sleeve from the right side of the hub.
56 Refer to Steps 27 through 30 for bearing inspection and replacement procedures.
57 Install the large spacer, sleeve, spacer washer and small spacer in the center of the hub.
58 Insert the bearings into position and press a new seal into each side of the hub. The lip of each seal should be lubricated with grease or oil before installation. Press the seal in 13/64 to 7/32-inch below the edge of the hub.
59 Install the spacers in the hub.
60 On 1982 models, place a washer in position on each side of the hub and secure the assembly with the two snap-rings.

1980 and later FXR (front hub)

61 Pry the seals out of both sides of the hub and remove the spacer from the left side of the hub.
62 Lift the bearings out of the hub, then slide the center spacer out. On later models there will be a spacer washer and thin spacer behind the right side bearing (shoulder on spacer washer should face the bearing).
63 Refer to Steps 27 through 30 for bearing inspection and replacement procedures.

64 Insert the long spacer into the hub and, where fitted, install the thin spacer and spacer washer (shoulder facing the bearing). Install the bearings.

65 Place the spacer in the left side of the hub and press the seals in. Be sure to lubricate the lips of the seals before installation. The seals should be flush with the edge of the hub when they're seated.

1980 and 1981 FLT and 1980 through 1983 FXR (rear hub)

66 Remove the spacer from the left side of the hub.

67 Pry the seal out of the right side of the hub, then remove the inner bearing race.

68 If the bearing must be removed, special tool no. HD-95760-69 must be used.

69 Clean the bearing and associated components and inspect them for wear and damage. If any component is defective, it must be replaced with a new one. If the bearing was removed from the hub, it must be pressed back into the hub using special tool no. HD-94440-81. When pressing the bearing into the hub, be sure to press only on the side of the bearing with the numbers or letters stamped on it.

70 Using the same special tool, press the seal into the hub. Be sure to lubricate the seal lip with oil before installation.

71 Remove the special tool from the hub but leave the inner bearing race inside the bearing and seal.

72 Install the spacer in the hub from the left side. The flat end of the spacer should go against the inner bearing race. Push the spacer into the hub as far as possible, using moderate thumb pressure. This should cause the inner bearing race to project slightly from the seal on the right side of the hub.

1982-on FLT and 1984-on FXR (rear hub)

73 Remove the spacers from the hub sides and pry out the oil seals. Lift the bearings out of the hub and withdraw the center long spacer. A thin spacer and spacer washer are fitted to later models; take note of which side of the hub these are fitted to when removing.

74 Refer to Steps 27 through 31 for bearing race replacement.

75 Install all components in the reverse of the removal procedure, noting that the thin spacer and spacer washer must be positioned in the correct hub side, as noted on removal (the flanged face of the spacer washer must face the bearing).

76 Early models with an enclosed drive chain will have the left side bearing race installed in the sprocket boss; refer to Chapter 6, Section 17 for details.

1985-on FX/Softail and all FXD Dyna models

Front hub

77 Where fitted, remove its retaining ring and lift off the chromed hub cover. Remove the spacer from the hub and pry out the oil seals on both sides. Lift out the bearings. On later models there will be a spacer washer and thin spacer on one hub side; take note of which side it is fitted. Remove the long center spacer.

78 Follow the procedure in Steps 27 through 31 to replace the bearing races. Refit all components in a reverse of the removal procedure, noting that the spacer washer on later models must be installed with its flanged shoulder facing the bearing.

Rear hub

79 These models have a spacer, seal, bearing and bearing race on each side of the hub, plus a long spacer in the hub center. Later models also have a spacer washer and thin spacer on one side of the hub (take note of which side before removing); the spacer washer must be installed with its shoulder facing the bearing.

80 Follow the procedure in Steps 27 through 31 for bearing race replacement.

30 Cylinder compression – check

1 Among other things, poor engine performance may be caused by leaking valves, incorrect valve clearances, a leaking head gasket or worn pistons, rings and/or cylinder walls. A cylinder compression check will help pinpoint these conditions and can also indicate the presence of excessive carbon deposits in the cylinder head.

2 The only tools required are a compression gauge and a spark plug wrench. Depending on the results of the initial test, a squirt-type oil can may also be needed. A compression gauge that screws into the spark plug hole is preferred over the type that requires hand pressure to maintain the seal at the plug hole.

3 Warm up the engine to normal operating temperature (ten or fifteen minutes of stop and go riding should be sufficient). Park the motorcycle and remove any dirt around the spark plugs with compressed air or a small brush, then remove the plugs. Work carefully, don't strip the spark plug hole threads and don't burn your hands. Use jumper wires to ground the spark plug wires on the cylinder heads.

4 Install the compression gauge in one of the spark plug holes. Make sure the choke is open and hold or block the throttle wide open. The transmission must be in Neutral.

5 Crank the engine over a minimum of five to seven revolutions and note the initial movement of the compression gauge needle as well as the final total gauge reading. Repeat the procedure on the remaining cylinder and compare the results to the Specifications.

6 If the compression in both cylinders built up quickly and evenly to the specified amount, you can assume the engine upper end is in reasonably good mechanical condition. Worn or sticking piston rings and worn cylinders will produce very little initial movement of the gauge needle, but compression will tend to build up gradually as the engine spins over. Valve and valve seat leakage, or head gasket leakage, is indicated by low initial compression which doesn't tend to build up.

7 To further confirm your findings, add about 1/2-ounce of engine oil to each cylinder by inserting the nozzle of a squirt-type oil can through the spark plug holes. The oil will tend to seal the piston rings if they're leaking. Repeat the test on both cylinders.

8 If the compression increases significantly after the addition of the oil, the piston rings and/or cylinders are definitely worn. If the compression doesn't increase, the pressure is leaking past the valves or the head gasket. Leakage past the valves may be caused by burned or cracked valve seats or faces, warped or bent valves or insufficient valve clearances.

9 If the compression readings are considerably higher than specified, the combustion chambers are probably coated with excessive carbon deposits. It's possible for carbon deposits to raise the compression enough to compensate for the effects of leakage past rings or valves. Refer to Chapter 2 to remove the cylinder heads and carefully decarbonize the combustion chambers.

31 Headlight aim – adjustment

1 An improperly adjusted headlight may cause problems for oncoming traffic or provide poor, unsafe illumination of the road ahead. Before adjusting the headlight, be sure to consult local traffic laws and regulations.

2 To set up the headlight, the machine should be placed on level ground at least 25-feet from a wall in its normal position (off the stand and with a rider – also a passenger if that's usually the case). On high beam, the height of the beam on the wall should be the same as the ground-to-headlight distance. This will ensure that oncoming drivers/riders won't be blinded.

All FL models, FXRT and FXRD

3 The headlight can be adjusted vertically by turning the screw at the top of the light. It may be necessary to remove the trim ring from the headlight assembly to gain access to the adjustment screws. The horizontal adjustment is made similarly, except the adjusting screw is at the side of the headlight.

FXWG, FXLR, FXSTC and FLSTC/F/N

4 Loosen the headlight assembly mounting bolt located below the lower triple clamp to adjust the beam horizontally.

5 Loosen the headlight assembly mounting bolt located above the lower triple clamp to adjust the beam vertically.

6 Tighten the two mounting bolts and recheck beam alignment.

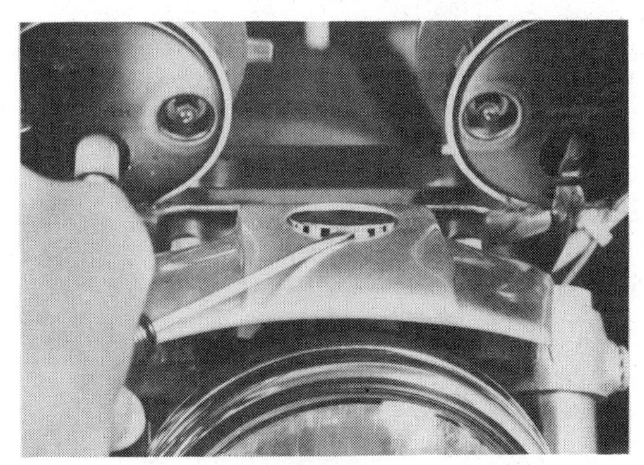

31.10a Pry the decorative plug out of the headlight housing . . .

31.10b . . . and loosen the mounting nut (FX models)

FXSTS model

7 The horizontal and vertical adjusters are located at the rear of the headlight shell, and are accessed from above. Loosen the through-bolt and nut which pass through the side of the mounting block to make vertical adjustment; note that the head lamp should be positioned as far forward as possible in its block so that it does not contact the fork springs.
8 Horizontal adjustment is made by loosening the Allen-head bolt which passes down through the top of the mounting block.
9 Tighten all adjustment bolts when complete and recheck the beam settings.

FXD and all other FXR and FX models

Refer to illustrations 31.10a and 31.10b

10 Pry the decorative plug out of the top of the headlight housing and loosen the mounting nut. Twist the headlight in the desired direction until the beam is aimed properly **(see illustrations)**.
11 Tighten the mounting nut and recheck the aim.
12 Snap the decorative plug into the top of the housing.

32 Sidestand – check and maintenance

1 The sidestand should be checked periodically for smooth operation and wear. Clean the stand and inspect it carefully for cracks and other damage. Also, be sure it's not bent.
2 The sidestand pivot must be in good condition and well lubricated so the stand will retract when the machine is raised to an upright position. This applies to the return spring also.
3 To lubricate the sidestand, first disengage the spring, then remove the sidestand pivot bolt. Apply a coat of grease to the pivot, then reinstall the stand.

33.3 Remove the rubber cap from the bleeder valve and attach the tubing to the valve fitting

33.6 Open the bleeder valve slightly with a wrench as pressure is applied to the brake lever or pedal – close the valve before releasing the brake

33 Brake hydraulic systems – bleeding and fluid replacement

1 Because the hydraulic systems used with both the front and rear disc brakes are virtually the same, the following procedures apply to both systems.

Bleeding

Refer to illustrations 33.3 and 33.6

2 Every time any part of the brake hydraulic system is disassembled or develops a leak, or when the fluid in the master cylinder reservoir runs low, air will enter the system and cause a decrease in performance. To eliminate the air, the system must be bled using the following procedure.
3 Before beginning, have an assistant on hand, as well as a supply of new brake fluid, an empty clear plastic container, a length of 3/16-inch plastic, rubber or vinyl tubing to fit over the bleeder valve and a wrench to open and close the bleeder valve **(see illustration)**.
4 Make sure the master cylinder reservoir is full of fluid and be sure to keep it at least half full during the entire operation. If, at any point, the reservoir runs low on fluid, the entire bleeding procedure must be repeated.
5 Loosen the bleeder valve slightly, then tighten it to a point where it's snug but can still be loosened quickly and easily.
6 Place one end of the tubing over the bleeder valve fitting and submerge the other end in brake fluid in the container **(see illustration)**.

1

7 Have your assistant apply the brakes a few times to get pressure in the system. On the last pump, have him hold the lever or pedal firmly depressed.
8 With the lever or pedal depressed, open the bleeder valve just enough to allow some fluid to leave the valve. Watch for air bubbles to exit the submerged end of the tube. When the fluid flow slows, close the valve again and have your assistant release the lever or pedal. If it's released before the valve is closed, air can be drawn back into the system. **Note:** *Avoid pulling the lever all the way back to the hand grip (front brake); this will cause overtravel of the piston and subsequent fluid leakage. This can be prevented by placing a 3/4-inch (20 mm) spacer between the hand grip and lever. Be careful when bleeding the rear brake also.*
9 Repeat Steps 7 and 8 until no air bubbles are seen in the fluid leaving the tube. Be sure to check the fluid level in the master cylinder frequently.
10 Completely tighten the bleeder valve.
11 Don't reuse old brake fluid – it attracts moisture which will deteriorate the brake system components. Also, don't mix different types of brake fluid.
12 Refill the master cylinder with fluid at the end of the operation.

Fluid replacement

13 Since brake fluid absorbs moisture and becomes contaminated in other ways while in use, it should be drained and the system filled with new fluid at the recommended intervals.
14 Before draining the old fluid, be sure to cover the painted surfaces of the gas tank and other components to prevent damage in the event brake fluid is spilled on them.
15 Remove the reservoir cap (leave the rubber diaphragm in place) and attach a piece of clear vinyl hose to the bleeder valve. Place the other end of the hose in a plastic container.
16 Loosen the bleeder valve and slowly pump the lever or pedal to drain the fluid from the system. In order to completely drain the brake caliper, it must be removed and turned upside-down (refer to Chapter 6). If you're working on the front brake system, repeat the procedure on the remaining caliper of a two caliper system.
17 After the system has been drained, close the bleeder valve(s) and add new, clean brake fluid of the recommended type to the reservoir. Slowly pump the lever or pedal with the bleeder valve(s) open (hoses attached) to fill the system. Add more fluid to the reservoir as the level drops.
18 Once the system is full, it must be bled to remove all air as described in the above Sub-section.
19 Make sure the fluid level in the reservoir is correct, then install and tighten the reservoir cap. **Warning:** *Don't ride the motorcycle unless you're absolutely sure the brake system is capable of performing as it's designed! If the lever or pedal is spongy, recheck everything and perform the bleeding procedure again.*

34 Compensating sprocket rubber dampers (belt drive models only) – lubrication

1 Refer to Chapter 2 for the procedure to follow for the removal of the primary drive belt and compensating sprocket.
2 With the compensating sprocket removed, it can be disassembled to lubricate the two sets of rubber dampers with Harley-Davidson POLY-OIL (part no. 99860-81).
3 Reassemble the compensating sprocket and install it with the primary drivebelt as described in Chapter 2.

35 Springer fork rocker bearings – adjustment

Caution: *This procedure requires the spring load to be taken off the rockers by disconnecting the spring forks. This can be done without removing the forks from the motorcycle.*
1 Refer to Chapter 5 and perform the first five Steps in springer fork removal.
2 Detach the spring fork legs from the rockers (see Chapter 5 if necessary). **Warning:** *Be sure to restrain the spring fork legs up against the rigid legs – if you don't, the spring pressure will snap them forward with extreme force, which could cause serious injuries!*
3 Loosen the jam nuts and bearing retainers on the rockers.
4 Move the RIGHT rocker UP against the rigid fork. Move the LEFT rocker BACK against the rigid fork.
5 Slowly tighten the bearing retainers until contact is felt, then tighten each one an additional (one) flat – mark the flats if necessary.
6 Hold each retainer so it can't move and tighten the jam nuts to 95 to 105 ft-lbs (129 to 142 Nm). **Note:** *Approximately 1/16-inch (1.6 mm) of the retainer will protrude past the jam nut when everything is installed and tightened correctly. The rockers may not feel like they're tightened equally, but it doesn't matter – they'll equalize within a few miles of riding. If it feels like there's metal-to-metal contact, the spherical bearings in the rocker(s) must be replaced.*
7 Reattach the spring fork legs to the rockers, then put everything back together.

Chapter 2 Engine, clutch and transmission

Contents

Specifications

Engine

General

	1200 cc	1340 cc
Type	Air-cooled, four-stroke, 45-degree V-twin	
Bore	3.438 in (87.3 mm)	3.498 in (88.8 mm)
Stroke	3.968 in (100.8 mm)	4.250 in (108.0 mm)
Displacement	73.66 cu in	81.6 cu in

Valves

Margin width limit	0.031 in (0.787 mm)
Seat width	
Pre-Evolution engine	0.050 to 0.090 in (1.270 to 2.286 mm)
Evolution engine	
Standard	0.040 to 0.062 in (1.016 to 1.575 mm)
Service limit	0.090 in (2.286 mm)

Valves

Valve stem protrusion from cylinder head boss		
Pre-Evolution engine (late 1978-on)	1.600 to 1.645 in (40.64 to 41.78 mm)	
Evolution engine		
Standard	1.990 to 2.024 in (50.55 to 51.41 mm)	
Service limit	2.034 in (51.66 mm)	
Valve-to-guide clearance		
Pre-Evolution engine	**Intake**	**Exhaust**
1970 through early 1981		
Standard	0.002 to 0.004 in (0.051 to 0.102 mm)	0.0035 to 0.0055 in (0.089 to 0.140 mm)
Service limit	0.002 to 0.006 in (0.051 to 0.152 mm)	0.0035 to 0.0075 in (0.089 to 0.191 mm)
Late 1981-on (with valve guide seal)		
Standard	0.0009 to 0.0026 in (0.023 to 0.066 mm)	0.0014 to 0.0031 in (0.036 to 0.079 mm)
Service limit	0.0009 to 0.0035 in (0.023 to 0.089 mm)	0.0014 to 0.0040 in (0.036 to 0.102 mm)
Evolution engine		
Standard	0.0008 to 0.0026 in (0.020 to 0.066 mm)	0.0015 to 0.0033 in (0.038 to 0.084 mm)
Service limit	0.0035 in (0.089 mm)	0.0040 in (0.102 mm)
Valve stem taper limit	0.0015 in (0.038 mm)	
Valve stem face (or tip) eccentricity limit	0.002 in (0.051 mm)	
Valve spring free length		
Pre-Evolution engine	**Inner**	**Outer**
1970 through early 1982	1.30 to 1.36 in (33 to 34.5 mm)	1.90 to 1.96 in (48 to 50 mm)
Late 1982-on	1.50 to 1.56 in (38 to 40 mm)	1.72 to 1.78 in (44 to 45 mm)
Evolution engine	1.926 to 1.996 in (48.920 to 50.698 mm)	2.105 to 2.177 in (53.467 to 55.296 mm)
Valve spring tension/length (late 1978-on)		
Pre-Evolution engine		
Late 1978 through early 1982		
Open	69 to 81 lbs @ 51/64 in (31 to 37 Kg @ 20.2 mm)	179 to 195 lbs @ 15/16 in (81 to 89 Kg @ 24 mm)
Closed	20 to 26 lbs @ 1-3/16 in (9 to 12 Kg @ 30 mm)	104 to 120 @ 1-3/8 in (47 to 54 Kg @ 35 mm)
Late 1982-on		
Open	77 to 87 lbs @ 3/4 in (35 to 40 Kg @ 19 mm)	182 to 204 lbs @ 15/16 in (83 to 93 Kg @ 24 mm)
Closed	32 to 40 lbs @ 1-3/16 in (15 to 18 Kg @ 30 mm)	76 to 94 lbs @ 1-3/8 in (35 to 43 Kg @ 35 mm)
Evolution engine		
Open	98 to 112 lbs @ 1.107 to 1.213 in (44 to 51 Kg @ 28 to 31 mm)	183 to 207 lbs @ 1.282 to 1.378 in (83 to 94 Kg @ 33 to 35 mm)
Closed	38 to 49 lbs @ 1.577 to 1.683 in (17 to 22 Kg @ 40 to 43 mm)	72 to 92 lbs @ 1.751 to 1.848 in (33 to 42 Kg @ 44 to 47 mm)

Rocker arms

Pre-Evolution engine	
Shaft-to-rocker arm bushing clearance	
Standard	0.0005 to 0.002 in (0.013 to 0.051 mm)
Service limit	0.0005 to 0.0035 in (0.013 to 0.089 mm)
End clearance	0.004 to 0.025 in (0.102 to 0.635 mm)
Evolution engine	
Shaft-to-rocker arm bushing clearance	
Standard	0.0005 to 0.002 in (0.013 to 0.051 mm)
Service limit	0.0035 in (0.089 mm)
Rocker arm-to-bushing clearance	0.002 to 0.004 in (0.051 to 0.107 mm)
Shaft-to-cover bore clearance	
Standard	0.0007 to 0.0022 in (0.018 to 0.056 mm)
Service limit	0.0035 in (0.089 mm)
End clearance	
Standard	0.003 to 0.013 in (0.076 to 0.330 mm)
Service limit	0.025 in (0.635 mm)

Cylinder head warpage limit (Evolution engine)	0.006 in (0.152 mm)

Cylinders (Evolution engine)

Bore	
Diameter (standard)	3.501 in (88.93 mm)
Taper limit	0.002 in (0.05 mm)
Out-of-round limit	0.003 in (0.08 mm)

Cylinders (Evolution engine) continued

Gasket surface warpage limit	
Top (cylinder head)	0.006 in (0.15 mm)
Base	0.008 in (0.20 mm)

Pistons and rings

Piston-to-bore clearance	
Pre-Evolution engine	
Standard	
1970 through early 1978	0.001 to 0.002 in (0.025 to 0.050 mm)
Late 1978-on	0.002 to 0.0025 in (0.050 to 0.063 mm)
Service limit	0.002 to 0.005 in (0.050 to 0.127 mm)
Evolution engine	
Standard	
KSG piston (late 1985-on)	0.00075 to 0.00175 in (0.019 to 0.044 mm)
All others	0.00055 to 0.00165 in (0.014 to 0.042 mm)
Service limit	0.0053 in (0.135 mm)
Compression ring side clearance	
Pre-Evolution engine	
Standard	0.004 to 0.005 in (0.102 to 0.127 mm)
Service limit	0.004 to 0.006 in (0.102 to 0.152 mm)
Evolution engine	
Standard	
Top ring	0.002 to 0.0045 in (0.050 to 0.114 mm)
Second ring	0.0016 to 0.0041 in (0.040 to 0.104 mm)
Service limit (both)	0.006 in (0.152 mm)
Oil control ring side clearance	
Pre-Evolution engine	
Standard	0.003 to 0.005 in (0.076 to 0.127 mm)
Service limit	0.003 to 0.006 in (0.076 to 0.152 mm)
Evolution engine	
Standard	0.0016 to 0.0076 in (0.041 to 0.193 mm)
Service limit	0.008 in (0.20 mm)
Ring end gap	
Compression rings	
Pre-Evolution engine	
Standard	0.010 to 0.020 in (0.254 to 0.508 mm)
Service limit	0.010 to 0.031 in (0.254 to 0.787 mm)
Evolution engine	
Standard	0.007 to 0.020 in (0.178 to 0.508 mm)
Service limit	0.030 in (0.762 mm)
Oil ring rail	
Pre-Evolution engine	0.010 to 0.045 in (0.254 to 1.143 mm)
Evolution engine	0.009 to 0.065 in (0.229 to 1.651 mm)

2

Connecting rods

Piston pin-to-connecting rod clearance	
Pre-Evolution engine	
Standard	0.0008 to 0.0012 in (0.020 to 0.030 mm)
Service limit	0.0008 to 0.0020 in (0.020 to 0.051 mm)
Evolution engine	
Standard	0.0003 to 0.0007 in (0.008 to 0.018 mm)
Service limit	0.001 in (0.025 mm)
Big end side play	0.005 to 0.030 in (0.127 to 0.762 mm)
Big end radial clearance	
Pre-Evolution engine	
Standard	0.001 to 0.0015 in (0.025 to 0.038 mm)
Service limit	0.001 to 0.0017 in (0.025 to 0.043 mm)
Evolution engine	0.0004 to 0.002 in (0.010 to 0.050 mm)

Tappets

Tappet-to-guide clearance	
Standard	0.001 to 0.002 in (0.025 to 0.050 mm)
Service limit	0.003 in (0.076 mm)
Guide-to-crankcase clearance (Evolution engine)	0.000 to 0.004 in (0 to 0.101 mm)
Roller radial fit	
Standard	0.0005 to 0.001 in (0.013 to 0.025 mm)
Service limit	0.0005 to 0.0015 in (0.013 to 0.038 mm)

Tappets (continued)

Roller end clearance	
Pre-Evolution engine	
Standard	0.008 to 0.010 in (0.203 to 0.254 mm)
Service limit	0.008 to 0.025 in (0.203 to 0.635 mm)
Evolution engine (service limit)	0.015 in (0.381 mm)

Gearcase

Breather gear end play	
Pre-Evolution engine	
1970 through early 1978	0.001 to 0.005 in (0.025 to 0.127 mm)
Late 1978 through 1983	0.001 to 0.016 in (0.025 to 0.406 mm)
Evolution engine	
Minimum	0.001 to 0.011 in (0.025 to 0.280 mm)
Maximum	0.016 in (0.406 mm)
Camshaft-to-bushing clearance	
Standard	0.0008 to 0.0018 in (0.020 to 0.046 mm)
Service limit	0.0030 in (0.076 mm)
Camshaft-to-bearing clearance	
Pre-Evolution engine	0.0005 to 0.0030 in (0.013 to 0.076 mm)
Evolution engine	
Standard	0.0005 to 0.0025 in (0.013 to 0.064 mm)
Service limit	0.005 in (0.127 mm)
Camshaft end play	
1970 through early 1978	0.001 to 0.005 in (0.025 to 0.127 mm)
Late 1978 through 1987	0.001 to 0.016 in (0.025 to 0.406 mm)
1988-on	0.001 to 0.050 in (no spacer) (0.025 to 1.270 mm)
Oil pump drive shaft-to-bushing clearance	
Pre-Evolution engine	
Standard	0.0008 to 0.0012 in (0.020 to 0.030 mm)
Service limit	0.0008 to 0.0025 in (0.020 to 0.064 mm)
Evolution engine	
Standard	0.0004 to 0.0025 in (0.010 to 0.064 mm)
Service limit	0.0035 in (0.089 mm)

Flywheel/crankshaft assembly

Maximum runout at rim	
Pre-Evolution engine	
1970 through early 1978	0.003 in (0.076 mm)
Late 1978-on	0.006 in (0.152 mm)
Evolution engine	0.015 in (0.381 mm)
Maximum runout at shaft	
Pre-Evolution engine	
1970 through early 1978	0.001 in (0.025 mm)
Late 1978-on	0.002 in (0.051 mm)
Evolution engine	0.003 in (0.076 mm)
Flywheel assembly end play	
Pre-Evolution engine	
1970 through 1981	0.001 to 0.006 in (0.025 to 0.152 mm)
1982-on	0.001 to 0.004 in (0.025 to 0.107 mm)
Evolution engine	0.001 to 0.006 in (0.025 to 0.152 mm)
Timing side (pinion shaft) bearing axial play	
Pre-Evolution engine	
Standard	0.0004 to 0.0008 in (0.010 to 0.020 mm)
Service limit	
1200 cc	0.0008 to 0.0020 in (0.020 to 0.051 mm)
1340 cc	0.0004 to 0.0020 in (0.010 to 0.051 mm)
Evolution engine	0.0002 to 0.0009 in (0.005 to 0.023 mm)
Gearcase cover bushing-to-mainshaft clearance	
Pre-Evolution engine	
Standard	0.0005 to 0.0012 in (0.013 to 0.030 mm)
Service limit	0.0005 to 0.0025 in (0.013 to 0.064 mm)
Evolution engine	
Standard	0.001 to 0.0025 in (0.025 to 0.064 mm)
Service limit	0.0035 in (0.089 mm)
Sprocket shaft bearing	
Inner race-to-shaft	0.0002 to 0.0015 in (0.005 to 0.038 mm)
Outer race-to-crankcase	0.0012 to 0.0032 in (0.030 to 0.081 mm)

Clutch

Type	
1970 through early 1984	Dry, multiple-disc with coil springs
Late 1984-on	Wet, multiple-disc with diaphragm spring

Dry clutch

Spring adjustment (from edge of spring collar-to-outer disc surface)	
Four-speed transmission	1-1/32 in (26.194 mm)
Five-speed transmission	7/8 to 1-1/32 in (22.225 to 26.194 mm)
Minimum friction disc lining thickness	1/32 in (0.794 mm)
Clutch spring	
Free length	1-45/64 in (26.103 mm)
Tension (late 1978-on)	30 to 38 lbs @ 1-1/4 in (14 to 17 Kg @ 32 mm)

Wet clutch

Steel plate warpage limit	
1984 through 1989	0.011 in (0.279 mm)
1990-on	0.006 in (0.152 mm)
Minimum steel plate thickness (Pre-1990 models only)	0.044 in (1.118 mm)
Minimum friction disc thickness	
Pre-1990 models	0.078 in (1.981 mm)
1990-on	0.661 in, 16.79 mm (stack of all eight friction discs – if measurement is less, replace friction discs and steel plates)

Primary drive

Primary chaincase vacuum (with vent hoses pinched closed)	25-inches (635 mm) of water minimum at 1500 rpm

Transmission

Four-speed

Mainshaft main drive gear	
End play	
1970 through 1981	0.0025 to 0.0135 in (0.064 to 0.343 mm)
1982 through 1984	0.010 to 0.025 in (0.254 to 0.635 mm)
1985-on	0.010 to 0.035 in (0.254 to 0.889 mm)
Bushing on mainshaft (loose)	
Standard	0.0018 to 0.0032 in (0.046 to 0.081 mm)
Service limit	0.004 in (0.102 mm)
Mainshaft	
Runout limit	0.003 in (0.076 mm)

First gear end bearing	Loose	Press fit
In housing		
Through 1984	0.0013 to 0.020 in (0.033 to 0.508 mm)	0.0001 in (0.002 mm)
1985-on	0.0013 to 0.020 in (0.033 to 0.508 mm)	0.0007 in (0.018 mm)
On shaft		
Through 1984	0.001 to 0.0015 in (0.025 to 0.038 mm)	0.0007 in (0.018 mm)
1985-on	0.0006 to 0.0015 in (0.015 to 0.038 mm)	0.0007 in (0.018 mm)
Housing in case		
Through 1984	0.0005 to 0.0009 in (0.013 to 0.023 mm)	0.0010 to 0.0015 in (0.025 to 0.038 mm)
1985-on	0.001 to 0.003 in (0.025 to 0.076 mm)	0.0015 in (0.038 mm)

Third gear (through 1984)	
End play	
Standard	0.000 to 0.017 in (0 to 0.432 mm)
Service limit	0.000 to 0.020 in (0 to 0.508 mm)
Gear on shaft (loose)	
Standard	0.0012 to 0.0023 in (0.030 to 0.058 mm)
Service limit	0.0012 to 0.0030 in (0.030 to 0.076 mm)
Bushing in gear	Press fit
Third gear (1985-on)	
End play	0.005 to 0.021 in (0.127 to 0.533 mm)
Gear on shaft (loose)	
Standard	0.0016 to 0.0021 in (0.041 to 0.053 mm)
Service limit	0.003 in (0.076 mm)
Shifter fork clutch gear spacing (late 1978-on)	0.100 to 0.110 in (2.540 to 2.794 mm)
Countershaft	
Runout limit	0.003 in (0.076 mm)

Transmission

Four-speed (continued)

Drive gear end bearing (loose)	
Standard	
1970 through early 1978	0.0005 to 0.0019 in (0.013 to 0.048 mm)
Late 1978-on	0.00025 to 0.002 in (0.006 to 0.051 mm)
Service limit	0.0005 to 0.002 in (0.013 to 0.051 mm)
First gear end bearing	
Standard	0.0005 to 0.0019 in (0.013 to 0.048 mm)
Service limit	0.0005 to 0.002 in (0.013 to 0.051 mm)
Gear end play	
Standard	
1970 through early 1978	0.007 to 0.012 in (0.178 to 0.305 mm)
Late 1978-on	0.004 to 0.012 in (0.102 to 0.305 mm)
Service limit	0.004 to 0.015 in (0.102 to 0.381 mm)
First gear	
Bushing on shaft (loose)	
Standard	0.000 to 0.0015 in (0 to 0.038 mm)
Service limit	0.000 to 0.0020 in (0 to 0.051 mm)
Bushing in gear (loose)	
Standard	0.0005 to 0.0025 in (0.013 to 0.064 mm)
Service limit	0.0005 to 0.0030 in (0.013 to 0.076 mm)
Second gear	
End play	
Standard	0.003 to 0.017 in (0.076 to 0.432 mm)
Service limit	0.003 to 0.020 in (0.076 to 0.508 mm)
Bushing on shaft (loose)	
Standard	0.000 to 0.0015 in (0 to 0.038 mm)
Service limit	0.000 to 0.002 in (0 to 0.051 mm)
Bushing in gear (loose)	
Standard	0.0005 to 0.002 in (0.013 to 0.051 mm)
Service limit	0.0005 to 0.0025 in (0.013 to 0.064 mm)
Shifter fork clutch gear spacing	
1970 through early 1978	
First and second	0.080 to 0.090 in (2.032 to 2.286 mm)
Third and fourth	0.100 to 0.110 in (2.540 to 2.794 mm)
Late 1978-on	0.080 to 0.090 in (2.032 to 2.286 mm)
Gear backlash	
Standard	0.003 to 0.006 in (0.076 to 0.152 mm)
Service limit	0.003 to 0.010 in (0.076 to 0.254 mm)
Shifter cam end play (through early 1979 models)	
Standard	0.005 to 0.0065 in (0.127 to 0.165 mm)
Service limit	0.005 to 0.007 in (0.127 to 0.178 mm)

Five-speed

Mainshaft	
Runout	0.000 to 0.003 in (0 to 0.076 mm)
End play	None
First gear end bearing (press fit)	
In housing	
Standard	0.0001 in (0.002 mm)
Service limit	0.0001 to 0.0005 in (0.002 to 0.013 mm)
On shaft	
Standard	0.0007 in (0.018 mm)
Service limit	0.0007 to 0.001 in (0.018 to 0.025 mm)
Housing in case	
Standard	0.0010 in (0.025 mm)
Service limit	0.0010 to 0.0015 in (0.025 to 0.038 mm)
First gear	
End play	0.0037 to 0.0339 in (0.094 to 0.861 mm)
Clearance	0.000 to 0.0080 in (0 to 0.203 mm)
Second gear	
End play	0.0037 to 0.0329 in (0.094 to 0.836 mm)
Clearance	0.000 to 0.0080 in (0 to 0.203 mm)
Third gear	
End play	0.0050 to 0.0420 in (0.127 to 1.067 mm)
Clearance	0.0003 to 0.0019 in (0.008 to 0.048 mm)
Fourth gear	
End play	0.0050 to 0.0310 in (0.127 to 0.787 mm)
Clearance	0.0003 to 0.0019 in (0.008 to 0.048 mm)

Specification		
Main drive gear (fifth)		
Bearing fit in transmission case	**Tight**	**Loose**
Through 1983	0.0009 in (0.023 mm)	0.0020 in (0.051 mm)
1984-on	0.0004 in (0.010 mm)	0.0001 in (0.002 mm)
Fit in bearing		
Through 1983	0.0020 in (0.051 mm)	0.0009 in (0.023 mm)
1984-on	0.0009 in (0.023 mm)	0.0001 in (0.002 mm)
Fit on mainshaft	0.0001 to 0.0009 in (0.002 to 0.023 mm)	
End play	None	
Shifter dog clearances	**Minimum**	**Maximum**
Second-fifth		
Through 1983	0.025 in (0.635 mm)	0.139 in (3.531 mm)
1984-on	0.035 in (0.899 mm)	0.139 in (3.531 mm)
Second-third		
Through 1983	0.021 in (0.533 mm)	0.164 in (4.166 mm)
1984-on	0.035 in (0.899 mm)	0.164 in (4.166 mm)
First-fourth		
Through 1983	0.030 in (0.762 mm)	0.152 in (3.861 mm)
1984-on	0.035 in (0.899 mm)	0.152 in (3.861 mm)
First-third (1984-on only)	0.035 in (0.899 mm)	0.157 in (3.988 mm)
Side door bearing-to-side door	0.0001 to 0.0014 in (0.002 to 0.036 mm)	
Side door bearing-to-mainshaft		
Tight	0.0007 in (0.178 mm)	
Loose	0.0001 in (0.002 mm)	
Countershaft		
Runout	0.000 to 0.003 in (0 to 0.076 mm)	
End play	None	
First gear		
End play	0.0050 to 0.0039 in (0.127 to 0.099 mm)	
Clearance	0.0003 to 0.0019 in (0.008 to 0.048 mm)	
Second gear		
End play	0.0050 to 0.0440 in (0.127 to 1.118 mm)	
Clearance	0.0003 to 0.0019 in (0.008 to 0.048 mm)	
Third gear		
End play	0.0037 to 0.0329 in (0.094 to 0.836 mm)	
Clearance	0.0000 to 0.0080 in (0 to 0.203 mm)	
Fourth gear		
End play	0.0050 to 0.0390 in (0.127 to 0.991 mm)	
Clearance	0.0000 to 0.0080 in (0 to 0.203 mm)	
Fifth gear		
End play	0.0050 to 0.0440 in (0.127 to 1.118 mm)	
Clearance	0.0000 to 0.0080 in (0 to 0.203 mm)	
Shifter clutch clearance		
Second-third	**Minimum**	**Maximum**
Through 1983	0.019 in (0.483 mm)	0.160 in (4.064 mm)
1984-on	0.035 in (0.889 mm)	0.164 in (4.166 mm)
First-third		
Through 1983	0.021 in (0.533 mm)	0.157 in (3.988 mm)
1984-on	0.035 in (0.889 mm)	0.157 in (3.988 mm)
Side door bearing-to-side door	0.0014 to 0.0001 in (0.036 to 0.002 mm)	
Side door bearing-to-countershaft		
Tight	0.0007 in (0.178 mm)	
Loose	0.0001 in (0.002 mm)	
Shifter cam assembly end play	0.001 to 0.004 in (0.025 to 0.102 mm)	
Transmission case outer surface-to-middle of center cam groove (through 1983 only)	3.043 in (77.29 mm)	
Outside machined surface of bearing support block-to-center cam groove nearest surface	1.992 to 2.002 in (50.59 to 50.85 mm)	
Shifter forks		
Taper	0.000 to 0.020 in (0 to 0.508 mm)	
End play		
Shifter fork-to-cam groove	0.0017 to 0.0019 in (0.043 to 0.048 mm)	
Shifter fork-to-gear groove	0.0010 to 0.0110 in (0.025 to 0.279 mm)	

Torque specifications

	Ft-lbs (unless otherwise indicated)	Nm
Primary chaincase-to-engine bolts		
FLT/FXR models through 1992	16 to 18	22 to 24
All others	18 to 22	24 to 30

2

Torque specifications (continued)

	Ft-lbs (unless otherwise indicated)	Nm
Primary chaincase-to-transmission		
1984 through 1992 FLT/FXR models		
5/16-inch bolts	13 to 16	18 to 22
3/8-inch bolts	21 to 27	28 to 37
All others		
Nuts	30 to 35	41 to 47
Bolts	18 to 22	24 to 30
Upper engine mounting bracket-to-head nuts		
Evolution engine	22 to 28	30 to 38
All others	35 to 40	47 to 54
Engine mount bolts/nuts	35 to 38	47 to 52
Rocker arm cover		
Pre-Evolution engine	12 to 15	16 to 20
Evolution engine		
1/4-inch bolts	10 to 13	14 to 18
5/16-inch bolts	15 to 18	20 to 24
Rocker arm shaft (acorn) locknut	12 to 18	16 to 24
Rocker arm shaft end cap (1980 and 1981 only)	8 to 12	11 to 16
Cylinder head bolts		
Pre-Evolution engine	55 to 75	75 to 102
Evolution engine (1984 FXST models only)		
Step 1	7 to 9	9 to 12
Step 2	15 to 17	20 to 23
Step 3	24 to 26	33 to 35
Evolution engine (all other models)		
Step 1		
Bolt 1 only	7	9
All others	7 to 9	9 to 12
Step 2	12 to 14	16 to 19
Step 3	Tighten an additional 90-degrees (1/4-turn)	
Cylinder barrel base nuts	32 to 40	43 to 54
Oil pump cover nuts/bolts		
Pre-Evolution engine		
Plastic gasket	45 to 50 in-lbs	5 to 6
White paper gasket	50 to 60 in-lbs	6 to 7
Black paper gasket	90 to 120 in-lbs	10 to 14
Evolution engine	90 to 120 in-lbs	10 to 14
Tappet guide bolts		
Pre-Evolution engine		
1970 through early 1978	10	14
Late 1978-on	65 to 105 in-lbs	7 to 12
Evolution engine		
Through 1990	90 to 120 in-lbs	10 to 14
1991-on	12 to 15	16 to 20
Crankshaft timing (pinion) gear nut	35 to 45	47 to 61
Flywheel sprocket shaft nut		
1970 through early 1978	400	540
Late 1978 through early 1981	300 to 440	400 to 596
Late 1981-on	290 to 320	393 to 434
Pinion shaft nut		
Late 1978 through early 1981	120 to 160	163 to 217
Late 1981-on	140 to 170	190 to 231
Crank pin nut		
1980 and early 1981	150 to 250	203 to 339
Late 1981-on	180 to 210	244 to 285
Crankcase stud nuts		
1970 through early 1978	22 to 26	30 to 35
Late 1978 through 1983	12 to 15	16 to 20
Crankcase bolts/nuts		
Pre-Evolution engine	22 to 26	30 to 35
Evolution engine		
Through 1990	15 to 19	20 to 26
1991-on (see text)	15 to 17	20 to 23
Gearcase cover screws	90 to 120 in-lbs	10 to 14
Tappet screen plug		
Pre-Evolution engine	90 to 160 in-lbs	10 to 18
Evolution engine	90 to 120 in-lbs	10 to 14

	Ft-lbs (unless otherwise indicated)	Nm
Primary chaincase cover screws		
1980 and later FLT/FXR models	7 to 9	9 to 12
1985 and later FX/Softail models	9 to 10	12 to 14
All others	18 to 22	24 to 30
1970 through early 1978 models only		
Kickstarter lever pinch bolt	25 to 30	34 to 41
Starter cover	13 to 16	18 to 22
Starter crank gear nut	50 to 60	68 to 81
Mainshaft bearing retaining plate	6 to 9	8 to 12
Four-speed models only		
Mainshaft ball bearing nut	50 to 60	68 to 81
Countershaft nut	55 to 65	75 to 88
Shifter fork nut		
1970 through early 1978	25	34
Late 1978-on	10 to 12	14 to 16
Shifter clutch nut (kickstart models only)		
1970 through 1975	50 to 60	68 to 81
1976-on	34 to 42	46 to 57
Mounting plate-to-transmission		
1970 through early 1978	18 to 22	24 to 30
Late 1978-on		
Nut	21 to 27	28 to 37
Bolt	30 to 33	41 to 45
Top cover		
Screws (through early 1979)	80 to 110	108 to 149
Bolts (late 1979-on)	13 to 16	18 to 22
Frame-to-transmission bolts (1985-on)	21 to 27	28 to 37
Clutch hub nut		
Through 1984	50 to 60	68 to 81
1985-on	35 to 50	47 to 68
Transmission sprocket nut		
1970 through early 1978	140 to 150	190 to 203
Late 1978 through early 1983	105 to 120	142 to 163
Late 1983-on	80 to 90	108 to 122
Sprocket nut lock screw (1985-on)	50 to 60 in-lbs	6 to 7
Compensating sprocket nut	80 to 100	108 to 137
Drain plug	15	20
Five-speed models only		
Starter drive-to-chaincase	10 to 12	14 to 16
Mounting bolts	35 to 38	47 to 52
Mounting bracket-to-transmission	13 to 16	18 to 22
Side door mounting screws	13 to 16	18 to 22
Shifter arm screw	18 to 22	24 to 30
Shifter arm adjusting screw locknut	20 to 24	27 to 33
Countershaft-to-side door	27 to 33	37 to 45
Mainshaft-to-side door	27 to 33	37 to 45
Compensating sprocket nut	150 to 165	203 to 224
Clutch hub nut		
Through 1989	50 to 60	68 to 81
1990-on	70 to 80	95 to 108
Side cover-to-side door		
Through 1990	7 to 9	9 to 12
1991-on	10 to 12	14 to 16
Support block	7 to 9	9 to 12
Transmission sprocket nut		
Through 1983	90 to 110	122 to 150
1984-on	110 to 120	150 to 163
Transmission sprocket Allen-head bolt – through 1991	50 to 60 in-lbs	6 to 7
Transmission sprocket lockplate bolts – 1992-on	7 to 9	9 to 12
Transmission drain plug		
FLT through 1992 and FXR	7	9
FLT 1993-on	14 to 30	19 to 40
FX/Softails and Dynas	0.16 to 0.18 in (4.0 to 4.5 mm) above surface of housing	
Oil pan-to-transmission bolts	7 to 9	9 to 12

1 General information

The engine used in the Harley-Davidson Glides is a large capacity, 45-degree V-twin, mounted inline with the frame. In this Chapter, the early engine (through 1984) with cast-iron heads and cylinders is referred to as the *pre-Evolution* engine, while the later engine (1985-on) with aluminum heads and cylinders, as well as redesigned rocker boxes, is called the *Evolution* engine – the term used by the factory. Actually, the Evolution engine was first installed in 1984 FXST models and the pre-Evolution engine was used in all other 1984 models, so there was some overlap of engine types and model years.

Aluminum alloy castings are used throughout the engine, with the above-mentioned exception of the cylinder barrels and heads on pre-Evolution engines, which were cast-iron. The engine has overhead valves actuated by hydraulic tappets and pushrods from the camshaft mounted in the gearcase at the right-hand side of the engine. The tappets on pre-Evolution engines are two-piece. The Evolution engine has one-piece tappets (with larger rollers than the pre-Evolution engine) and hollow, one-piece pushrods that feed oil to the rocker arm shafts, eliminating the need for external oil lines. A gear-type oil pump, driven off the crankshaft, provides pressurized oil to the main engine components.

An electric starter is standard on most models. The early FX models were equipped with a kickstarter, but they can be equipped with an electric starter as well.

A four-speed transmission was offered in all years covered by this manual and a five-speed was available beginning in 1980. A multi-plate, dry clutch with coil springs was used on all 1970 through early 1984 models, while a wet multi-plate clutch (with a large diaphragm spring) was introduced in late 1984 and modified for the 1990 model year. Access to the transmission is available after the primary drive components and clutch have been removed.

2 Repair operations possible with the engine in the frame

It's not necessary to remove the engine from the frame unless the crankcases have to be separated to gain access to the crankshaft, connecting rods or bearings. If only the cylinder heads and related valvetrain components (including the camshaft and tappets), or the pistons, rings or barrels require attention, the work can be done with the engine in the frame, provided the various disassembly procedures outlined in this Chapter are modified slightly. However, if major work or a complete overhaul is necessary, remove the engine from the frame.

Operations that can be completed with the engine in the frame include removal and installation of the . . .

Alternator
Starter motor
Transmission and related components
Clutch and primary drive components
Cylinder heads and cylinder barrels
Camshaft and drive gear
Oil pump and related components
Tappets and pushrods
Rocker arms and shafts
Valves, springs, seals and guides
Rocker box gaskets
Cylinder head and base gaskets

This does not, however, apply to the main crankcase assembly, which is split vertically and cannot be separated until the engine is out of the frame. Conversely, the crankcase assembly must be bolted together before it can be reinstalled in the frame.

3 Major engine repair – general information

1 It's not always easy to determine when or if an engine should be completely overhauled, as a number of factors must be considered.

2 High mileage isn't necessarily an indication an overhaul is needed, while low mileage, on the other hand, doesn't preclude the need for an overhaul. Frequency of servicing is probably the single most important consideration. An engine that has regular and frequent oil and filter changes, as well as other required maintenance, will most likely give many miles of reliable service. Conversely, a neglected engine, or one that hasn't been broken in properly, may require an overhaul very early in its life.

3 Exhaust smoke and excessive oil consumption are both indications that piston rings and/or valve guides need attention. Make sure oil leaks aren't responsible before deciding the rings or guides are bad. Refer to Chapter 1 and perform a cylinder compression check to determine for certain the nature and extent of the work required.

4 If the engine is making obvious knocking or rumbling noises, the engine bearings are probably at fault. Main bearings are generally more durable than rod bearings or bushings, so begin with the latter when trying to make a diagnosis. Certain knocking noises may be caused by piston slap as well as bearings. In most cases, valves, piston rings, piston pins, bushings and bearings will all need attention at the same time.

5 Loss of power, rough running, excessive valve train noise and high fuel consumption rates may also point to the need for an overhaul, especially if they're all present at the same time. If a complete tune-up doesn't remedy the situation, major mechanical work is the only solution. To diagnose and repair noisy tappets and valvetrain components, first check the oil pressure with the engine running (2000 rpm) at normal operating temperature. If it's above 50 or below 5 psi, check the oil pump, oil hoses and crankcase passages for restrictions. If oil is reaching the tappets, they may be dirty or defective (see Section 20). Inspect the pushrods and tappet blocks to make sure everything is installed correctly and look for unusual wear patterns. Check the camshaft lobes for unusual or excessive wear (see Section 21). Make sure the camshaft and pinion gears are correctly matched and not worn or damaged. Inspect the rocker arms and shafts for wear and binding. Check the valves to make sure they fit correctly in the guides and look for scuffing and galling on the stems. The valve seats must be tight in the head and the valves must seat properly.

6 An engine overhaul generally involves restoring the internal parts to the specifications of a new engine. During an overhaul the piston rings are replaced and the cylinders are bored and/or honed. If a rebore is done, then new pistons are also required. All major engine bearings are generally replaced with new ones and, if necessary, the crankshaft is also replaced. Generally the valves are serviced as well, since they're usually in less than perfect condition at this point. While the engine is being overhauled, other components such as the carburetor and the starter motor can be rebuilt also. The end result should be a like-new engine that will give as many trouble-free miles as the original.

7 Before beginning the engine overhaul, read through all of the related procedures to familiarize yourself with the scope and requirements of the job. Overhauling an engine isn't all that difficult, but it is time consuming. Plan on the motorcycle being tied up for a minimum of four or five weeks. Check on the availability of parts and make sure any necessary special tools, equipment and supplies are obtained in advance.

8 Most work can be done with typical shop hand tools, although a number of precision measuring tools are required for inspecting parts to determine if they must be replaced. Often a dealer service department or motorcycle repair shop will handle the inspection of parts and offer advice concerning reconditioning and replacement. If special tools or equipment are needed for a particular job, let a dealer or repair shop handle it. As a general rule, time is the primary cost of an overhaul so it doesn't pay to install worn or substandard parts.

9 As a final note, to ensure maximum life and minimum trouble from a rebuilt engine, everything must be assembled with care in a spotlessly clean environment.

4 Engine removal

Refer to illustrations 4.15a, 4.15b, 4.17, 4.24, 4.25, 4.26, 4.30, 4.34a, 4.34b and 4.35

1 Because of its size and weight, at least two people must be available for this task, preferably with a third person to steady the cycle while the

4.15a Loosen the set screws to remove the throttle cable . . .

4.15b . . . and the choke cable (early model shown)

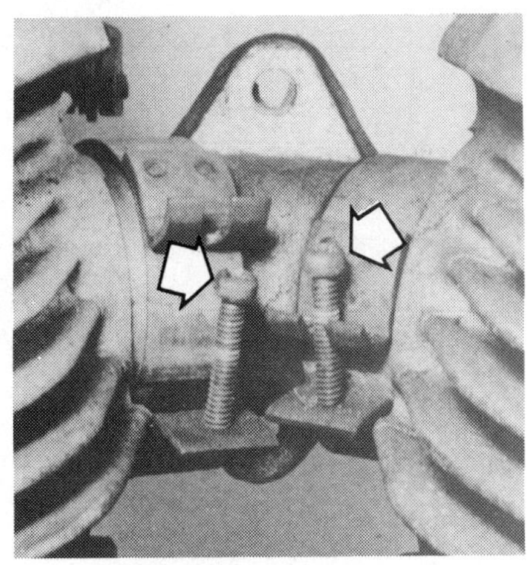

4.17 Where applicable, loosen the two clamp screws and pull the intake manifold out of the heads

engine is lifted in and out. An engine hoist is recommended, if available. If the cylinder heads and barrels are removed first, one person may manage to lift the engine out of the frame.

2 Since the motorcycle isn't normally equipped with a centerstand, the initial disassembly should be done with the machine resting on the side-stand, unless some form of rigid stand is available (like the ones used in many repair shops). Since the lower frame tubes are close together, the machine won't be too stable if it's supported on them. If the machine is on the sidestand, be sure it's on level ground and the wheels are chocked so it won't roll.

3 Disconnect the negative cable from the battery.

4 Turn the fuel valve to the Off position, then disconnect the fuel line from the carburetor. Drain the fuel from the fuel tank(s) into an approved gasoline container.

5 Refer to Chapter 3 and remove the fuel tank. **Note:** *On FLT and FXR models (1985-on), rocker box, tappet block and camshaft repairs can be accomplished at this point without continuing the engine removal procedure.*

6 On models so equipped, detach the buddy seat by removing the pivot bolt from the front of the seat frame.

7 Place a drain pan under the crankcase and remove the primary chaincase (wet clutch models) and transmission drain plugs, allowing the oil to drain. This task is easier if the engine is warm so that the oil flows more freely.

8 Drain the engine oil tank by removing the drain plug and directing the oil via a metal chute into a container. On 1993 and later FLT and 1991 and later Dyna models (which have an oil pan on the bottom of the transmission), remove the drain plug and allow the oil to drain into the pan along with the transmission and primary oil.

9 When all of the oil has drained, replace and tighten all the drain plugs after checking the sealing washers to make sure they're clean and in good condition. Note that the chaincase plug is magnetic; any metal particles that have collected should be cleaned off. Tighten the drain plugs now, otherwise this point may be overlooked during reassembly.

10 Remove the pivot pin securing the footboards (if equipped), by unscrewing the thin nut and driving the pin out. Remove the bracket supporting the rear of the right-hand footboard. It's held in place with a single bolt.

11 On models with a kickstarter, loosen the pinch bolt and pull the kickstarter off the shaft.

12 The upper cylinder head bracket must be removed from the engine (on models fitted with a stabilizer bar **don't** loosen the stabilizer jam nuts). Pay attention to the installed locations of the washers and be sure to reinstall them in the same positions during installation. It may be helpful to draw a sketch to refresh your memory if the motorcycle will be apart for an extended period of time. On models where the bracket also retains the ignition switch, tape the switch/bracket to the upper frame tube so disconnection of the wires is unnecessary.

13 Label the spark plug wires and disconnect them from the spark plugs. Unscrew the spark plugs from the cylinder heads.

14 Remove the air cleaner assembly from the engine. Begin by unscrewing the bolt(s) securing the cover to the assembly. Lift the filter element out, then remove the bolts securing the air cleaner backplate to the mounting bracket(s) and the carburetor. On some models, the bolts to the carburetor are safety wired together. On these models, you'll have to remove the safety wire first. Remove the backplate support bracket. On some models, rubber hoses are attached to the backplate – remove them as well.

15 Disconnect the throttle cable(s) and the choke control cable from the carburetor. On early models, the cables can be disconnected by loosening the set screws at the carburetor **(see illustrations)**. Note that they have specially protected ends to prevent the cable from fraying or spreading when pressure is applied. On later models, the cables must be disconnected at the throttle first. **Note:** *On models with an Evolution engine, detach the VOES (Vacuum Operated Electrical Switch) hose from the carburetor.*

16 Remove the bolts or nuts securing the carburetor to the intake manifold and separate the carburetor from the engine. It may be easier to simply loosen the hose clamps and pull the intake manifold out of both compliance fittings to remove the carburetor and manifold as an assembly. **Note:** *On 1990 and later models, pull the carburetor out of the manifold sealing ring and unbolt the manifold from the heads.* The float bowl will probably be full of gasoline so be sure to drain it. Store the carburetor in a safe place, away from sparks and open flames.

17 Loosen the clamps securing the intake manifold to the cylinder head compliance fittings and pull the manifold off the engine, if not already done **(see illustration)**. Stuff clean rags into the intake ports in the heads to prevent dirt or debris from entering.

18 Remove the single Allen-head bolt or nuts that secure the exhaust pipe flanges at each exhaust port. If equipped, remove the chrome pipe guards by completely unscrewing the hidden clamps. On Electra Glide models, the left-hand exhaust pipe is secured by a single bolt which passes into the clutch housing. Loosen all the muffler-to-pipe and pipe-to-pipe clamps. Disconnect the muffler either by removing the bracket bolts or by removing the muffler and the bracket from the frame as a complete unit. In some cases it may be possible to remove the exhaust system without first separating the exhaust pipes from the mufflers. A rubber hammer can be used to separate the pipes at the cylinder heads or at the pipe joints. If the joints have rusted together and defy removal in the normal manner, penetrating oil may help. Allow plenty of time for the oil to soak in before reattempting removal.

19 On 1985 and later FLT/FXR models, remove the large center bolt from the front engine mount and the bolt from the outer end of the front stabilizer. Remove both front engine mount bolts and detach the mounting plate and stabilizer assembly. **Note:** *With the external components removed from the engine to this extent, it's possible to remove the cylinder*

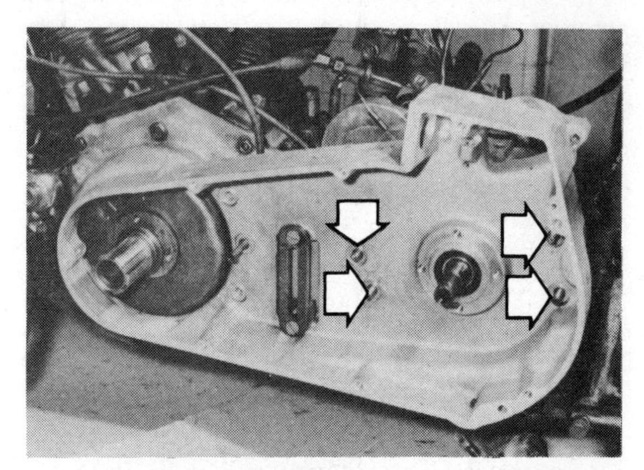

4.24 Remove the nuts from the inner primary chaincase-to-transmission studs (arrows) (some models have bolts)

4.25 Unplug the rectifier/regulator at the rubber plug on the crankcase

heads and barrels (see Section 6). If more extensive repairs are required, continue with the remaining engine removal steps. On 1985 and later FX/ Softail models, an access hole has been provided in the frame to remove the left rear rocker cover bolt. A rolled up paper tube should be inserted into the hole first and pushed down until it touches the engine to prevent dropping the bolt into the frame as it's removed. Also, you may have to compress the rear intake valve spring to provide enough clearance for removal of the lower rocker cover (use special tool no. HD-34641 to compress the spring). On some other models, the rear bolts in the rear cylinder rocker cover can't be removed separately – loosen them all the way, then detach the box from the head with the bolts in the holes.

20 On later FLT/FXR models (1984-on), remove the left footboard and rear bracket or the footrest brackets. On FXR models, remove the shift lever.

21 Remove the screws securing the primary chaincase cover and separate the cover from the engine. If the cover is stuck, don't pry it loose. Instead, tap around the edges with a soft-face hammer. If necessary, remove the shift lever from the shaft to allow the cover to be removed. The pedal is secured on a splined shaft with a pinch bolt. **Note:** *On FX/Softail models (1985-on), remove the compensating sprocket shaft nut so the sprocket can come off the shaft as the engine is removed. On FLT/FXR models, remove the adjuster bolt as well.*

22 Remove the clutch and sprocket assemblies as described in Section 35. Disconnect the final drive chain/belt as described in Chapter 5.

23 Remove the bolts securing the primary chaincase to the engine and gearbox, accessible from the front and rear face of the chaincase; they will be locked in place with tab washers or safety wire.

24 Loosen the four bolts or nuts attaching the inner primary chaincase to the transmission **(see illustration)**.

25 Unplug the rectifier/regulator (left-hand side of the engine) and the breaker point or sensor plate primary wires (right-hand side of the engine). The regulator/rectifier is connected to the alternator by a socket which is a push fit in the front, left-hand side of the crankcase. Carefully ease the socket out of position **(see illustration)**. Removal of the rectifier/regulator isn't necessary, but since it's in close proximity to the engine, it may be accidentally damaged. The unit is retained by four nuts and bolts.

26 Disconnect its wire(s), and unscrew the oil pressure switch from oil pump cover (early models) or crankcase (later models) **(see illustration)**. **Note**: *If the motorcycle has an Evolution engine, the oil lines must be disconnected from the pump and crankcase as well.*

27 On models with an automatic chain oiler, disconnect the chain oiler hose from the oil pump. Label the remaining oil lines and disconnect them from the oil pump or the back of the primary chaincase and oil tank. It may help to draw a sketch of the oil line connections to help during reassembly.

28 Remove the crankcase breather line from the oil pump.

29 The alternator rotor (magnet ring) must be removed on all early models except FLT and FXR. Refer to Chapter 7 for the removal procedure.

30 The rear brake master cylinder assembly on FL models should be disconnected from the frame and pivoted down, out of the way. On FX models, the right-hand footrest and the brake pedal assembly must be removed **(see illustration)**. **Note:** *If you're working on an FX model with an Evolution engine, that doesn't have forward foot controls, remove the right-hand footrest, the brake pedal and the master cylinder assembly.*

31 On late 1978 through 1983 four-speed models, the starter and housing must be removed along with the primary chaincase. Label the wires to the starter to assist in reassembly. Refer to Chapter 7, if necessary, for the

4.26 Remove the nut securing the wire to the oil pressure sending unit

4.30 Remove the right-hand footrest and bracket

4.34a The engine is retained by two bolts at the rear and . . .

4.34b . . . two bolts at the front

starter motor removal procedure.
32 Remove the speedometer drive cable from the crankcase on early models.
33 On models with an Evolution engine, detach the clutch cable from the engine bracket or detach the bracket from the engine.
34 Remove the two rear and the two front engine mounting bolts **(see illustrations)**. Make sure all electrical wires and control cables have been disconnected from the engine and are out of the way. **Note:** *On FXRT models, the front cylinder upper rocker cover must be removed, along with the two bolts from the lower fairing support bracket. Raise the right side of the fairing about one-inch and support it with wood blocks.*
35 The engine is very heavy and there's little room for maneuvering it out of the frame. An engine hoist would be very helpful, but it's not absolutely necessary **(see illustration)**. The engine should be lifted out of the frame to the right side.

4.35 Lift the engine out of the right side of the frame

5 Engine disassembly and reassembly – general information

Refer to illustration 5.2

1 Before beginning work on the engine, the external surfaces should be cleaned thoroughly. A motorcycle engine has very little protection from road grit and other foreign material, which sooner or later will find its way into the disassembled engine if this simple precaution isn't followed.
2 In addition to the precision measuring tools mentioned earlier, you will need a torque wrench, a valve spring compressor, oil line brushes **(see illustration)**, a motorcycle piston ring removal and installation tool and a piston ring compressor (low-profile type). Some new, clean engine oil, some engine assembly lube or moly-base grease and several types of Loctite will also be required.
3 An aerosol engine degreaser can be used effectively, especially if it's allowed to penetrate the film of grease and oil before it's washed away. When rinsing it off, make sure the water can't get into the internal areas of the engine – plug or cover all openings.
4 If the engine hasn't been disassembled before, either purchase or borrow an impact screwdriver. This will prevent damage to the screws used for engine assembly, most of which will be very tight. If an impact screwdriver isn't available, you can use a screwdriver bit attached to an extension and a sliding T-handle as a substitute.
5 Disassemble the engine on a clean surface and have a supply of clean, lint-free rags available.
6 Never use force to remove any stubborn part unless specific mention is made of it in the text. There's usually a good reason why a part is difficult to remove, often because the disassembly operation has been tackled in the wrong sequence.
7 Disassembly will be easier if a simple engine stand is constructed that will accept the engine mounting bolts. This arrangement will permit the engine to be clamped rigidly to the workbench, leaving both hands free for disassembly.

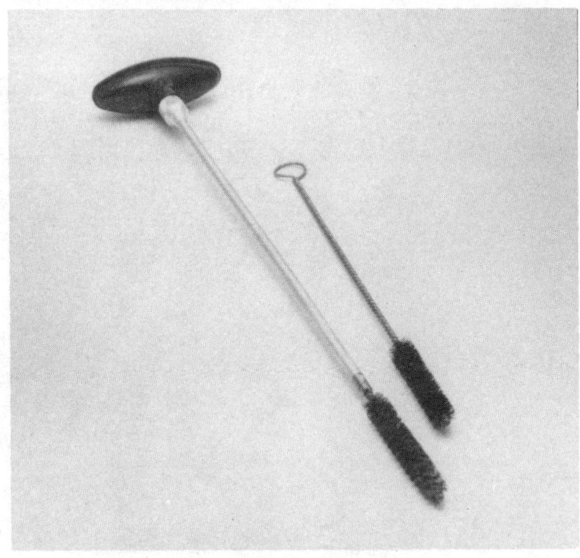

5.2 A selection of brushes is required for cleaning holes and passages in the engine components

8 When disassembling the engine, keep "mated" parts together (including gears, cylinders, pistons, etc. that have been in contact with each other during engine operation). These "mated" parts must be reused or replaced as an assembly.

6 Cylinder heads and barrels – removal

1 Begin by removing the intake manifold (if it's still in place). It's secured by two clamps. Loosen both clamp screws and pull the intake manifold out of the head fittings. Note that on early models there's an O-ring inside each port, to ensure an airtight joint.

Pre-Evolution engine

Refer to illustrations 6.2a, 6.2b, 6.3, 6.4a, 6.4b, 6.5 and 6.7

2 Remove the rocker boxes. Begin by unscrewing the fittings and removing the small diameter external oil lines, which are easily damaged **(see illustrations)**.

3 Each rocker box is secured by several nuts **(see illustration)**, which should be loosened in 1/4-turn increments. Make sure the valves of the cylinder involved are closed, to avoid stress on the casting from an open valve. Mark the rocker boxes so they can't be interchanged and remove them.

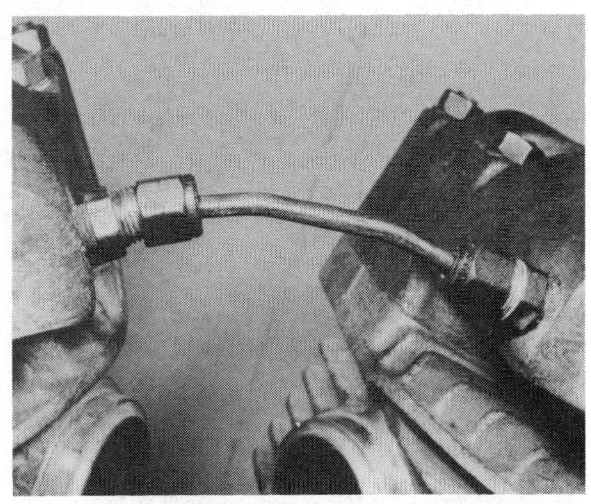

6.2a Detach the oil line between the rocker boxes and . . .

6.2b . . . the line between the crankcase and head (pre-Evolution engine)

1 Head steady bolt
2 Washer
3 Head steady bracket
4 Washer
5 Washer
6 Cotter pin
7 Castellated nut
8 Nut
9 End cap
10 O-ring
11 Rocker arm shaft
12 Bushing
13 Intake rocker arm
14 Exhaust rocker arm
15 Spacer
16 Rocker box
17 Sealing washer
18 Acorn nut
19 Nut
20 Plain washer
21 Union
22 Sleeve
23 Union nut
24 Crossover oil line
25 Spark plug Helicoil insert
26 Valve spring keepers
27 Retaining plate
28 Inner spring
29 Outer spring
30 Spring seat
31 Valve guide
32 O-ring
33 Cylinder head
34 Stud
35 Stud
36 Exhaust valve seat
37 Exhaust valve
38 Intake valve seat
39 Intake valve
40 Head gasket
41 Stud
42 Cylinder barrel
43 Plain washer
44 Cylinder head bolt
45 Cylinder base gasket
46 Cylinder barrel nut
47 Lock washer
48 Stud
49 Union
50 Sleeve
51 Union nut
52 Oil feed line
53 Cam follower
54 Follower roller foot
55 Gasket
56 Guide block
57 Bolt
58 Cork seal
59 Pushrod tube
60 Pushrod tube seal
61 Plain washer
62 Spring
63 Spring cover
64 Keeper plate
65 Pushrod adjuster
66 Locknut
67 Pushrod
68 Pushrod tube seal
69 Cylinder head gasket

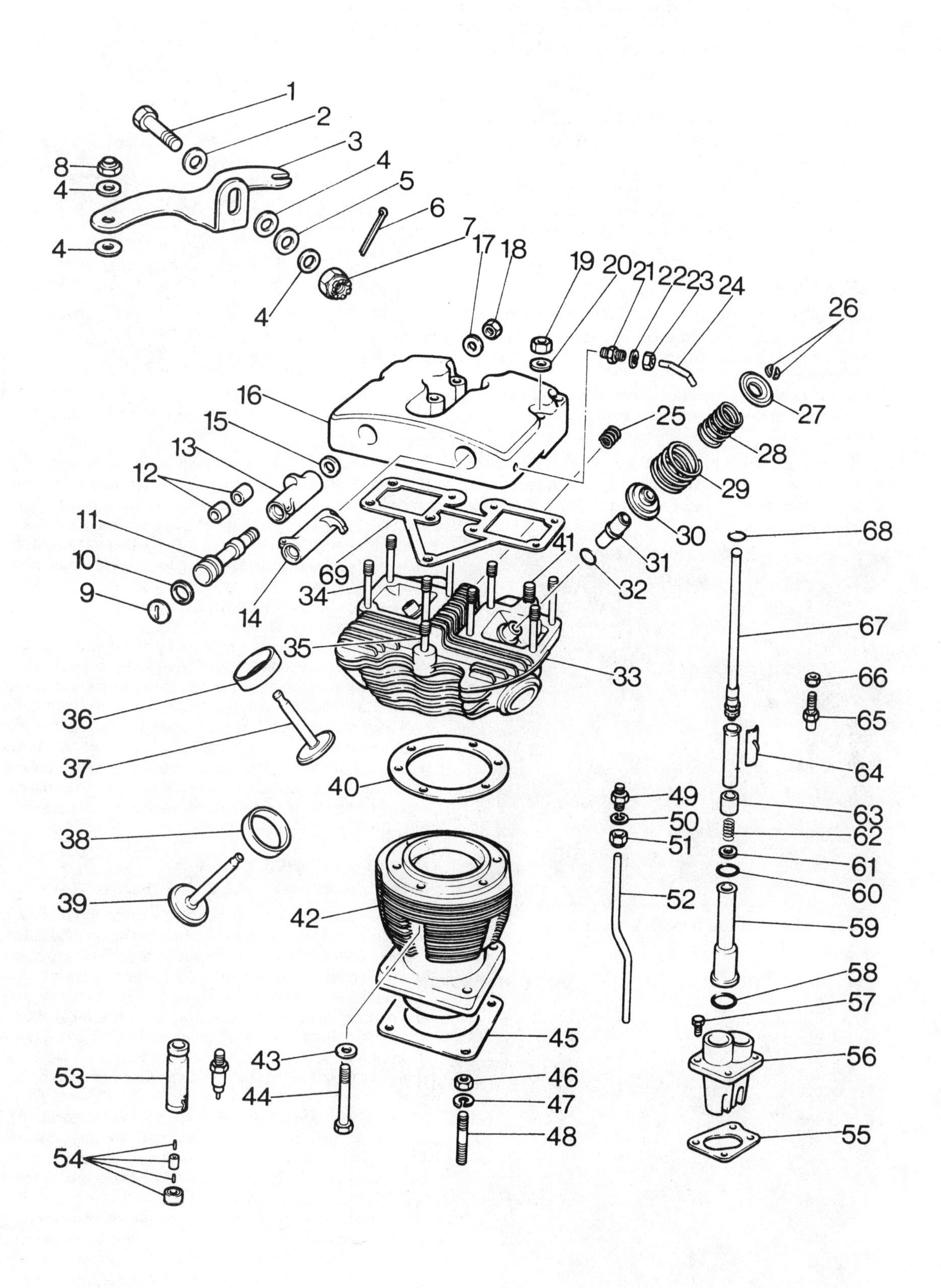

6.3 Cylinder head, barrel and rocker box components – exploded view (pre-Evolution engine)

2

6.4a Remove the pushrods and mark them so they can be returned to their original locations

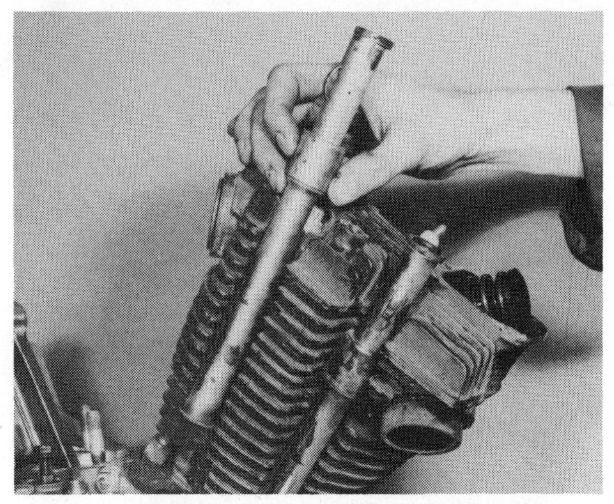

6.4b Once the rocker box has been detached, the pushrod tube can be removed as a complete unit

4 The pushrods should be lifted out of the tubes and clearly marked to make sure they're replaced in their original locations **(see illustration)**. The pushrod tubes can then be detached, each as a complete assembly **(see illustration)**.

6.5 The head bolts thread into the bottom of the cylinder head

6.7 Stuff clean rags into the crankcase opening before removing the barrel

5 Each cylinder head is retained by five bolts, which will be very tight **(see illustration)**. Loosen them in 1/4-turn increments, following a criss-cross pattern, until they can be removed by hand. If the cylinder head isn't free after the bolts have been removed, tap it lightly with a soft-face hammer in an attempt to break the seal. Don't attempt to pry it off, or fins will be broken. There's no point in marking the cylinder head, since they cannot be interchanged.

6 On late 1981 and early 1982 models, remove the clamp and oil hose from the fitting on each cylinder.

7 Each cylinder barrel is retained by four nuts around the base flange. When the nuts are removed, the cylinder barrel can be detached **(see illustration)**. Be sure to catch the piston as it emerges from the bore. If it's allowed to fall out, either the piston or the piston rings may be damaged. If only a top-end overhaul is required, pad the mouth of the crankcase with a clean rag as each barrel is removed. This will prevent particles of broken piston ring or other foreign matter from dropping into the crankcase as each piston emerges from its bore. Otherwise you would have to strip the engine completely in order to retrieve the broken parts.

Evolution engine

Refer to illustrations 6.8, 6.9, 6.10, 6.11 and 6.13

8 Push down on the front pushrod tube spring retainer **(see illustration)**, then remove the keeper and pull the upper tube out of the recess in the underside of the cylinder head (it's a good idea to start with the front head). Repeat the procedure for the remaining pushrod tube.

9 Remove the Allen-head screws and washers from the upper rocker arm cover, then detach the cover **(see illustration)**. If the engine is in the frame, the bolts on the rear cylinder head are difficult to reach – modify an Allen wrench so it'll fit into the bolt heads and still clear the frame. If the cover is stuck, tap it gently with a soft-face hammer to break the gasket seal.

10 Remove the middle rocker arm cover **(see illustration)**. Remove and discard all rocker cover gaskets and use new ones during reassembly.

11 Turn the crankshaft until both valves in the affected head are closed, then remove the large bolts retaining the lower rocker arm cover **(see illustration)**.

12 Carefully tap the rocker arm shafts out of the lower rocker arm cover (from the left side), then lift out the rocker arms and remove the pushrods and tubes (the pushrods must not be interchanged or turned end-for-end, so make sure they're marked!). Make sure the parts are marked or stored in marked containers – they must be returned to their original locations during reassembly.

13 Remove the lower rocker arm cover-to-cylinder head bolts **(see illustration)**. There are two Allen-head bolts and three hex-head bolts.

14 Detach the lower rocker arm cover and gaskets. Discard the gaskets and use new ones during reassembly.

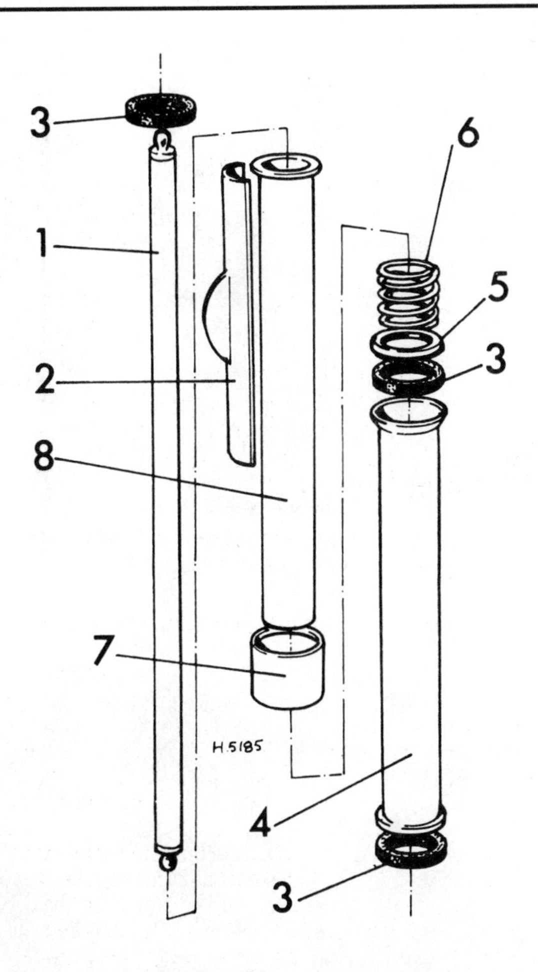

6.8 **Pushrod tube components – exploded view (both engines)**

1	*Pushrod*	5	*Washer*
2	*Keeper*	6	*Spring*
3	*O-ring/seal*	7	*Spring retainer*
4	*Lower tube*	8	*Upper tube*

6.9 Remove the four Allen-head screws (arrows) and detach the upper rocker arm cover, . . .

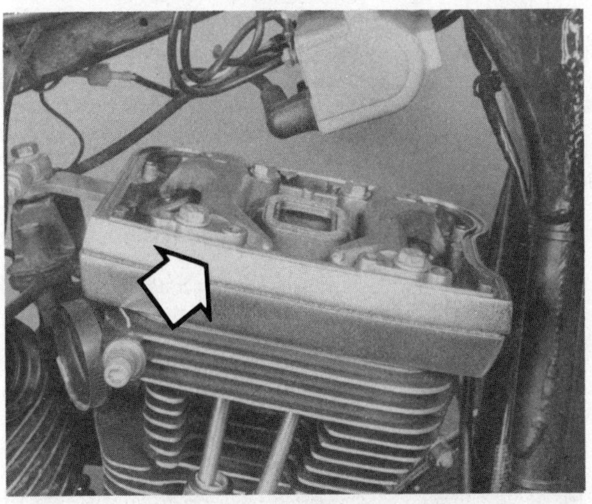

6.10 . . . then lift off the middle rocker arm cover and gasket (arrow) (Evolution engine)

6.11 Remove the four large bolts (arrows) (the two on the right side retain the rocker arm shafts), then make sure the valves are completely closed and drive out the shafts from the left side, which will release the rocker arms

6.13 Lift out the rocker arms (make sure they're marked so they can be reinstalled in their original locations), then remove the three hex-head bolts and two Allen-head bolts (arrows) and detach the lower rocker arm cover from the cylinder head

7.3 Don't let the tappets fall into the case as the guide is removed

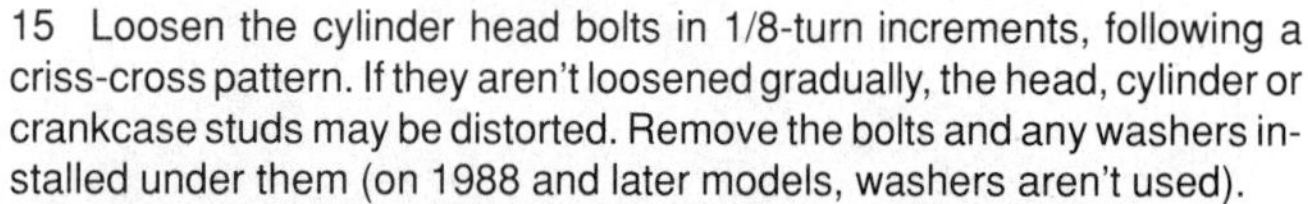

15 Loosen the cylinder head bolts in 1/8-turn increments, following a criss-cross pattern. If they aren't loosened gradually, the head, cylinder or crankcase studs may be distorted. Remove the bolts and any washers installed under them (on 1988 and later models, washers aren't used).

16 Carefully lift off the cylinder head, then remove the gasket and O-rings around the dowel sleeves. Discard the gasket and O-rings and use new ones during installation.

17 Mark the cylinder (front or rear), then raise it enough to place a clean shop rag under the piston. This will prevent debris (such as broken piston rings) from falling into the crankcase.

18 Remove the cylinder. Be very careful not to scratch or bend the crankcase studs and don't allow the piston to fall and be damaged. Once the cylinder is off, slip six-inch sections of rubber hose over the studs to protect them and the piston. If the studs are damaged in any way, they could fail during engine operation.

Both engines

19 Wear safety glasses when removing the piston pin circlips. Remove one circlip from each piston and discard it – circlips should never be re-used. To prevent the circlip from flying out, place your thumb over it as it's removed from the groove with an awl.

20 Press the pin out and remove the piston. If the piston pins are a tight fit, they can be freed by warming the piston crown to expand the metal. To do this, place a clean rag in boiling water, wring it out and wrap it around the piston crown. Mark each piston inside the skirt, to ensure it's installed facing the right direction in the correct cylinder during reassembly. **Note:** *The pistons used in the Evolution engine are very hard and should be handled with care. If they get scratched or gouged, they could score the cylinder(s) during engine operation.*

7 Tappets and guides – removal

Refer to illustration 7.3

Note: *The hydraulic tappets/cam followers are installed in pairs in two separate tappet guides, secured at the top of the case by four small bolts. On pre-Evolution engines, each tappet is a two-piece assembly (a tappet and separate cam follower) and they may fall apart as they're removed. Evolution engines have one-piece tappets.*

1 On late 1981 and early 1982 models, remove the clamp from the oil hose and pull the oil hose off the fitting.

2 Use a scribe and mark the position of the tappet guide on the case to ensure proper alignment during installation.

3 Remove the bolts securing the guide, then fashion a U-shaped wire from a large paper clip, engage the wire in the tappets to hold them in place and lift the tappets and guide out as an assembly **(see illustration)**. It may be necessary to tap around the edge of the guide with a soft-face hammer to break the gasket seal. Be careful not to drop the tappets out the bottom of the guide, into the case.

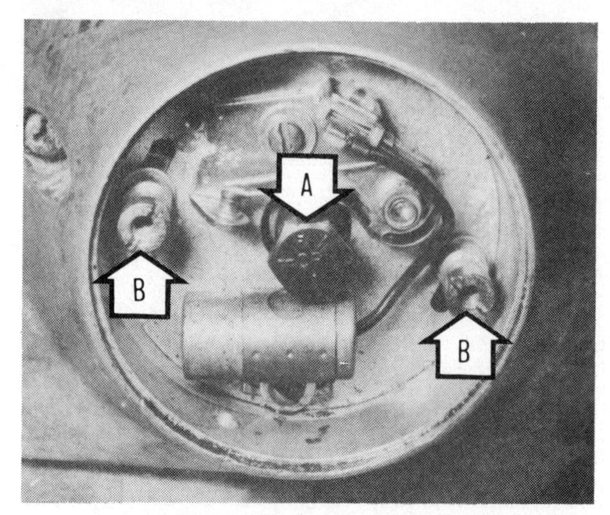

8.2 Remove the bolt (A) and the two screws (B), then lift out the contact breaker baseplate

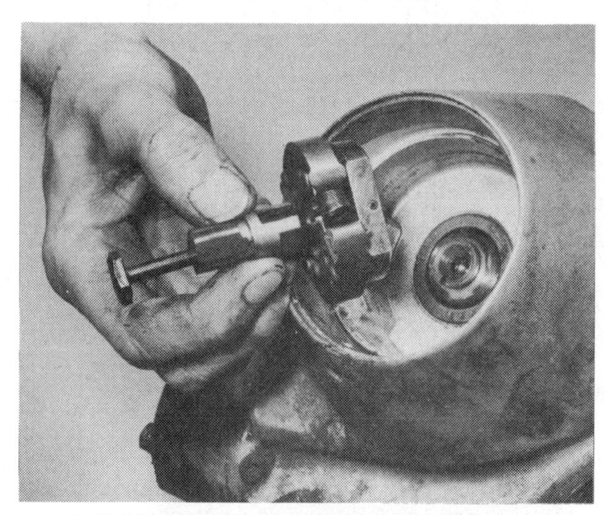

8.4 Remove the contact breaker point cam and the centrifugal advance unit by pulling them out

4 Label the components or store them so they will be reinstalled in the same locations.

5 Remove the pushrod tube seals and washers from the top of each tappet guide.

6 Remove the tappets from the bottom of the guide if necessary.

7 Remove all traces of the old gasket and sealant from the guide and crankcase.

8 Repeat the procedure for the remaining guide.

8 Ignition components – removal

1970 through early 1978 models

Refer to illustrations 8.2 and 8.4

1 Remove the screws securing the breaker point cover to the gearcase cover. Detach the point cover and the gasket.

2 Scribe a line across the baseplate and gearcase. The scribed line will ensure the ignition timing is close to the required setting after reassembly. Remove the center bolt that retains the contact breaker point cam and the two screws that secure the contact breaker baseplate assembly **(see illustration)**.

8.10 On later models with electronic ignition, drill out the rivet heads and remove the outer cover, . . .

8.11 . . . then remove the screws (arrows) and detach the inner cover and gasket to get at the sensor plate

8.13 Mark the sensor plate, then remove the screws (arrows) . . .

8.14 . . . and pull out the sensor plate to gain access to the rotor – remove the bolt (arrow) and detach the rotor

3 Lift out the baseplate.

4 Pull off the contact breaker point cam and the centrifugal advance unit **(see illustration)**.

Late 1978 and 1979 models

5 Remove the outer cover from the gearcase cover by removing the two screws or drilling the heads off the two rivets.

6 Lift the ignition module out of the timing case – be careful, the wires are still attached – and set it aside.

7 Remove the screws securing the timer plate, then remove the bolt from the center of the trigger rotor.

8 Remove the small screws and detach the sensor from the timer plate and the electrical lead from the module to the timer plate.

9 Lift the timer plate out of position and pull the trigger rotor off the advance assembly. The advance assembly can now be removed from the gearcase cover.

1980 and later models

Refer to illustrations 8.10, 8.11, 8.13 and 8.14

10 Remove the outer cover from the gearcase cover by drilling the heads off the two rivets **(see illustration)**.

11 Remove the two screws securing the inner cover, then detach the cover and gasket **(see illustration)**.

12 Unplug the wire harness connector (follow the wires to the connector).

13 Mark the sensor plate and gearcase (timing) cover with a scribe or permanent felt-tip marker to ensure the ignition timing can be returned to its original setting, then remove the screws and detach the sensor plate **(see illustration)**. The sensor plate can stay with the gearcase cover if you don't want to try to pull the wires through the hole and thread them back in again later. Just be very careful not to damage the sensor as the gearcase cover is handled (wrap a clean shop rag around it and secure it with tape to protect it).

14 Remove the rotor bolt and detach the rotor **(see illustration)**.

9 Gearcase cover, camshaft and timing gears – removal

Note: *If the gearcase is being removed with the top-end of the engine installed, valve train pressure must be released to alleviate loading on the camshaft.*

Refer to illustrations 9.1, 9.3, 9.5, 9.6, 9.8, 9.9, 9.10a and 9.10b

1 Unscrew the tappet oil screen plug and lift out the O-ring, the spring and the oil screen from the top of the gearcase **(see illustration on next page)**.

2 Remove the ignition components from the gearcase cover as described in the previous Section. Remove the tappets and guides as described in Section 7.

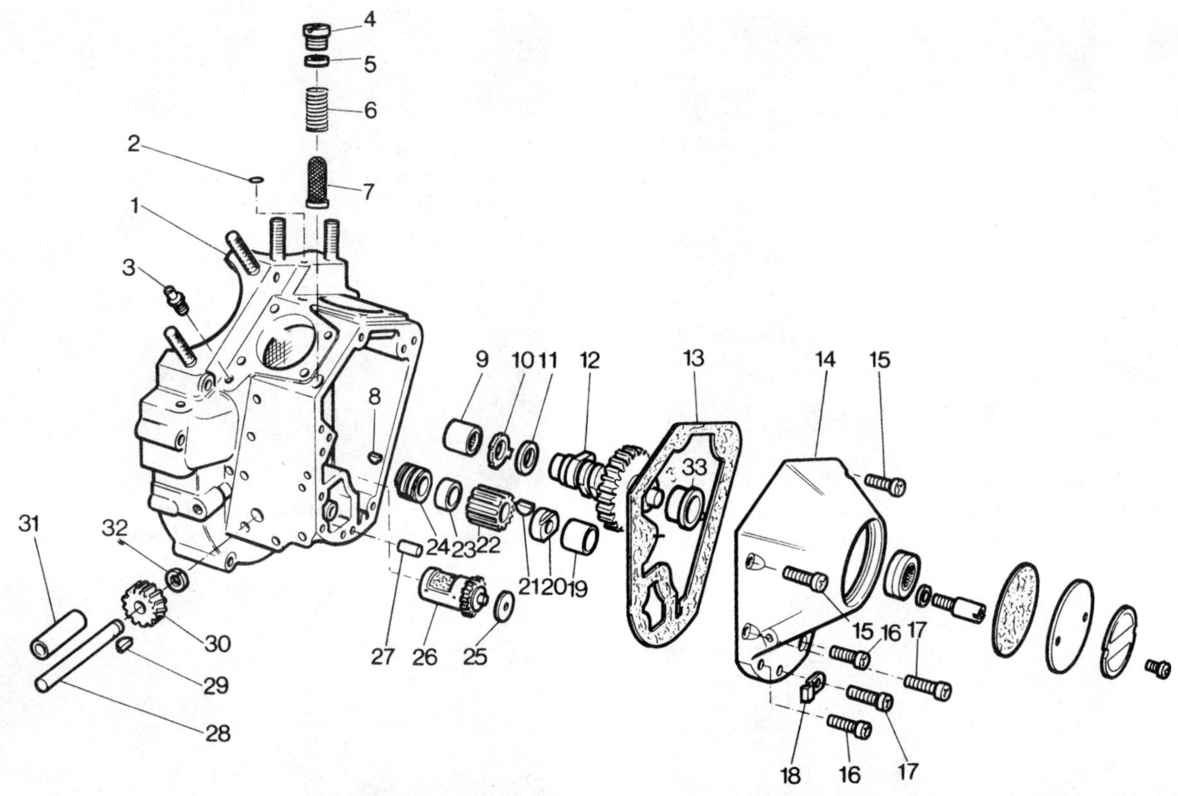

9.1 Gearcase and cover components – exploded view

1	*Right crankcase half*	12	*Camshaft*	23	*Timing gear spacer**
2	*Rubber plug*	13	*Gasket*	24	*Oil pump drive gear*
3	*Rocker oil feed union (pre-Evolution engine only)*	14	*Gearcase (timing) cover*	25	*Spacer*
4	*Plug*	15	*Screw*	26	*Timed breather*
5	*O-ring*	16	*Screw*	27	*Dowel pin*
6	*Spring*	17	*Screw*	28	*Oil pump drive shaft*
7	*Hydraulic tappet oil screen*	18	*Cable guide*	29	*Woodruff key*
8	*Woodruff key*	19	*Mainshaft (crankshaft) support bushing*	30	*Oil pump driven gear*
9	*Crankshaft bearing*	20	*Special nut (plain nut 1993-on)*	31	*Bushing*
10	*Thrust washer*	21	*Woodruff key*	32	*Circlip*
11	*Camshaft end play control shim**	22	*Timing gear*	33	*Camshaft outer bushing*

** Components not fitted from 1993-on*

9.3 The gearcase cover screws are different lengths – be sure to keep track of where they go

9.4a Service tool necessary for gearcase removal – 1993 and later

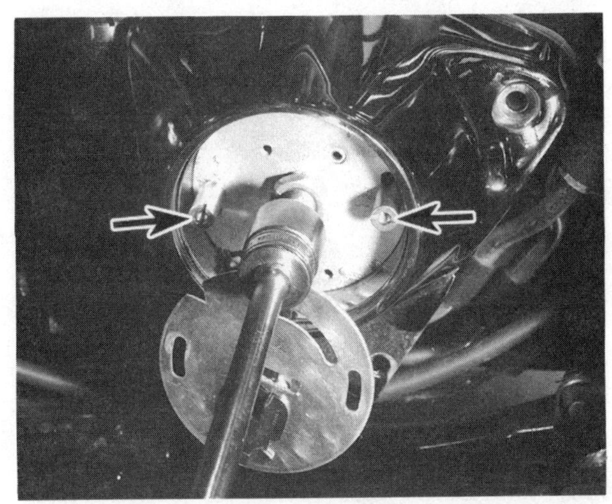

9.4b Tool in position – note use of sensor plate screws (arrows)

9.5 Remove the crankcase breather gear and spacer from the gearcase bore

9.6 Carefully remove the camshaft, shim (where fitted) and thrust washer from the gearcase

9.8 Unscrew the special nut (or plain nut) securing the timing gear

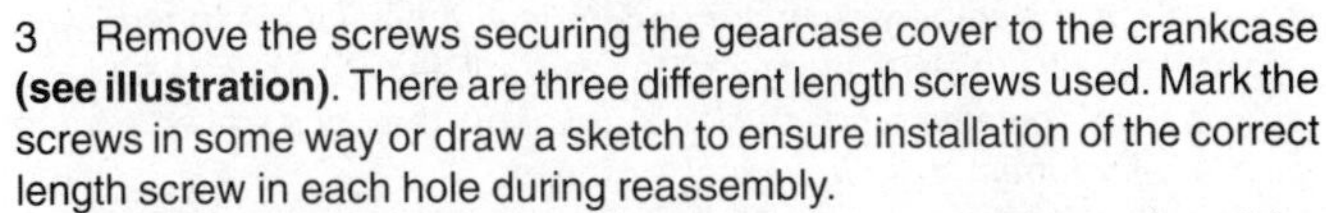

3 Remove the screws securing the gearcase cover to the crankcase **(see illustration)**. There are three different length screws used. Mark the screws in some way or draw a sketch to ensure installation of the correct length screw in each hole during reassembly.

4 If care is exercised, it may be possible to tap around the edge of the gearcase cover with a soft-faced hammer to break the gasket seal, but it is recommended that a puller be used in order to remove the cover safely. A puller can be obtained through Harley-Davidson dealers **(see illustrations)**, or could be fabricated by using the ignition sensor plate as a template. The puller is inserted in the same way as the sensor plate and the center bolt tightened down onto the camshaft end, thus pulling the gearcase off as it is tightened. **Caution:** *On no account lever the gearcase off – such action will destroy its finish and could incur breakage.*

5 Remove the breather gear and spacer from the left side of the gearcase **(see illustration)**.

6 Carefully slide the camshaft out of the gearcase, along with the thrust washer and shim **(see illustration)**.

7 Lock the crankshaft to keep it from turning by inserting a tight fitting rod through the small end of one of the connecting rods. Place two pieces of wood across the opening in the crankcase. Carefully turn the crankshaft until the rod bears down on the two pieces of wood. If the engine isn't disassembled, apply the rear brake with the transmission in gear to keep the crankshaft from turning.

8 Unscrew the nut securing the timing gear to the crankshaft. On models through 1992 it has a left-hand thread (turn clockwise to loosen); the nut has only two flats and must be removed with an open-end wrench **(see illustration)**. On 1993 and later models a plain 6-sided nut is used, with a conventional right-hand thread.

9 On models through 1992 the timing gear is pressed onto the crankshaft, so a puller will be required to draw the gear off the shaft **(see illustration)**. On 1993 and later models the gear is a slip fit on its splines.

9.9 A two-jaw puller can be used to remove the timing gear from the end of the crankshaft – models through 1992

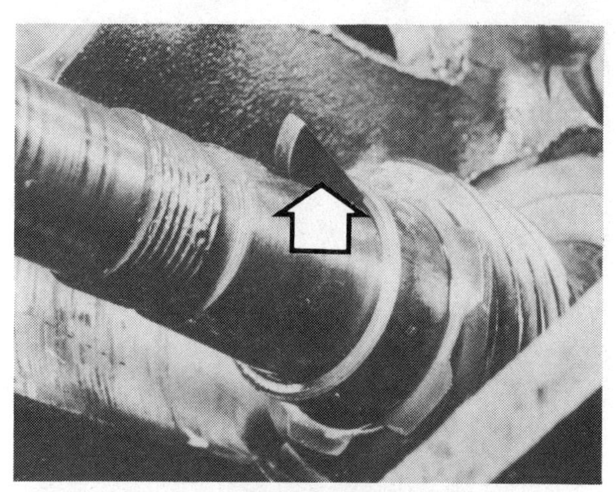

9.10a Pry the Woodruff key out of the crankshaft and store it in a safe place

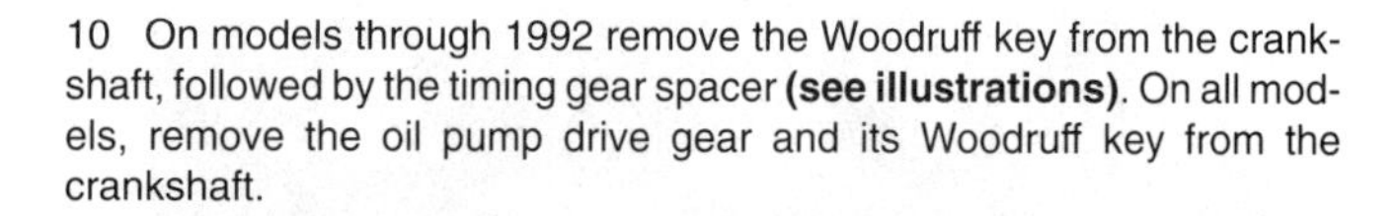

10 On models through 1992 remove the Woodruff key from the crankshaft, followed by the timing gear spacer **(see illustrations)**. On all models, remove the oil pump drive gear and its Woodruff key from the crankshaft.

10 Oil pump – removal

Note: *If the engine is in the frame follow the procedure given in Chapter 3, Section 11.*

Refer to illustration 10.1

1 The oil pump driven gear is located on the drive shaft by a Woodruff key and is secured by a small circlip. Removal of the circlip requires care and a steady hand as access is obscured by the crankcase wall. If possible, use a pair of external snap-ring pliers with 90-degree offset jaws. If this type of tool isn't available, it's possible to spring the clip out of position with a small screwdriver. Don't lose the clip as it comes off the shaft **(see illustration)**.

2 Remove the bolts/nuts holding the pump to the case and pull the oil pump assembly off, bringing with it the pump drive shaft. The oil pump driven gear and Woodruff key will fall into the case. The oil pump outer cover will probably slip out of place when the four mounting bolts are removed. Make sure when this happens, and also as the main body of the pump is removed, that the pump gears don't fall out. It's important to keep the gears in place or replace them in their original positions.

3 Refer to Chapter 3 for the oil pump overhaul procedure.

10.1 Use snap-ring pliers to remove the circlip from the oil pump drive shaft (which will free the driven gear)

9.10b Slide the timing gear spacer (models through 1992) and oil pump drive gear off the shaft

11 Alternator – removal

Refer to illustrations 11.1, 11.2, 11.3 and 11.4

Note: *If the engine is in the frame, disconnect the negative battery cable from the battery, then remove the primary drive components to gain access to the alternator rotor. On 1989 and later FX/Softail models, you don't have to remove the inner primary chaincase to detach the alternator.*

1 If the engine has been removed from the frame, support it on blocks so the alternator is facing up, away from the workbench. Pry the splined shaft extension off the crankshaft **(see illustration)** (not all models).

2 The alternator rotor on most models has an internally splined boss and is pressed onto the crankshaft. The rotor must be removed from the crankshaft with a puller. A Harley-Davidson special tool can be used or a puller with three bolts that will screw into the threaded holes in the alternator rotor boss can be used. Place the plate over the end of the crankshaft and screw the bolts into the rotor boss. Tighten the bolts evenly to draw the rotor off the crankshaft **(see illustration)**. **Note:** *The rotors on some Evolution engines (late 1985 through 1987 models) are a slip-fit on the crankshaft and have large spacers on each side. After the compensating sprocket nut and primary drive components are removed, the spacers and rotor will slip off the shaft. Later engines with splined rotor hubs also have large spacer washers on each side – note how they're installed and be sure to return them to their original locations.*

11.1 Use a pry bar to pry the drive sprocket shaft extension off the end of the crankshaft

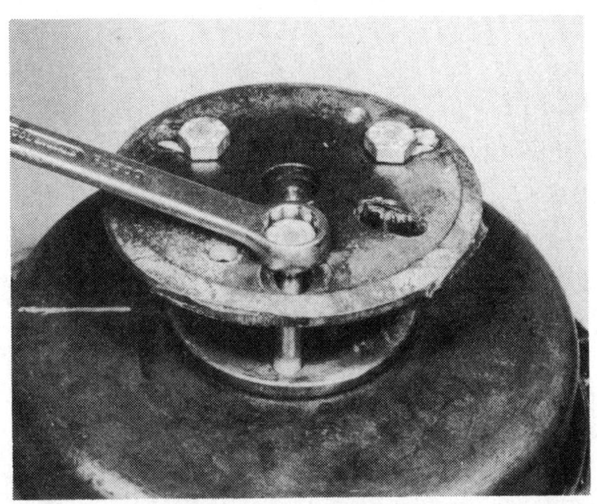

11.2 A puller must be used to remove the alternator rotor from the shaft (some later models don't require a puller to remove the rotor)

11.3 Remove the four alternator stator screws after releasing the lock plate tabs

3 Bend back the tabs of the lock plates that secure the four stator mounting screws, then remove the screws **(see illustration)**. Later model Evolution engines don't have lock plates and the screws have Torx heads. Discard the Torx screws if they're removed and install new ones during reassembly.
4 Remove the two countersunk screws securing the bridge piece that holds the alternator wires in place **(see illustration)**. Remove the bridge piece.
5 The rubber plug, which forms the ends of the alternator wires, is a tight fit in the crankcase. Select a number of blunt tools to push the plug in, toward the alternator. Don't use any sharp instruments – they may pierce the plug and damage the wires or plug pins.
6 Lift the alternator stator out.

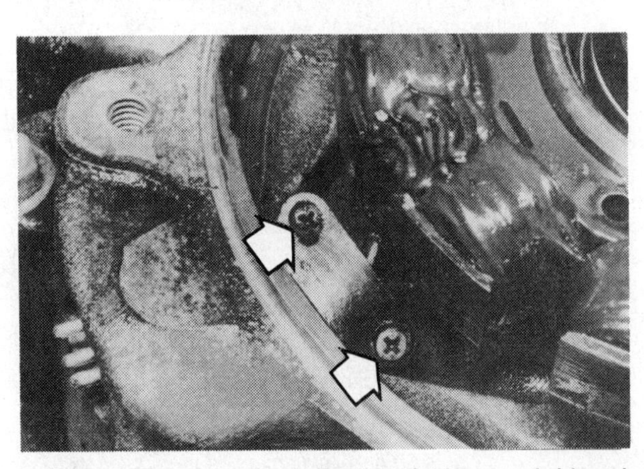

11.4 Remove the screws securing the wire harness bridge piece

12 Splitting the crankcases

Refer to illustration 12.4

1 Check the end play of the crankshaft/flywheel assembly before the crankcases are split. Attach a dial indicator to the case with the stem in contact with the end of the crankshaft. Pull the crankshaft to the drive side of the engine as far as possible.
2 Zero the dial indicator, then push the crankshaft completely over to the timing side. The total end play should be within the tolerance given in the Specifications section of this chapter. Greater play indicates the need for new drive side main bearings. If the play is greater than the specified amount, but the drive side main bearings are in good condition, shims (early models) or a thicker bearing spacer ring (later models) may be installed during reassembly.
3 Remove the nut from one end of each crankcase stud, and remove the bolts. Tap the studs out of the crankcase with a drift punch, having marked them so that they can be returned to their original holes in the casing on reassembly.
4 Tap around the edge of the right-hand crankcase with a soft-face hammer until the gasket seal is broken. Don't pry the two cases apart by inserting screwdrivers or other tools between the mating surfaces or damage will result, causing oil leaks. After the seal has been broken, lift the right-hand crankcase half off the timing side main bearing **(see illustration on next page)**.
5 The drive side crankshaft is a press fit in two tapered roller main bearings and the oil seal collar. Because of this, the flywheel assembly must be pressed out of the left crankcase half, leaving the oil seal, oil seal collar and outer (sprocket side) tapered roller bearing in the crankcase and the flywheel side (inner) tapered roller bearing on the crankshaft. Removal of the flywheel assembly requires a press, which will enable a controlled amount of pressure to be applied to the crankshaft end. Many automotive and motorcycle repair shops use this type of press and will carry out the operation for a nominal fee.
6 After crankshaft removal, position a large drift punch on the inner race of the sprocket side (outer) tapered roller bearing and drive the bearing, oil seal collar and oil seal out from the inside of the crankcase half. A piece of pipe with an outside diameter the same size as the bearing inner race will also work.

13 Inspection and repair – general information

1 Before examining the parts of the disassembled engine for wear, clean them thoroughly. Use solvent to remove all traces of old oil and sludge that accumulated in the engine.
2 Examine all castings for cracks and other damage, especially the crankcase castings. If a crack is discovered, it'll require professional repair or replacement of the part.
3 Carefully examine each part to determine the extent of wear, using the figures listed in the Specifications Section of this Chapter. If there's any question or doubt, be safe and replace the component. Notes included in the following text will indicate what type of wear can be expected and whether the part concerned can be reused.
4 Use a clean lint-free rag for cleaning and drying the various components. This will reduce the risk of small particles obstructing the internal oilways, causing the lubrication system to fail.

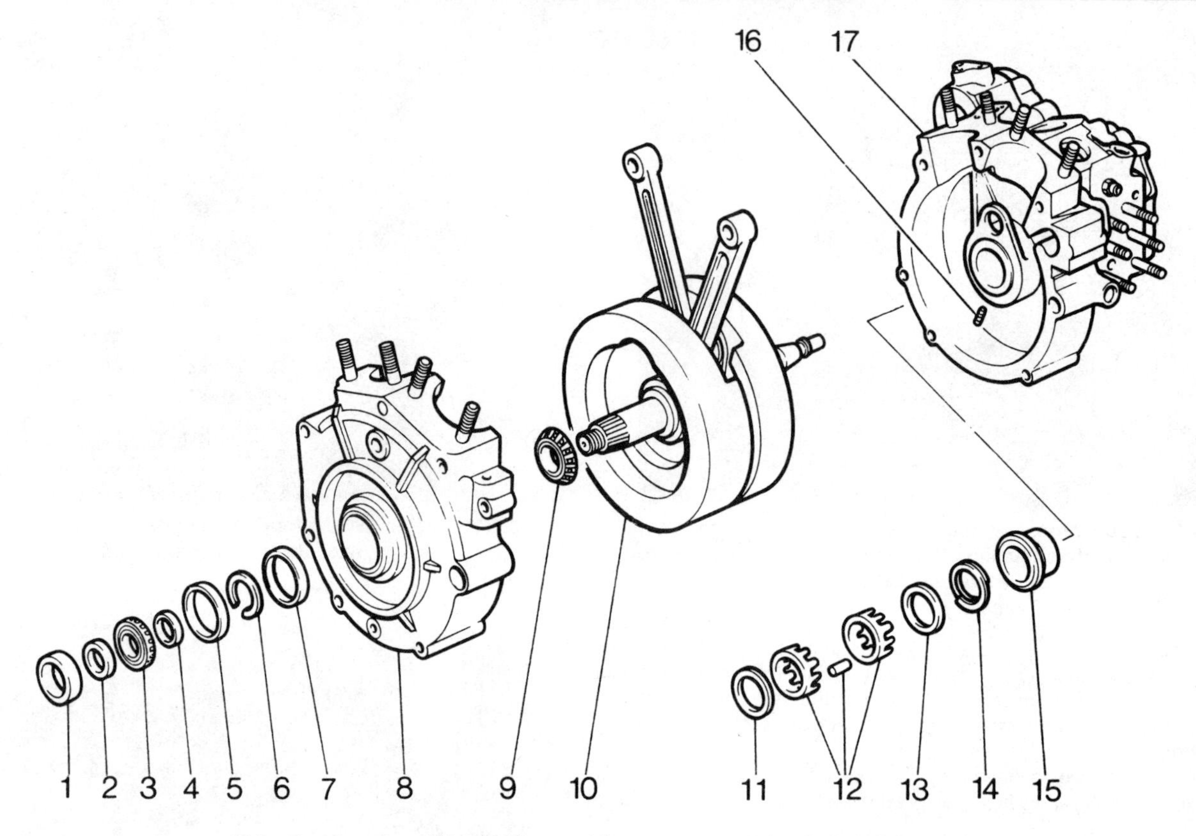

12.4 Crankcase and flywheel assembly components – exploded view

1 *Oil seal*
2 *Oil seal collar*
3 *Sprocket side (outer) tapered roller main bearing/inner race*
4 *Inner race spacer ring*
5 *Sprocket side main bearing outer race*
6 *Circlip*
7 *Flywheel side main bearing outer race*
8 *Left side crankcase*
9 *Flywheel side (inner) tapered roller main bearing/inner race*
10 *Flywheel assembly (crankshaft)*
11 *Washer*
12 *Timing side main bearing rollers and cage*
13 *End washer*
14 *Spiral lock ring*
15 *Timing side main bearing outer race*
16 *Timing side main bearing outer race set screw*
17 *Right side crankcase*

14.8a Timing side main bearings are secured by a double spiral lock ring

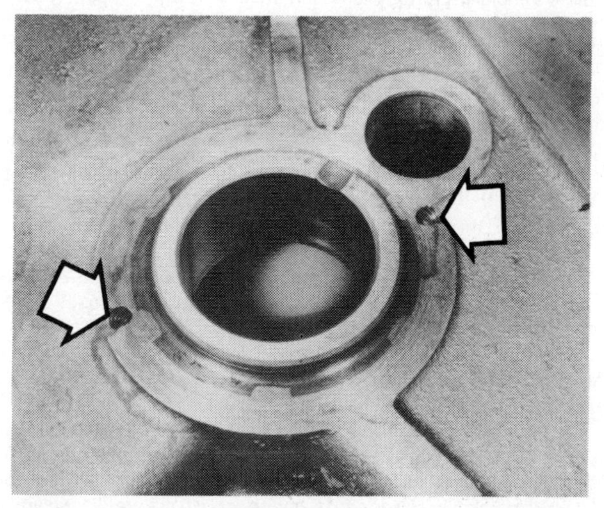

14.8b The timing side main bearing outer race is located by two set screws (arrows)

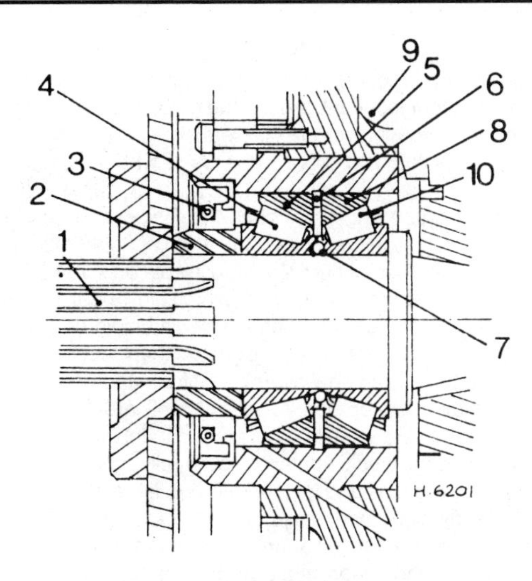

14.9 Drive side main bearing configuration

1 *Drive side flywheel (crankshaft)*
2 *Oil seal collar*
3 *Oil seal*
4 *Outer tapered roller main bearing – rollers and inner race*
5 *Sprocket side bearing – outer race*
6 *Outer race spacing circlip*
7 *Inner race spacer ring*
8 *Flywheel side bearing outer race*
9 *Left (drive side) crankcase half*
10 *Inner tapered roller main bearing – rollers and inner race*

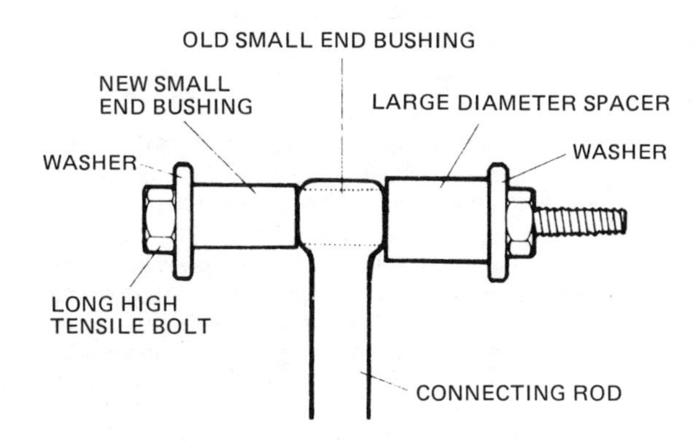

14.10 The connecting rod small end bushings can be removed and installed with a home-made tool like this one – the new bushing will be drawn in, forcing the old one out, as the nut is tightened

5 Above all, work in clean, well-lit surroundings so problems don't pass undetected. Failure to detect damage or signs of advanced wear may necessitate another complete teardown at a later date, due to the premature failure of the part concerned.

14 Crankshaft and connecting rod bearings – inspection

Refer to illustrations 14.8a, 14.8b, 14.9 and 14.10

1 The crankshaft assembly consists of a pair of heavy flywheels joined together by a single crankpin and big-end assembly, which serves both cylinders. An ingenious overlapping arrangement permits the two connecting rods to share the common big-end assembly without fouling each other during rotation. The front connecting rod fits inside the forked rear rod.

2 It isn't possible to separate the flywheel assembly without access to a large press or to realign the components to a high enough standard of accuracy without a lathe. As a result, if repairs are required, take the flywheel assembly to a Harley-Davidson dealer.

3 Failure of the big-end bearing is usually accompanied by a pronounced knock from within the crankcase, which gets progressively worse. Vibration will also be experienced. There should be no vertical play whatsoever in either of the connecting rods after the old oil has been washed out of the big-end bearing. If even a small amount of play is evident, it must not be ignored; the bearing should be replaced. Don't run the engine with a worn big-end bearing – the connecting rods or crankpin will break and cause extensive damage.

4 Don't confuse big-end wear with side play in the bearing (a small amount of side play is acceptable). It should be maintained within the limits in the Specifications. The side play should be checked with a feeler gauge.

5 Worn main bearings are evidenced by an audible rumble from the lower end of the engine and vibration that's particularly noticeable through the footrests. Wear of the drive side tapered roller bearings can be determined by measuring the flywheel assembly end play as described in Section 12. Again, if any play is evident, the main bearings must be replaced.

6 While the engine is apart, examine the main bearings for signs of wear and damage after the oil has been washed out. The usual indications are chipped or broken rollers, scuff marks in the hardened surface of the rollers and roughness as the bearing is revolved. If a bearing fails soon after a complete overhaul, another teardown of the engine will be necessary, so don't reinstall questionable parts.

7 Two different approaches are possible when attending to worn timing side main bearings. The bearings can be replaced outright as complete units or oversize rollers can be installed to restore the original clearances. The latter method requires honing of the inner and outer races in conjunction with selection of the correct oversize rollers. Special tools and expertise are required to carry out this operation, which should be left to a Harley-Davidson dealer.

8 To remove the timing side main bearing inner races and cages from the mainshaft, pry the spiral lock ring off the shaft and remove the end washer **(see illustration)**. Slide the bearing cages off the shaft, being careful not to let the rollers fall out of the cages. The bearing rollers run directly on the mainshaft. If the bearing track is worn, the only remedy is to replace the mainshaft, which requires separation of the flywheels. As noted in Step 2, this is a job for a dealer service department. The timing side main bearing outer race is a press fit in the crankcase and is secured by two screws threaded into the inside of the case **(see illustration)**. To prevent damage to the housing, the crankcase should be heated to 275 to 300-degrees F before the race is removed. The case should be heated in a similar manner when installing the new bearing outer race. Even after replacement of the timing side main bearings, the outer race must be honed and new rollers installed to give the correct running clearance.

9 The drive side main bearing outer races can be removed and installed with the crankcase cold. Drive the outer race out and the inner race in with a brass punch. To remove the outer race spacing circlip **(no. 6 in illustration 14.9),** rotate the circlip in the groove until one end is directly below the oilway, drilled at an angle from the inside of the case through to the center of the bearing housing. Pass a 1/8-inch pin punch or similar tool through the oilway to dislodge the end of the circlip, then pry the clip out with a screwdriver. **Note:** *The outer race spacing circlip must be installed so that the casing oilway lies between the opening in the circlip.* The drive side main bearings must be replaced as a set, together with the circlip and the inner race spacer ring. These components shouldn't be replaced individually. The drive side main bearing inner race can be pried off the mainshaft. Make sure the main bearing outer and inner races are kept in pairs and the components aren't interchanged.

10 The connecting rod small end bearings are plain bushings. When replacement is needed, the new bushing can be pressed in to displace the old bushing **(see illustration)** (be sure to align the oil hole in the bushing with the oil hole in the connecting rod). It'll also need reaming to size.

2

15 Cylinder barrels – inspection

Note: *The cylinders in Evolution engines usually have four faint polish marks running the length of the bore near the stud holes. The marks are about 0.375-inch (10 mm) wide and appear as the engine accumulates running time. They're normal and shouldn't cause concern.*

1 The usual indication of badly worn cylinder barrels is excessive oil consumption, accompanied by blue smoke from the exhaust and possibly piston slap (a metallic rattle that occurs when there's little or no load on the engine). If the top of the cylinder bore is examined carefully, you probably will find a ridge on the thrust side, denoting the limit of travel of the upper piston ring. The depth of this ridge will vary according to the amount of wear that has taken place.

2 Using a telescoping gauge and outside micrometer, or an inside micrometer, measure the bore about 1/2-inch (13 mm) below the limit of top ring travel, first from back-to-front in the barrel, then from side-to-side. This will show whether the bore has worn out-of-round. Repeat the measurements just above the lower limit of oil ring travel and also approximately half-way between these two points. If the bore hasn't worn more than 0.002-inch (0.05 mm), isn't out-of-round or badly scored, the cylinder can be reused.

3 If the bore wear exceeds 0.002-inch (0.05 mm) or the bore is tapered, out-of-round or damaged in any way, a rebore is required. Oversize pistons are available in various rebore sizes.

4 If the cylinder barrel is already at the maximum rebore size, you'll have to install a replacement cylinder barrel. It's not possible to go beyond the maximum bore limit without seriously weakening the barrel.

5 Have the barrels honed at a dealer service department and install new piston rings regardless of what is done with the pistons.

16 Pistons – inspection

1 Before the inspection process can be carried out, the pistons must be cleaned and the old piston rings removed.

2 Using a piston ring installation tool, carefully remove the rings from the pistons. Don't nick or gouge the pistons in the process. Make a note of how the rings are installed so they'll be reinstalled properly.

3 Scrape all traces of carbon from the tops of the pistons. A hand-held wire brush or a piece of fine emery cloth can be used once the majority of the deposits have been scraped away. Do not, under any circumstances, use a wire brush mounted in a drill motor to remove deposits from the pistons; the piston material is soft and will be eroded away by the wire brush.

4 Use a piston ring groove cleaning tool to remove any carbon deposits from the ring grooves. If a tool isn't available, a piece broken off an old ring will do the job. Be very careful to remove only the carbon deposits. Don't remove any metal and don't nick or gouge the sides of the ring grooves.

5 Once the deposits have been removed, clean the pistons with solvent and dry them thoroughly. Make sure the oil return holes in the back sides of the oil ring grooves are clear.

6 If the pistons aren't damaged or worn excessively and if the cylinders aren't rebored, new pistons won't be necessary. Normal piston wear appears as even, vertical wear on the thrust surfaces of the piston and slight looseness of the top ring in its groove. New piston rings, on the other hand, should always be used when an engine is rebuilt.

7 Carefully inspect each piston for cracks around the skirt, at the pin bosses and at the ring lands.

8 Look for scoring and scuffing on the thrust faces of the skirt, holes in the piston crown and burned areas at the edge of the crown. If the skirt is scored or scuffed, the engine may have been suffering from overheating and/or abnormal combustion, which caused excessively high operating temperatures. The oil pump should be checked thoroughly. A hole in the piston crown, an extreme to be sure, is an indication that abnormal combustion (pre-ignition) was occurring. Burned areas at the edge of the piston crown are usually evidence of spark knock (detonation). If any of the above problems exist, the causes must be corrected or the damage will occur again.

9 Measure the piston ring side clearance by laying a new piston ring in the ring groove and slipping a feeler gauge in beside it. Check the clearance at three or four locations around the groove. Be sure to use the correct ring for each groove; they are different. If the clearance is greater than specified, new pistons will have to be used when the engine is reassembled.

10 On pre-Evolution engines, check the piston-to-bore clearance by measuring the bore and the piston diameter. Make sure the pistons and cylinders are correctly matched. Measure the piston across the skirt on the thrust faces at a 90-degree angle to the piston pin, about 3/8-inch (10 mm) up from the bottom of the skirt. Subtract the piston diameter from the bore diameter to obtain the clearance. If it's greater than specified, the cylinders will have to be bored and new oversize pistons and rings installed.

11 If the appropriate precision measuring tools aren't available, the piston-to-cylinder clearances can be obtained, though not quite as accurately, using feeler gauge stock. Feeler gauge stock comes in 12-inch lengths and various thicknesses and is generally available at auto parts stores.

12 To check the clearance, select a 0.004-inch feeler gauge and slip it into the cylinder along with the appropriate piston. The cylinder should be upside-down and the piston must be positioned exactly as it normally would be. Place the feeler gauge between the piston and cylinder on one of the thrust faces (90-degrees to the piston pin bore).

13 The piston should slip through the cylinder (with the feeler gauge in place) with moderate pressure (don't try to force it). If it falls through, or slides through easily, the clearance is excessive and a new piston will be required. If it won't fit into the cylinder, the clearance is less than 0.004-inch and progressively thinner feeler gauges will have to be used to determine the clearance. If the piston binds at the lower end of the cylinder and is loose toward the top, the cylinder is tapered, and if tight spots are encountered as the piston/feeler gauge is rotated in the cylinder, the cylinder is out-of-round.

14 Repeat the procedure for the remaining piston and cylinder. Be sure to have the cylinders and pistons checked by a Harley-Davidson dealer service department or a motorcycle repair shop to confirm your findings before purchasing new parts. ***Note:*** *The pistons and cylinders for Evolution engines* ***must*** *be checked and matched by a dealer service department.*

17 Valves, valve seats and valve guides – servicing

1 Because of the complex nature of this job and the special tools and equipment required, servicing of the valves, the valve seats and the valve guides (commonly known as a valve job) is best left to a professional.

2 The home mechanic can, however, remove and disassemble the heads, do the initial cleaning and inspection, then reassemble and deliver the heads to a dealer service department, motorcycle repair shop or even an automotive machine shop for the actual valve servicing.

3 The dealer service department will remove the valves and springs, recondition or replace the valves and valve seats, replace the valve guides, check and replace the valve springs, spring retainers and keepers (as necessary), replace the valve seals (if used) with new ones and reassemble the valve components.

4 After the valve job has been performed, the heads will be in like-new condition. When the heads are returned, be sure to clean them again very thoroughly before installation on the engine to remove any metal particles or abrasive grit that may still be present from the valve service operations. Use compressed air, if available, to blow out all the holes and passages.

18 Rocker arms/shafts and pushrods – inspection

Refer to illustrations 18.2a, 18.2b, 18.5, 18.6, 18.7 and 18.8

1 It's unlikely that excessive wear will occur in either the rocker arms or the shafts unless the flow of oil has been impeded or the machine has covered a lot of miles. A clicking noise from the rocker area is the usual symptom of wear in the rocker arms, and shouldn't be confused with a somewhat similar noise caused by excessive valve clearances.

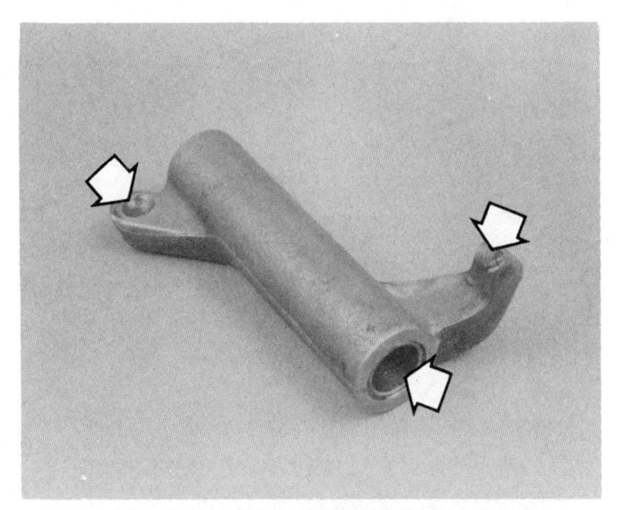

18.2a Check the rocker arm bushings, the pushrod recesses and the valve stem pads for wear and damage (arrows)

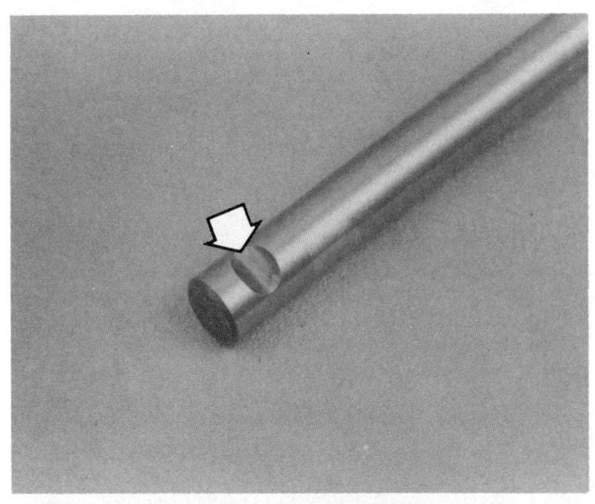

18.2b Check the rocker arm shafts for wear at the areas where the rocker arm bushings ride – the Evolution engine rocker arm shafts have a cutout in one end (arrow) for the bolt to pass through

18.5 Check the pushrod ends for wear and make sure the oil holes are clear

18.6 On pre-Evolution engines, remove the rocker arm shaft end cap and . . .

2 Check the rocker arm bushings and the shafts for evidence of excessive wear and damage **(see illustrations)**.

3 Measure the outside diameter of the shaft (where it rides in the bushings) and the inside diameter of the rocker arm bushings. Subtract the shaft diameter from the bushing diameter to obtain the clearance. If it's excessive, install new parts or have the bushings in the rocker arm(s) replaced by a dealer service department.

4 Check each rocker arm where it contacts the pushrod and valve stem. If cracks, scuffing or breakthrough in the case hardened surface are evident, install a new rocker arm.

5 Check the ends of the pushrods where they ride in the tappets and the rocker arms **(see illustration)**. Check for distortion by rolling them on a flat surface such as a pane of glass. If wear or distortion is evident, new pushrods should be installed.

6 To disassemble a rocker box on pre-Evolution engines, unscrew the rocker arm shaft end caps found on the right-hand side of each rocker box. Early models have caps with a screwdriver slot, later models a hex-shaped socket to accept an Allen wrench **(see illustration)**. Remove both caps and the O-ring behind each one (if equipped). Discard the O-ring.

7 Unscrew the acorn nuts from the other end of the rocker box and tap the shaft out of position from the acorn nut end **(see illustration)**.

18.7 . . . unscrew the acorn nut from the other end

18.8 Lift out the end play control spacer (arrow)

19.6a Compress the springs with a valve spring compressor, then remove the keepers (arrows) with a magnet or needle-nose pliers

19.6b Remove the retainer, the valve springs . . .

19.6c . . . and the spring seat, then pull off the valve stem seals (if used) with a pair of pliers

8 Remove the end play control spacer and the rocker arm **(see illustration)**. This may take a certain amount of manipulation. The dismantled assembly will include the rocker arm, a spacer and the shaft itself. Mark all the parts before tapping out the second shaft so they're eventually replaced in their original positions.

9 Replacement of the rocker arm bushings should be done by a dealer service department, since the new ones may have to be reamed to size to fit the shafts.

19 Cylinder head and valves – disassembly, inspection and reassembly

1 As was mentioned in Section 17, valve servicing and valve guide replacement should be left to a dealer service department or motorcycle repair shop. However, disassembly, cleaning and inspection of the valves and related components can be done (if the necessary special tools are available) by the home mechanic. This way no expense is incurred if the inspection reveals that service work isn't required at this time.

2 To properly disassemble the valve components without the risk of damaging them, a valve spring compressor is absolutely necessary. If the special tool isn't available, have a dealer service department or motorcycle repair shop handle the entire process of disassembly, inspection, service or repair (if required) and reassembly of the valves

Disassembly

Refer to illustrations 19.6a, 19.6b and 19.6c

3 Before the valves are removed, scrape away any traces of gasket material from the head gasket sealing surface. If you're working on an Evolution engine, be careful – don't nick or gouge the soft aluminum head. Gasket removing solvents, which work very well, are available at most motorcycle shops and accessory stores.

4 Carefully scrape all carbon deposits out of the combustion chamber. A hand held wire brush or a piece of fine emery cloth can be used once the majority of deposits have been scraped away.

5 Before proceeding, arrange to label and store the valves along with their related components so they can be kept separate and reinstalled in the same valve guides they are removed from.

6 Compress the valve springs on the first valve with a spring compressor, then remove the keepers **(see illustration)** and the retainer from the valve assembly. Don't compress the springs any more than absolutely necessary. Carefully release the valve spring compressor and remove the valve spring retainer, the springs, the spring seat and the valve from the head **(see illustrations)**. If the valve binds in the guide (won't pull through) push it back into the head and deburr the area around the keeper groove with a very fine file or whetstone.

7 Repeat the procedure for the remaining valves. Remember to keep the parts for each valve together so they can be reinstalled in the same location.

8 Once the valves have been removed and labeled, pull off the valve stem seals (1981 and later models) with pliers and discard them. The old seals should never be reused.

9 Next, clean the cylinder heads with solvent and dry them thoroughly. Compressed air will speed the drying process and ensure that all holes and recessed areas are clean.

10 Clean all of the valve springs, keepers, retainers and spring seats with

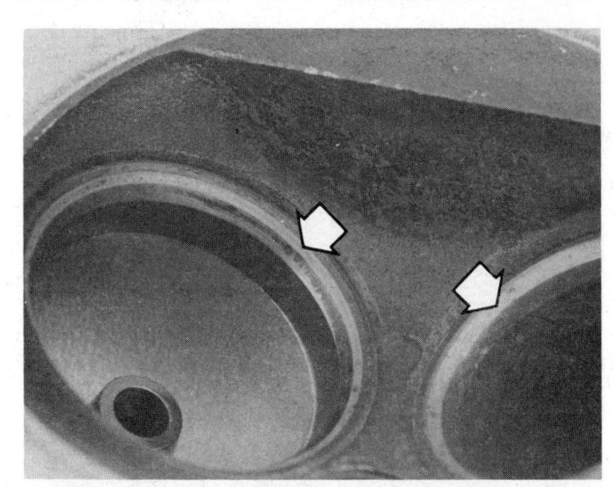

19.14a Check the valve seats (arrows) in each head – look for pits, cracks and burned areas

19.14b Use a ruler to measure the width of each valve seat

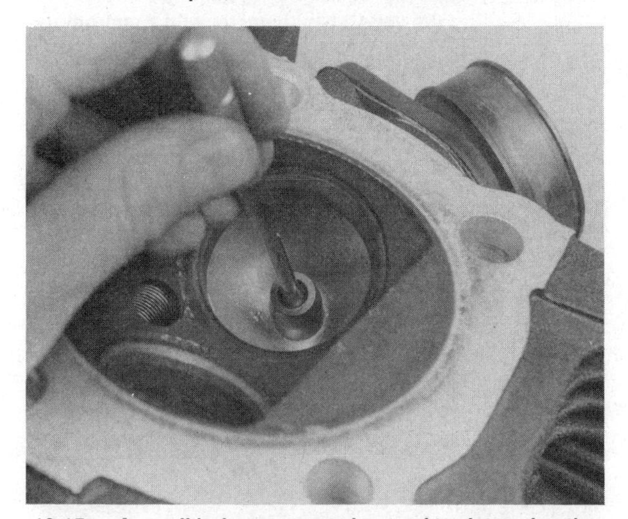

19.15a A small hole gauge can be used to determine the inside diameter of the valve guide (s)

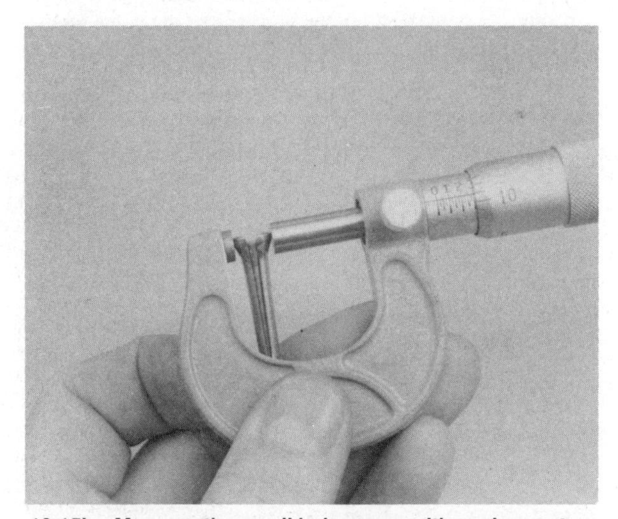

19.15b Measure the small hole gauge with a micrometer to obtain the actual size of the guide (s)

solvent and dry them thoroughly. Do the parts from one valve at a time so no mixing of parts between valves occurs.

11 Scrape off any deposits that may have formed on the valves, then use a motorized wire brush to remove deposits from the valve heads and stems. Again, make sure the valves don't get mixed up.

Inspection

Refer to illustrations 19.14a, 19.14b, 19.15a, 19.15b, 19.16a, 19.16b, 19.17, 19.18a and 19.18b

12 Inspect the heads very carefully for cracks and other damage. If cracks are found a new head is in order.

13 Using a straightedge and feeler gauge, check the head gasket mating surface for warpage (Evolution engine only). Lay the straightedge diagonally (corner-to-corner), intersecting the head bolt holes, and try to slip a 0.006-inch feeler gauge under it at each location. If the feeler gauge can be inserted between the head and the straightedge, the head is warped and must be replaced with a new one.

14 Examine the valve seats in each of the combustion chambers **(see illustration)**. If they're pitted, cracked or burned, the heads will require valve service that's beyond the scope of the home mechanic. Measure the valve seat width and compare it to the Specifications **(see illustration)**. If it's not within the specified range, or if it varies around its circumference, valve seat service is required.

15 Clean the valve guides to remove any carbon buildup, then measure the inside diameters of the guides (at both ends and the center of the guide) with a small hole gauge and a 0-to-1 inch micrometer **(see illustrations)**. Record the measurements for future reference. These measurements, along with the valve stem diameter measurements, will enable you to compute the valve-to-guide clearance. This clearance, when compared to the Specifications, will be one factor that will determine the extent of valve service work required. The guides are measured at the ends and at the center to determine if they're worn in a bell-mouth pattern (more wear at the ends). If they are, guide replacement is an absolute must.

16 Carefully inspect each valve face for cracks, pits and burned spots **(see illustration)**. Check the valve stem and the keeper groove area for

19.16a Check the valve face and margin for wear and cracks

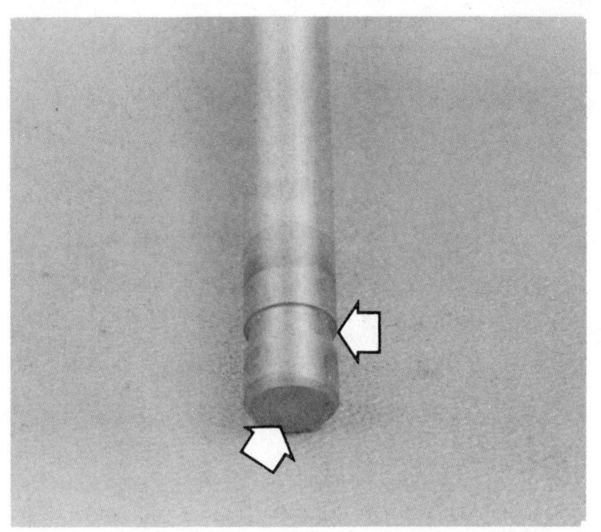

19.16b Look for wear on the very end of the valve stem and make sure the keeper groove isn't distorted in any way

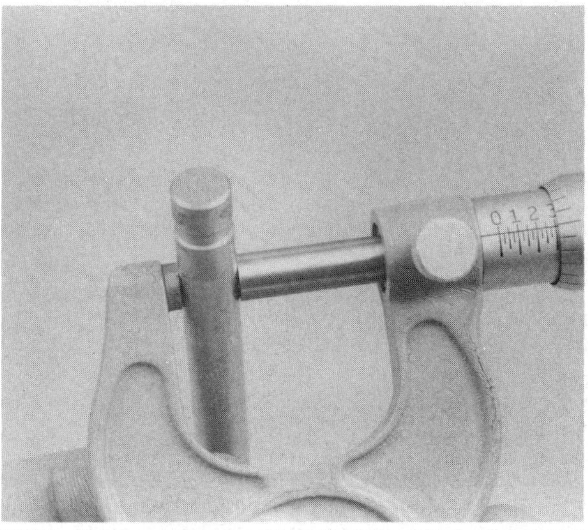

19.17 Measure the valve stem diameter with a micrometer

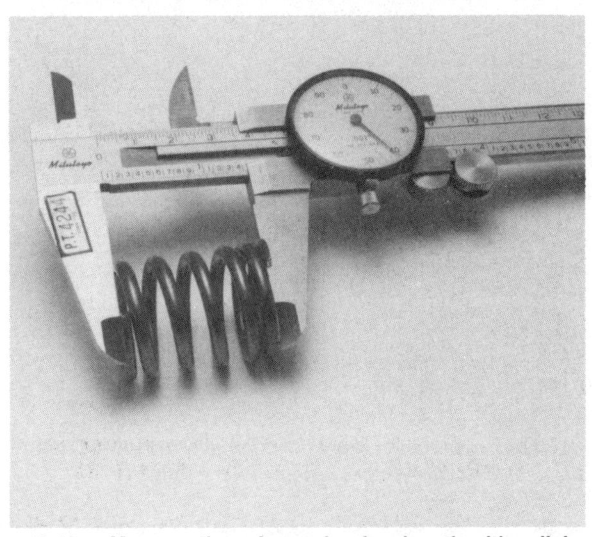

19.18a Measure the valve spring free length with a dial or vernier caliper

19.18b Check each valve spring for squareness with an accurate square

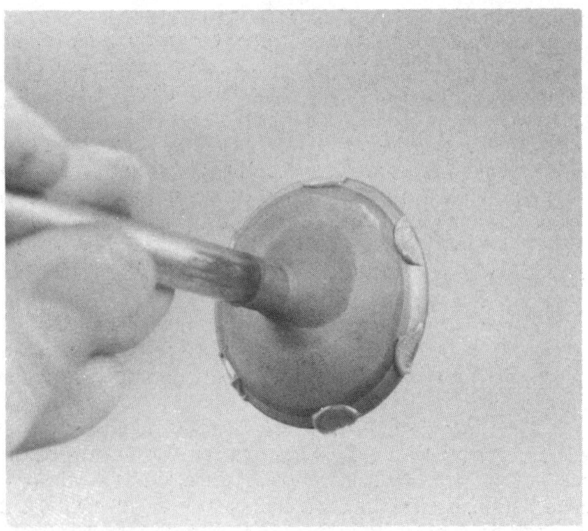

19.22 Apply the lapping compound very sparingly, in small dabs, to the valve face only

19.23a Rotate the lapping tool or hose back-and-forth between the palms of your hands

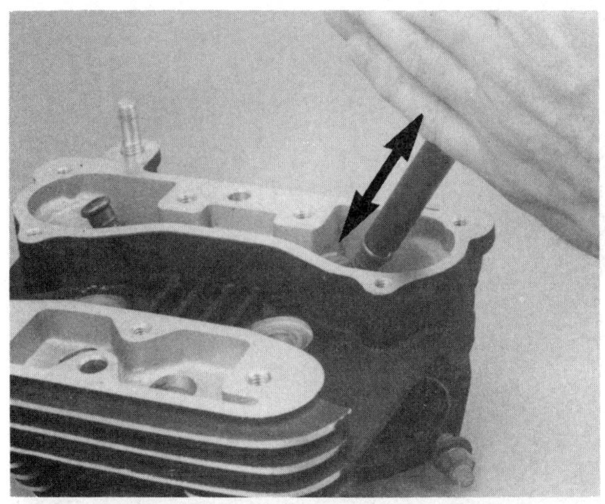

19.23b Lift the tool and valve periodically to redistribute the lapping compound on the valve face and seat

19.24a After lapping, the valve face should exhibit a uniform, unbroken contact pattern (arrow) . . .

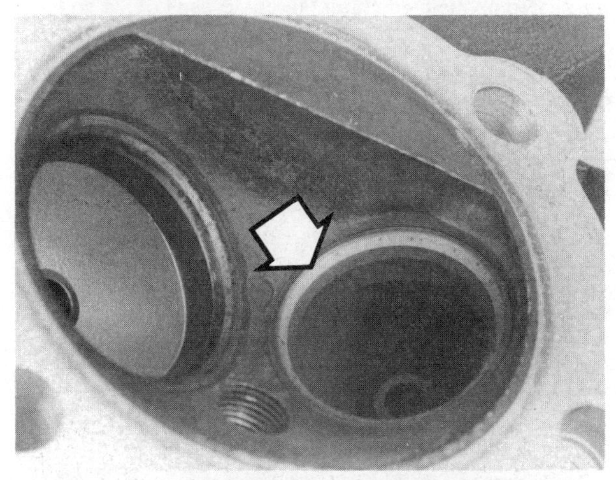

19.24b . . . and the seat should be the specified width (arrow), with a smooth, unbroken appearance

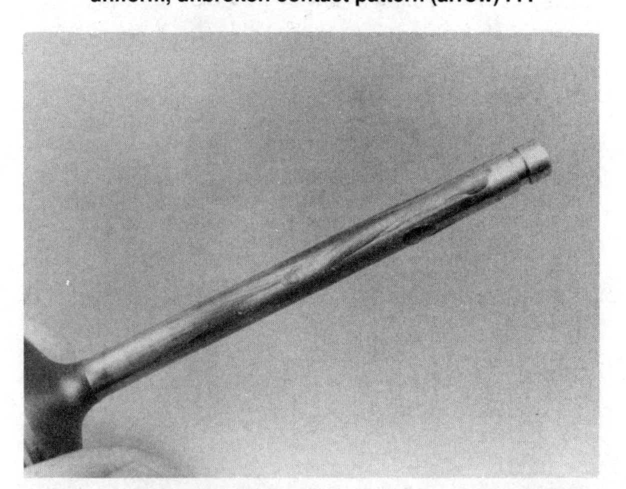

19.27 Lubricate the valve stem with moly-base grease before installing the valve in the cylinder head

cracks **(see illustration)**. Rotate the valve and check for any obvious indication that it's bent. Check the end of the stem for pitting and excessive wear. The presence of any of the above conditions indicates the need for valve servicing.

17 Measure the valve stem diameter **(see illustration)**. By subtracting the stem diameter from the valve guide diameter, the valve-to-guide clearance is obtained. If the valve-to-guide clearance is greater than specified, the guides will have to be replaced and new valves may have to be installed, depending on the condition of the old ones. Insert each valve into its guide and hold it tightly against the seat, then measure the distance from the cylinder head spring seat surface to the very end of the valve stem. If it's greater than specified, the valve and/or seat must be replaced.

18 Check the end of each valve spring for wear and pitting. Measure the free length and compare it to the Specifications **(see illustration)**. Any springs that are shorter than specified have sagged and shouldn't be re-used. Stand the spring on a flat surface and check it for squareness **(see illustration)**.

19 Check the spring retainers and keepers for obvious wear and cracks. Questionable parts shouldn't be reused – extensive damage will occur in the event of failure during engine operation.

20 If the inspection indicates that no service work is required,the valve components can be reinstalled in the heads.

Reassembly

Refer to illustrations 19.22, 19.23a, 19.23b, 19.24a, 19.24b, 19.27, 19.28, 19.30, 19.31a and 19.31b

21 Before installing the valves in the head, they should be lapped to ensure a positive seal between the valves and seats. This procedure requires fine valve lapping compound (available at auto parts stores) and a valve lapping tool. If a lapping tool isn't available, a piece of rubber or plastic hose can be slipped over the valve stem (after the valve has been installed in the guide) and used to turn the valve.

22 Apply a small amount of fine lapping compound to the valve face **(see illustration)**, then slip the valve into the guide. **Note:** *Make sure the valve is installed in the correct guide and be careful not to get any lapping compound on the valve stem.*

23 Attach the lapping tool (or hose) to the valve and rotate the tool between the palms of your hands. Use a back-and-forth motion rather than a circular motion **(see illustration)**. Lift the valve off the seat at regular intervals to distribute the lapping compound properly **(see illustration)**.

24 Continue the lapping procedure until the valve face and seat contact area is of uniform width and unbroken around the entire circumference of the valve face and seat **(see illustrations)**.

25 Carefully remove the valve from the guide and wipe off all traces of lapping compound. Use solvent to clean the valve and wipe the seat area thoroughly with a solvent-soaked cloth. Repeat the procedure for the remaining valves.

26 Once all of the valves have been lapped, check for proper valve sealing by pouring a small amount of solvent into each of the head ports with the valves in place and held tightly against the seats. If the solvent leaks past the valve(s) into the combustion chamber area, repeat the lapping procedure, then reinstall the valve(s) and repeat the check. Repeat the procedure until a satisfactory seal is obtained.

27 Coat the first valve stem with grease (preferably moly-based) then

19.28 Lay the spring seat in the head, over the guide, . . .

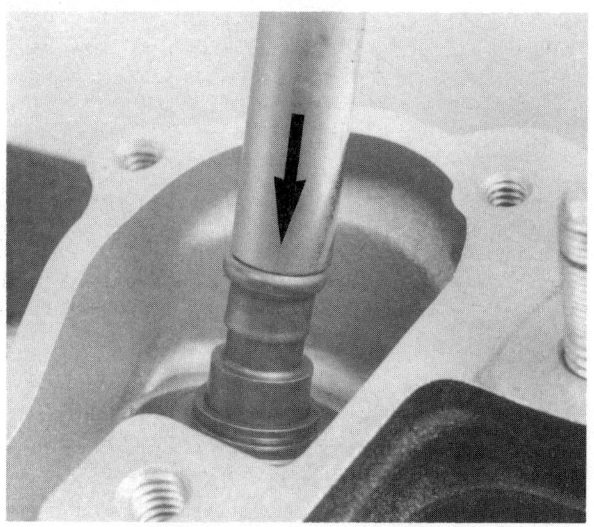

19.30 . . . THEN install the valve guide seal and make sure it's seated on the guide (don't forget to use Loctite on the guide before installing the seal)

19.31a Install the inner and outer valve springs, then position the retainer and compress the springs so the keepers can be installed

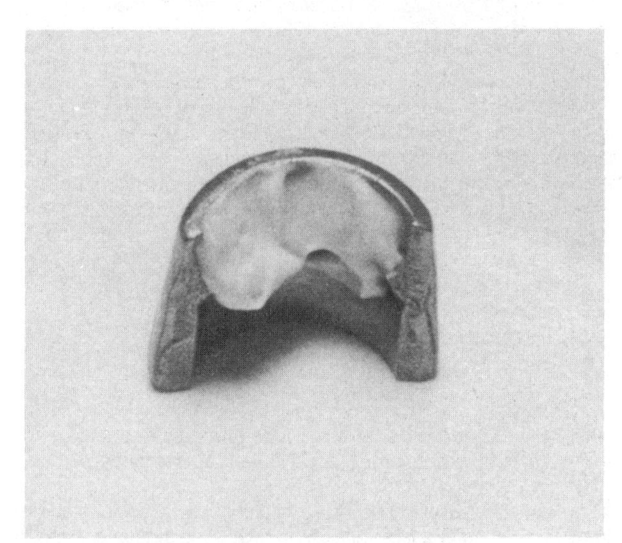

19.31b A small dab of grease will help hold the keepers in place on the valve while the spring is released

slide it into the guide **(see illustration)**.

28 Lay the spring seat in place in the cylinder head **(see illustration)**.

29 On models equipped with valve stem seals, place a protective sleeve over the valve stem so the keeper area of the stem is covered (if the seal is installed without the sleeve, the seal will be damaged). Apply a small amount of RC 620 Loctite (green) to the outside edge of the guide near the top.

30 Slide a new valve seal into position over the valve stem. Use a hammer and deep socket to install the seal **(see illustration)**. You'll be able to feel when the seal seats completely on the guide (don't continue hammering on it after it seats – you may damage it). Don't twist or cock it or it won't seal properly and DO NOT remove the valve from the guide once the seal is in place.

31 Position the valve springs and retainer over the valve stem and attach the valve spring compressor **(see illustration)**. Depress the valve springs only as far as absolutely necessary to slip the keepers into place. Make certain the keepers are securely locked in their retaining grooves. Apply a small amount of grease to the keepers to hold them in place as the pressure is released from the springs **(see illustration)**.

32 Repeat the procedure for the remaining valve in the first head and both valves in the other head.

33 Support the cylinder head on blocks so the valves can't contact the workbench top and gently tap each of the valve stems with a soft-face hammer. This will help seat the keepers in the grooves.

20 Hydraulic tappets – inspection

1 Before separating the components of both pairs of hydraulic tappet assemblies, note that the various components must not be interchanged with one another. Mixing up the parts must be avoided due to the very small clearances involved.

2 Remove each cam follower/tappet from the guide block and separate the follower from the tappet (if possible). The two component parts of the tappet can then be separated (early models only).

3 Clean the components thoroughly with solvent and allow them to dry. Blow out the oil holes with compressed air and use a length of fine wire to ensure the oil holes in the tappet guide blocks are open.

4 Use precision measuring tools to check the clearance between each tappet and guide block. If the clearance exceeds the specified limit, the components should be replaced with new ones. A worn cam roller can be replaced by pressing out the spindle pin. Press a new pin into place and

peen over or stake both ends to secure the pin in place.
5 The tappet must be a perfect fit, with no visible score marks in the bore it fits into or the body of the tappet. On early models, to test the two components, insert the tappet head into the cylinder and, without obscuring the oil passage, push the head down completely for six seconds and then release it. If the tappet head bounces back, the tappet assembly is in good condition. If the tappet doesn't bounce back, repeat the test with a finger covering the oil hole in the cylinder. If the piston now returns quickly a badly seating ball valve in the cylinder is indicated. Unless this is caused by a particle of dirt on the cylinder ball valve, the tappet must be replaced.
6 Check the socket that makes contact with the pushrod end. If wear is evident, the tappets must be replaced; the pushrods will probably require replacement also.
7 Soak the tappets in clean engine oil and keep them covered until engine reassembly.

21 Camshaft, gears and gearcase bearings – inspection

Refer to illustration 21.2

1 The timing gears are unlikely to require attention unless the engine has very high mileage or there has been a lubrication failure. Wear will be evident in the form of excessive backlash between the individual gears, with a characteristic "clacking" noise. If the gears are worn, the timing gear and camshaft must be replaced as a matched set. The dealer where you purchase the parts will make sure they're compatible.
2 The cam lobes should be checked for wear and damage in the form of pit marks, scuffing or flaking of the case hardened surfaces **(see illustration)**. Wear will be particularly evident on the flanks of the lobes, at the point where they begin to lift the tappets. If there's any doubt about the condition of the cam lobes, the camshaft should be replaced as a precaution.

21.2 Check the camshaft lobes, gear teeth and bearing journals (arrows) for wear, damage and heat discoloration

3 If the cam is worn or damaged, more than likely the tappets will require attention too. Check the rollers (which, if worn, can be replaced with a roller kit). Make sure the tappet guide itself isn't worn, causing the tappet to tilt and create additional mechanical noise. The pushrods should be examined and replaced if the ends are worn or damaged or if they're bent (see Section 18).
4 The needle roller bearings in the gearcase should be replaced if they don't turn smoothly or if damage or wear is evident. This also applies to all bushings. Replacement of the bearings should be done by a dealer service department (especially in the case of bushings, which may have to be reamed to size after installation). Don't forget the gearcase (timing) cover bushings. They're pegged in position and require expert attention when replacement is necessary.

22 Crankcase castings – inspection

1 After thorough cleaning, the crankcase castings should be examined for cracks and other signs of damage that may ultimately cause failure. Minor cracks can be repaired by welding, but if more extensive damage is apparent, replacement is recommended.
2 Note that crankcases are always supplied as a matched pair and should never be replaced any other way. This is important – the crankcases are line bored in pairs, so the bearing housings are aligned correctly. Replacement of only one half will result in a mismatch, which may cause the crankshaft to run out of line and absorb a surprising amount of power.
3 Make sure the mating surfaces are undamaged, otherwise oil leaks will be inevitable after reassembly. If there's any doubt about their ability to seal, use a liquid gasket sealer during reassembly.
4 Check the bearing housings to make sure they're undamaged. If they have worn as the result of a bearing rotating, it's possible to repair using a special bearing sealant such as red Loctite during reassembly. This can be used successfully only if the amount of wear is small.
5 Now is the opportunity to retap any of the threads, if they require attention. Most damage is caused by over tightening the drain plugs. If necessary, threaded holes can be repaired by installing Helicoil thread inserts, which will permit the original drain plug to be reused. Most dealers can perform this type of repair.

23 Oil seals – inspection and replacement

1 Even after very careful examination, it's difficult to determine whether an oil seal can be reused, especially if its been disturbed during disassembly. Because an oil seal failure will require another teardown at a later date, replace all oil seals during an overhaul as a precautionary measure.
2 Oil seals are very easily damaged during reassembly. Always be very careful when installing shafts and apply grease to the seal lips.
3 The most important oil seals on the engine are the ones on the drive side of the crankshaft assembly and in the gearcase cover, sealing the camshaft journal.

24 Engine reassembly – general information

1 Before reassembly is started, the various engine components should be cleaned thoroughly and placed close to the work area.
2 Make sure all traces of old gaskets have been removed and that the mating surfaces are clean and undamaged. One of the best ways to remove old gasket sealer is to apply aerosol gasket remover, available at most auto parts stores. This acts as a solvent and will ensure the sealer is removed without scraping and the accompanying risk of damage.
3 Gather up all necessary tools and have an oil can filled with clean engine oil available. Make sure all the new gaskets, oil seals and replacement parts are on hand; there's nothing more frustrating than having to stop in the middle of a reassembly sequence because a vital gasket or part has been overlooked.
4 Make sure the reassembly area is clean and well lit, with adequate working space. Refer to the torque specifications (when given). Many smaller bolts are easily sheared off if overtightened. Always use the correct size screwdriver bit for Phillips-head screws. Remember, if the heads are badly damaged during reassembly, the screws may be almost impossible to loosen later.

25 Crankcases – reassembly

Refer to illustrations 25.1, 25.2a, 25.2b, 25.3a, 25.3b, 25.4a and 25.4b

1 Install the timing side main bearing cages and rollers on the right-hand main shaft and the flywheel side (inner) tapered roller main bearing

25.1 Install the flywheel side (inner) tapered roller main bearing and spacer ring on the left-hand mainshaft

25.2a Support the crankshaft securely on blocks of wood, . . .

25.2b . . . then carefully lower the left side crankcase over the bearing

25.3a Slip the outer (sprocket side) tapered roller main bearing onto the mainshaft . . .

and inner race spacer ring on the left-hand main shaft **(see illustration)**. The tapered roller main bearing inner race should be driven into place with a length of pipe or the appropriate special tool.

2 Support the flywheel assembly securely on wooden blocks so the right-hand main shaft is facing down **(see illustration)**. Lubricate the tapered roller main bearing with clean engine oil and position the left side crankcase over the shaft **(see illustration)**. Make sure the connecting rods don't damage the crankcase mouths when the crankcase is lowered

25.3b . . . and drive it into place with a piece of pipe and a hammer – don't continue to pound on it once it's seated

25.4a Install the oil seal collar on the mainshaft

25.4b Grease the lip of the oil seal before installing it (fitted direction for wet-clutch models shown)

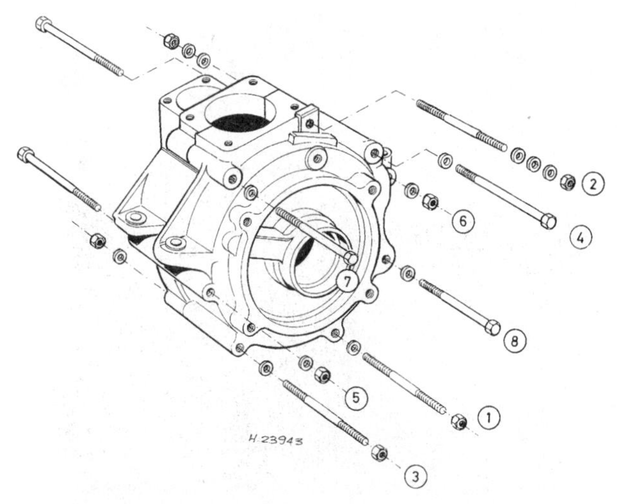

25.7 Crankcase stud and bolt positions – 1991 and later models (circled numbers indicate tightening sequence)

26.3 Secure the pump gear with the circlip (arrow)

into position.

3 Install the outer tapered roller main bearing in a manner similar to that used for installing the inner bearing **(see illustrations)**. Make sure the outer bearing is completely seated against the inner race spacer ring and make sure there's a small amount of end play in the flywheel assembly. If there is no end play, the tapered bearings are under preload. This condition must be rectified by pressing the flywheel assembly out of the bearing and installing one or more shims between the bearing spacer and one of the bearing inner races (early models) or by fitting a different thickness bearing spacer ring (later models). Shims are available in a variety of sizes. Adjust the end play to the specified limit.

4 Place the oil seal collar on the shaft so the widest diameter faces in **(see illustration)**. Drive the collar onto the shaft, using a length of pipe with an inside diameter slightly larger than the main shaft. Lubricate the sealing lip of the crankshaft oil seal and slide it over the main shaft and oil seal collar with the spring side facing in on dry-clutch models or facing out on wet-clutch models **(see illustration)**. Be careful to keep it square with the bore and tap it completely into the case. Make sure the flywheel assembly rotates freely in the crankcase half.

5 Invert the flywheel assembly, again blocking it securely on the workbench. Make sure the timing side main bearing spiral lock ring is in position and lubricate the bearing with clean engine oil.

6 Clean the mating surfaces of both crankcase halves and apply a non-hardening gasket sealant to the left-hand crankcase mating surface. Position the right-hand crankcase half over the mainshaft and lower it into position so the crankcase stud holes align.

7 On all models through 1990, insert the six crankcase studs and the two bolts, and install the stud washers and nuts. Note that the upper stud (between the crankcase mouths) may retain the speedometer drive cable clip or horn bracket on the left-hand side. On 1991 and later models, insert the three studs in their holes and secure their nuts lightly, then fit the bolts **(see illustration)**.

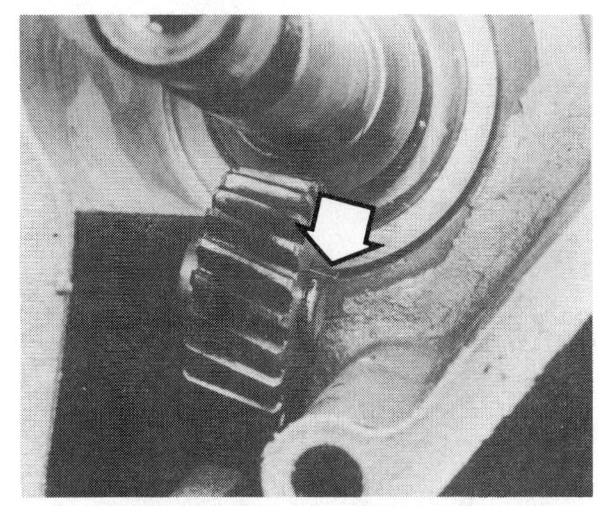

26.2 Position the oil pump driven gear, THEN install the Woodruff key (arrow)

8 Before tightening the crankcase nuts/bolts ensure that the cylinder barrel surfaces are exactly aligned; tap the casting gently to bring into alignment if necessary.

9 On all models through 1990 tighten the fasteners evenly in a criss-cross pattern to 15 to 19 ft-lbs (20 to 26 Nm). On 1991 and later models, all fasteners should be tightened evenly to 10 ft-lbs (14 Nm) following the sequence shown in illustration 25.7, and then after the cylinder barrels and heads have been installed, tightened to 15 to 17 ft-lbs (20 to 23 Nm). **Note:** *At all times during tightening make sure that the flywheel assembly is able to revolve freely.*

10 Before continuing, pad the crankcase mouths with clean rags to prevent dirt and other foreign matter from entering.

26 Oil pump – installation

Refer to illustrations 26.2 and 26.3

Note: *Use new gaskets and oil seals, but DO NOT use any gasket sealant when installing the oil pump.*

2

1 Lubricate the oil pump driveshaft and slide the shaft through the bushing in the timing case so the oil pump is up against the milled surface of the case. As the shaft is inserted, the driven gear must be installed. Due to lack of space, the gear cannot be installed once the shaft is inserted.

2 Align the gear keyway and shaft keyway and insert the small Woodruff key to secure the two components together **(see illustration)**.

3 Install the gear retaining circlip **(see illustration)**.

4 Install the mounting nuts/bolts. Tighten them evenly in a criss-cross pattern. Keep checking to make sure the pump revolves freely – uneven bolt tension may cause misalignment. A defective gasket may cause the gears to bind by interfering with them during rotation.

5 Don't overtighten the oil pump fasteners or distortion will occur, causing oil leaks.

6 Apply thread locking compound to the crankcase breather union and screw it into the crankcase to the rear of the oil pump. Reconnect the union with the chain oiler take off at the oil pump. Secure the union hose with new hose clamps (not all models).

27 Camshaft and timing gears – installation

Refer to illustrations 27.1, 27.2, 27.3, 27.6, 27.7, 27.8, 27.9 and 27.11

1 Place the inner Woodruff key in the keyway on the crankshaft and

27.1 Slide the oil pump drive gear and timing gear spacer (models through 1992) into position, . . .

27.2 . . . then install the timing gear

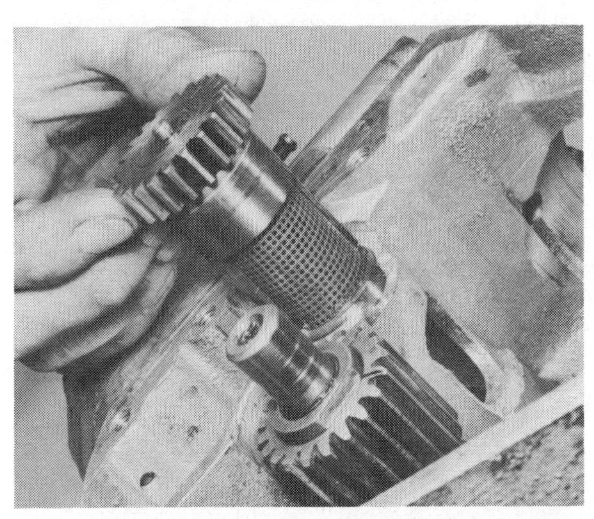

27.3 Insert the timed breather into the crankcase bore

27.6 Insert the needle bearing and position the thrust washer as shown before installing the camshaft (and shim)

27.7 Align the valve timing marks on the gears exactly as shown here (arrows)

27.8 Don't forget to install the timed breather spacer (apply a dab of grease to hold it in place while the cover is installed)

27.9 Install a new cover gasket over the dowel sleeves, then lubricate the cover bushings and seal and install the cover

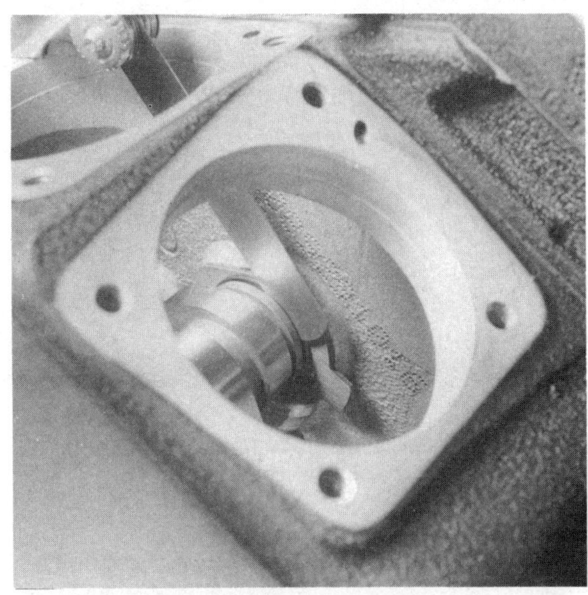
27.11 Check the camshaft end play with feeler gauges before permanently bolting the gearcase cover in place

slide the oil pump drive gear into position **(see illustration). Note:** *On early 1985 through 1992 Evolution engines, the chamfer on the oil pump drive gear must face IN. After the timing gear nut is tightened, make sure the timing gear spacer has noticeable end play.*

2 Install the spacer and outer Woodruff key (models through 1992) and replace the timing gear with the timing marks facing out **(see illustration)**. Lock the crankshaft, then install and tighten the timing gear nut. **Note:** *If the timing gear is replaced with a new one, the camshaft must be replaced as well because the gears must be matched for the correct clearance. When you purchase the new parts, the dealer should be able to supply them in matched sets.*

3 Slide the timed breather into position in the timing case **(see illustration)**. Before proceeding with reassembly, the timed breather end play should be checked and, if necessary, adjusted. Install a new timing case gasket and position the timed breather spacer on the end of the breather. Place a straightedge across the gasket and measure the gap between the straightedge and the spacer with a feeler gauge. Subtract 0.006-inch (0.152 mm) from the measurement to find the end play, which should be within the specified limits. If the end play is incorrect, substitute a thrust washer of a different thickness. The washers are available in 0.110, 0.115, 0.120, and 0.125-inch (2.80, 2.92, 3.04 and 3.17 mm) sizes.

4 Lubricate the camshaft needle bearing and the plain bushings in the gearcase cover.

5 Rotate the flywheel until the left-hand mark on the timing gear face is at 12 o'clock.

6 Install the special thrust washer in the rear of the crankcase so the camshaft needle bearing is covered **(see illustration)**.

7 Place the shim (not used on 1988 and later model Evolution engines) on the inner end of the camshaft and install the camshaft so the timing marks align **(see illustration)**.

8 Place the timed breather spacer in position **(see illustration)**.

9 Lubricate the lip of the camshaft seal and install the gearcase cover, together with a new gasket **(see illustration)**.

10 Install the cover screws and tighten them evenly, in a criss-cross pattern.

11 The camshaft end play must be checked **(see illustration)**. If it's outside the specified limit, it must be adjusted by installing different shims. Shims are available in increments of 0.005-inch (0.127 mm) from 0.050 to 0.095-inch (1.27 to 2.41 mm).

28 Ignition components – installation

1970 through early 1978 models

1 Installation of the ignition components and the advance unit is the reverse of the removal procedure described in Section 8 of this Chapter.

2 The centrifugal advance unit on all but early 1978 models will locate on the protruding end of the camshaft.

3 Install the contact breaker point baseplate, aligning the scribed marks made during disassembly so the timing is approximately correct.

Late 1978 and 1979 models

4 Be sure to seat the advance assembly for the late 1978 and 1979 ignition system squarely on the end of the camshaft. Install the trigger rotor with its flat side next to the cam stop roll pin on the advance assembly base. It must also engage both flyweights on the advance unit.

1980 and later models

5 When installing the rotor bolt, the threads should be coated with thread locking compound. If the sensor plate was completely removed on these models, it may be necessary to install new sockets, wire pins and body receptacle.

6 Adjust the timing as described in Chapter 1.

7 When installing the outer cover, special rivets must be used. These rivets are specially designed so the end doesn't fall off in the timing compartment. The use of regular rivets could damage the ignition commponents. The recommended rivets are part no. 8699.

29 Tappets and guides – installation

Refer to illustrations 29.3 and 29.4

1 Squirt oil through the opening in the top of the timing case so the timing gears and camshaft lobes are liberally coated with it.

2 Place new gaskets in position over the openings in the case.

3 Lubricate the components of each tappet and guide block.Note that the tappets must be installed with the machined faces and oil holes facing

29.3 **The oil hole in the tappet (arrow) MUST face in (pre-Evolution engine only)**

29.4 **The guide blocks are marked FRONT (cylinder) and REAR (cylinder) to avoid mix-ups**

in **(see illustration)** (pre-Evolution engines only).

4 To prevent the tappets from falling into the gearcase during reassembly, pinch the two tappets together as the guide block is lowered into position. **Note:** *The guide blocks are marked FRONT (cylinder) and REAR to avoid mix-ups during reassembly* **(see illustration)**.

5 Install and tighten the four bolts that retain each tappet guide block. You'll have to use a 12-point socket on some models to tighten the bolts. **Note:** *On Evolution engines, a tappet guide alignment tool must be threaded into the bolt hole nearest the tappet oil feed hole, then the other three bolts can be installed and tightened. Remove the guide tool, install the fourth bolt and tighten it securely.*

6 Replace the hydraulic tappet oil screen and spring in the opening in the top of the case. The closed end of the screen must face up. The sealing washer on the plug should be replaced with a new one before the plug is installed.

7 On late 1981 and early 1982 models, attach the oil hose and secure it in position with a hose clamp.

30 Piston rings – installation

1 Before installing the new piston rings, the ring end gaps must be checked.

2 Lay out the pistons and the new ring sets so the rings will be matched with the same piston and cylinder during the end gap measurement procedure and engine assembly.

3 Insert the top (no. 1) ring into the first cylinder and square it up with the cylinder walls by pushing it in with the top of the piston, The ring should be at least 1-inch below the top edge of the cylinder. To measure the end gap, slip a feeler gauge between the ends of the ring and compare the measurement to the Specifications.

4 If the gap is larger or smaller than specified, double-check to make sure you have the correct rings before proceeding.

5 If the gap is too small, it must be enlarged or the ring ends may come in contact with each other during engine operation, which can cause serious damage. The end gap can be increased by filing the ring ends very carefully with a fine file. Mount the ring in a vise equipped with soft jaws, holding it as close to the gap as possible, When performing this operation, file only from the outside in.

6 Excess end gap isn't critical unless it's greater than the service limit shown in the Specifications. Again, double-check to make sure you have the correct rings for the engine.

7 Repeat the procedure for each ring that will be installed in the first cylinder and for each ring in the remaining cylinder. Remember to keep the rings, pistons and cylinders matched up.

8 Once the ring end gaps have been checked/corrected, the rings can be installed on the pistons. Be sure to install them with the gaps staggered about 120-degrees.

9 The oil control ring (lowest on the piston) is installed first. Models from 1970 through early 1978 have a full width slotted ring with a spring expander. Late 1978 and later models have rings composed of three separate components (two side rails and a spring expander). Place the spring expander and the slotted oil ring in the lower groove of the piston in the same position they were in before disassembly. On late 1978 and later models, slip the spring expander into the groove, then install the upper side rail. Do not use a piston ring installation tool on the oil ring side rails as they may be damaged. Instead, place one end of the side rail into the groove between the spring expander and the ring land. Hold it firmly in place and slide a finger around the piston while pushing the rail into the groove. Next, install the lower side rail in the same manner.

10 After the oil ring components have been installed, check to make sure the components can be turned smoothly in the ring groove.

11 The no. 2 (middle) ring is installed next. Later model engines (1340 cc engines beginning with crankcase number 1279-345-165 and 1200 cc engines beginning with crankcase number 179-023-001) use a taper face type of compression ring for the no. 2 ring. The new type-ring is identifiable by the black color on the outside edge of the ring instead of the chrome edge. The ring should be installed so the stamped word TOP (and the dot or dash) are facing up.

12 To avoid breaking the ring, use a piston ring installation tool. Fit the ring into the middle groove on the piston. Don't expand the ring any more than necessary to slide it into place.

13 Install the top ring in the same manner as the middle ring.

14 Repeat the procedure for the other piston and rings. Be very careful not to confuse the top and middle rings on later models.

15 After all of the rings are installed, make sure they move freely in the grooves, but don't change the positions of the ring gaps.

31 Pistons – installation

Refer to illustrations 31.2, 31.4, 31.5 and 31.8

1 Prior to installing the pistons on the connecting rods, install the rings on the pistons as described in the previous Section.

2 Rotate the crankshaft until the rod journal is at TDC, then position the pistons over the rods, facing in the direction marked during disassembly. If new pistons are being installed, be sure they're installed in the cylinder the rings were measured in **(see illustration)**.

3 Place a new base gasket over the studs for each cylinder. Place clean rags around the openings in the crankcase to keep dirt and debris out. Install one new circlip in each piston to retain the piston pin. Make sure it's seated in the groove.

4 Coat the small end of the connecting rod with moly-base grease. Slide the piston pin through the side of the piston opposite the circlip. Align the piston and connecting rod and push the piston pin through the rod to the other side of the piston **(see illustration)**.

5 Install the other circlip on the other side of the piston. Be sure the circlips are completely seated in the grooves. The ends of the circlips should

31.2 The nub on the piston must be on the right-hand side of the engine (pre-Evolution engine)

31.4 Align the piston and connecting rod and insert the piston pin

be facing up **(see illustration)**.

6 Assemble the other piston and connecting rod in the same way.

7 Coat the pistons, rings and grooves and the cylinder walls with clean engine oil. Make sure the ring end gaps are still spaced about 120-degrees apart.

8 Compress the piston rings with a ring compressor **(see illustration)** and insert the piston into the cylinder bore, carefully lowering the barrel until the piston rings enter the bore. Remove the rag from the crankcase opening and lower the barrel until it's seated. On pre-Evolution engines, install and tighten the nuts and flat washers.

9 Install the remaining barrel over the other piston in the same manner.

10 If possible, two persons should be employed during this operation. One can insert the piston and rings while the other supports the barrel and lowers it into position.

11 Connect the oil hoses to the fittings on the base of the cylinder on late 1981 and early 1982 models. If the fittings were removed, they should be installed using new O-rings.

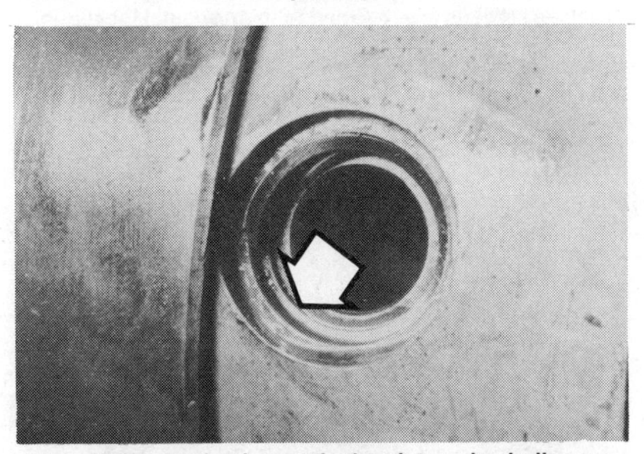

31.5 Make absolutely certain the piston pin circlips are seated in the piston grooves – if they fly out during engine operation, extensive damage will occur

32 Cylinder heads and rocker boxes – installation

Pre-Evolution engine

Refer to illustrations 32.2 and 32.5

1 If the rocker boxes have been disassembled, they should now be reassembled, keeping the parts in their original locations. Oil the rocker shafts prior to insertion and install new sealing washers behind the acorn nuts. Use new O-rings behind the caps on the other end.

2 Place a new rocker box gasket over the studs in the cylinder head. To prevent the heel of each rocker arm from contacting the valve springs when the rocker box is attached to the cylinder head, place the inverted rocker box on the workbench and lower the cylinder head into position **(see illustration)**. Hold the two components together, invert them and in-

31.8 A piston ring compressor attached to the piston makes installation of the barrel much easier

32.2 Carefully attach the cylinder head to the rocker box (pre-Evolution engine)

32.5 Install each pushrod and tube separately – make sure the pushrods are returned to their original locations

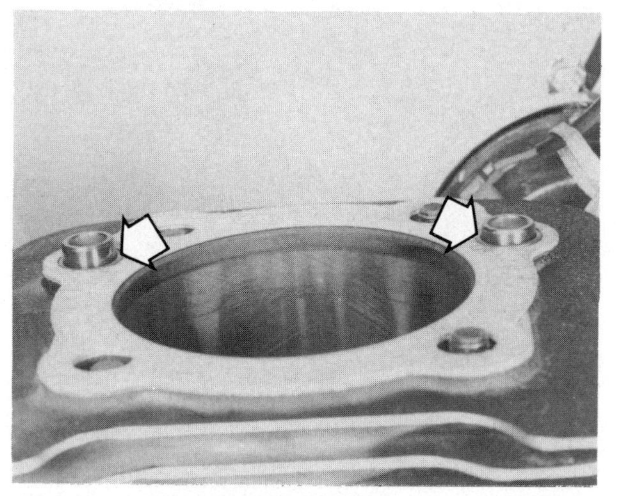

32.10 On Evolution engines, position the O-rings (arrows) over the dowel sleeves first, then install the head gasket and make sure it's aligned properly

32.12 Clean and lubricate the threads and the underside of each head bolt (arrows) prior to installation

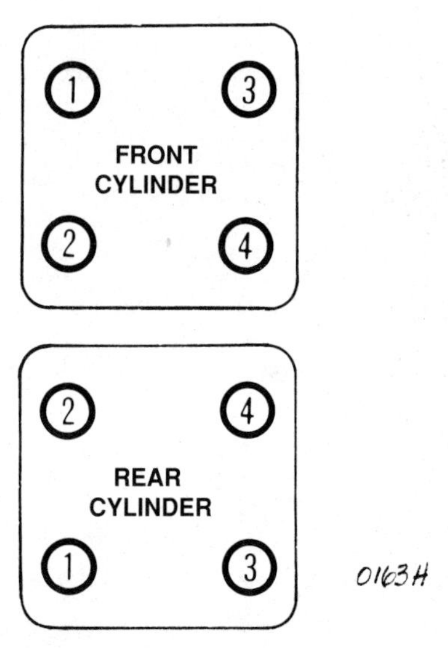

32.14 Evolution engine cylinder head bolt TIGHTENING sequence

stall the nuts. Tighten the nuts in 1/4-turn increments, following a criss-cross pattern, to the recommended torque.

3 Install the cylinder heads. Use new cylinder head gaskets (no sealant) and tighten the bolts in a criss-cross pattern in three steps.

4 Separate the components of each pushrod tube by removing the keeper from the upper portion of each tube. Replace the tube seals with new ones. Reassemble each pushrod tube, without the keeper, in the retracted position and insert them into the guide block recesses.

5 The pushrods should be installed in the following sequence. Rotate the crankshaft until the front cylinder exhaust tappet is just beginning to move. There's now sufficient clearance to install the rear cylinder exhaust pushrod. If necessary, loosen the pushrod adjustment locknut and screw in the adjuster to shorten the pushrod. Rotate the crankshaft until the front cylinder intake tappet is just beginning to move and install the rear cylinder intake pushrod. Repeat the procedure to install the front cylinder pushrods **(see illustration)**.

6 Rotate the crankshaft until one valve is completely open. The counterpart of this valve on the other cylinder is now completely closed and the tappet clearance can be checked accurately. Loosen the pushrod adjuster locknut and screw in the adjuster until there's perceptible play. Slowly unscrew the adjuster until the play is just eliminated. Mark the adjuster with a piece of chalk and then screw it down exactly four (4) turns. If the hydraulic tappet is dry (such as during reassembly), turn the adjusting screw down until the tappet is completely compressed. Mark the position, then turn the adjuster up exactly 1-3/4 turns on 1970 through early 1978 models, or exactly 1-1/2 turns on late 1978 through 1983 models. Tighten the locknut. Extend the pushrod tube so the upper and lower ends of the tube are located against the seals and then install the keeper.

7 Repeat the procedure for the remaining three pushrods.

8 Install the rocker oil feed line that runs from the crankcase to the rear rocker box and also the rocker box connection oil feed line. Use new rubber seals at each union.

9 Install the intake manifold (make sure the clamps are completely tight to prevent air leaks). Install new O-rings.

Evolution engine

Refer to illustrations 32.10, 32.12, 32.14, 32.16, 32.17, 32.18, 32.25a and 32.25b

Note: *Before beginning this procedure, clean the cylinder head bolt threads, then lubricate them with clean engine oil and thread the bolts onto the studs to make sure they don't bind.*

10 Install new O-rings over the dowel sleeves, then position the new head gasket on the cylinder barrel **(see illustration)**. No sealant is required on the head gaskets.

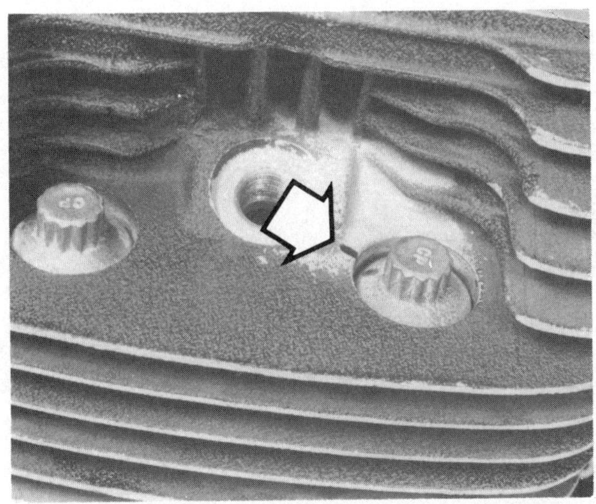

32.16 After the head bolts are all at the specified torque, make a mark on each bolt head flange and extend it onto the head (arrow)

32.17 Turn each cylinder head bolt an additional 90-degrees (1/4-turn) in an uninterrupted motion – this is very important and must be done exactly as described to prevent head gasket leaks (see text for 1984 FXST models)

32.18 Position the new rocker cover gaskets on the cylinder head (arrows) with the sealant beads facing up

32.25a Position new gaskets (arrows) in the lower rocker arm cover grooves, . . .

11 Make sure the bolt holes in the head are clean, then carefully lower it onto the cylinder – the dowel sleeves must enter the holes and align the head.

12 Lubricate the threads and the underside of each bolt head with engine oil **(see Illustration)**.

13 Slip the washers over the bolts (where applicable), then install the bolts finger-tight. **Caution:** *The procedure for tightening the cylinder head bolts is extremely critical to prevent gasket leaks, stud failure and cylinder or head distortion.*

14 Refer to the accompanying illustration and tighten bolt number one to 7 ft-lbs (9 Nm). Tighten the remaining bolts to 7 to 9 ft-lbs (9 to 12 Nm) following the sequence shown.

15 Tighten each bolt to 12 to 14 ft-lbs following the sequence in illustration 32.14. **Note:** *On 1984 FXST models, tighten the bolts to 15 to 17 ft-lbs. (20 to 23 Nm)*

16 Mark each bolt and the cylinder head with a felt-tip pen **(see illustration)**.

17 Turn each bolt, in the recommended sequence, an additional 90-degrees (1/4-turn) **(see illustration)**. **Note:** *On 1984 FXST models, don't turn the bolts 90-degrees. Instead, tighten them to 24 to 26 ft-lbs (33 to 35 Nm).*

18 Install the rocker cover gaskets with the sealant beads facing UP **(see illustration)**.

19 Install new seals and assemble the pushrod tubes, then slip the pushrods into the tubes and install them in their original locations. Make sure they're seated in the tappets. **Note:** *The pushrods are color coded to ensure correct installation (rear exhaust – purple; rear intake – blue; front intake – yellow; front exhaust – green). Don't turn the pushrods end-for-end; they must be mated with the rocker arm or tappet just like they were originally.*

20 Make sure the tappets for the cylinder being reassembled are on the cam lobe base circles (turn the crankshaft if necessary to reposition them).

21 Lubricate the rocker arm faces with moly-base grease, assemble the rocker arms and shafts in the lower rocker arm cover and slip all the bolts through the holes. The cutouts in the shafts must be positioned so the bolts will pass through the cover and the cutouts to retain the shafts.

22 Position the lower rocker arm cover on the head and thread all the bolts into place finger-tight. Make sure the pushrods are engaged in the rocker arm recesses.

23 Tighten the bolts in 1/4-turn increments, following a criss-cross pattern, to the specified torque. This will allow the tappets to bleed down slowly.

24 Make sure the pushrods spin freely.

25 Position new gaskets in the lower rocker arm cover **(see illustration)**,

32.25b ... then install the middle rocker arm cover and a new gasket (arrow) – use "Gaska-cinch" to hold the gaskets in place in the cover grooves if necessary

32.26 Rubber umbrella valve location in rear cylinder – 1993 and later models

35.5 The crankshaft must be locked in place to loosen the compensating sprocket nut

then install the middle rocker arm cover and a new gasket **(see illustration)**.

26 On 1993 and later models, check the breather passage from the rubber umbrella valve in the inner corner of each middle rocker cover to the bleed bolt hole on the side of the head – it should be unrestricted. If the rubber valve is damaged replace it **(see illustration)**.

27 Install the upper rocker arm cover and screws. Tighten the screws after making sure the middle rocker arm cover is positioned evenly on all sides.

33 Alternator – installation

1 Place the alternator stator over the left end of the crankshaft so it rests in approximately the correct position.

2 Insert the wiring plug into the recess in the crankcase and push it carefully into position until the ridge molded on the plug is seated against the inside edge of the case. Place the bridge piece over the wiring and secure it with the two screws.

3 Align the alternator stator so the four retaining bolts can be inserted. Lay the lock plates in position and install the bolts. Tighten the bolts evenly and then secure them by bending over the edges of the lock plates. Some Evolution engines have Torx-head bolts, which must not be reused – always install new bolts.

4 Install the alternator rotor on the splines of the crankshaft and carefully tap it into place. **Note:** *On 1986 and later Evolution engines, install the small diameter spacer on the shaft, followed by the alternator rotor, then the large diameter spacer.*

5 Install the primary drive components.

34 Engine – installation

1 As was evident during engine removal, there's very little clearance available for maneuvering the engine into the frame. You should have three persons available for this operation – two to lift the engine and locate it correctly, while the third person steadies the frame. Lift the engine in from the right-hand side of the machine.

2 When the engine is in position, install the two front and two rear engine mounting bolts, being careful not to damage the threads as they're tapped into place. Move the engine as necessary to align the bolt holes. Install and tighten the nuts. Replace the cylinder head support bracket and tighten the retaining nuts and the main bolt. Note the washer positioned between the frame lug and bracket – it shouldn't be left off or the stress will strain the bracket and frame.

3 Reverse the removal procedure for the remaining components.

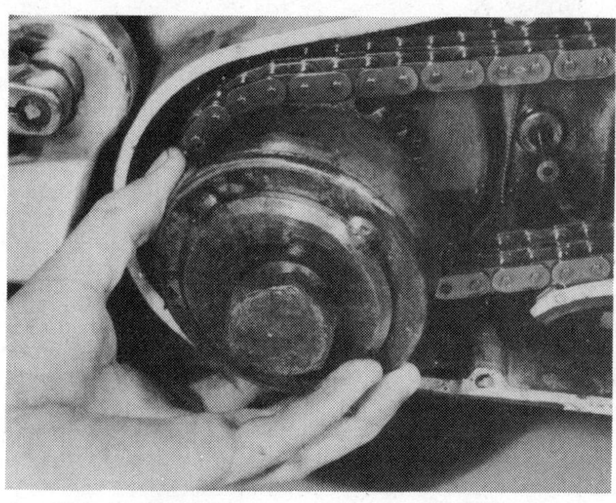

35.6a Remove the nut and cover, followed by ...

35.6b ... the collar (not used on all models) and the sliding cam

35 Primary drive and clutch components – removal

Primary drive components

Refer to illustrations 35.5, 35.6a, 35.6b, 35.6c, 35.7 and 35.8

1 Shift the transmission into gear. On some models the gear shift lever must be removed in order to remove the primary chaincase cover. On these models, the shift lever is secured to the splined shaft by a pinch bolt.

2 Remove the screws securing the primary chaincase cover. The screws should be loosened a little at a time in a criss-cross pattern.

3 Tap around the edge of the cover with a soft-face hammer to break the gasket seal. Detach the cover from the chaincase.

4 Lock the engine so the crankshaft can't turn. In order to do this, the transmission must be in gear and the drive sprocket must be held with a chain wrench or some other tool. Another way to do this is to apply the rear brake while the transmission is in gear or remove the spark plugs and bring the piston to the TDC position. Back the crankshaft up 1/8 of a revolution and fill the combustion chamber with nylon rope inserted through the spark plug hole. Be sure the end of the rope is still outside the engine.

5 Loosen the compensating sprocket nut. Some early models are equipped with a nut that requires a special peg wrench for removal and installation. Harley-Davidson special tool no. 94557-55 is recommended, but it's unlikely to be readily available. If the special tool isn't available, attach a U-bolt and use a large screwdriver or pry bar as a lever. On other models the nut is very tight and will require a large wrench with a long handle in order to apply enough leverage to loosen it **(see illustration)**.

6 On primary chain models, unscrew the nut and remove the cover, the collar and the sliding cam **(see illustrations)**. The compensating sprocket and the shaft extension must remain in place until the primary chain is ready to come off (after the clutch is disassembled).

7 On belt primary drive models, remove the lock washer and outer plate **(see illustration)**. The compensating sprocket assembly can remain in place until the belt is removed (after the clutch is disassembled). **Note:** *A set screw secures the outer plate to the compensating sprocket. There's*

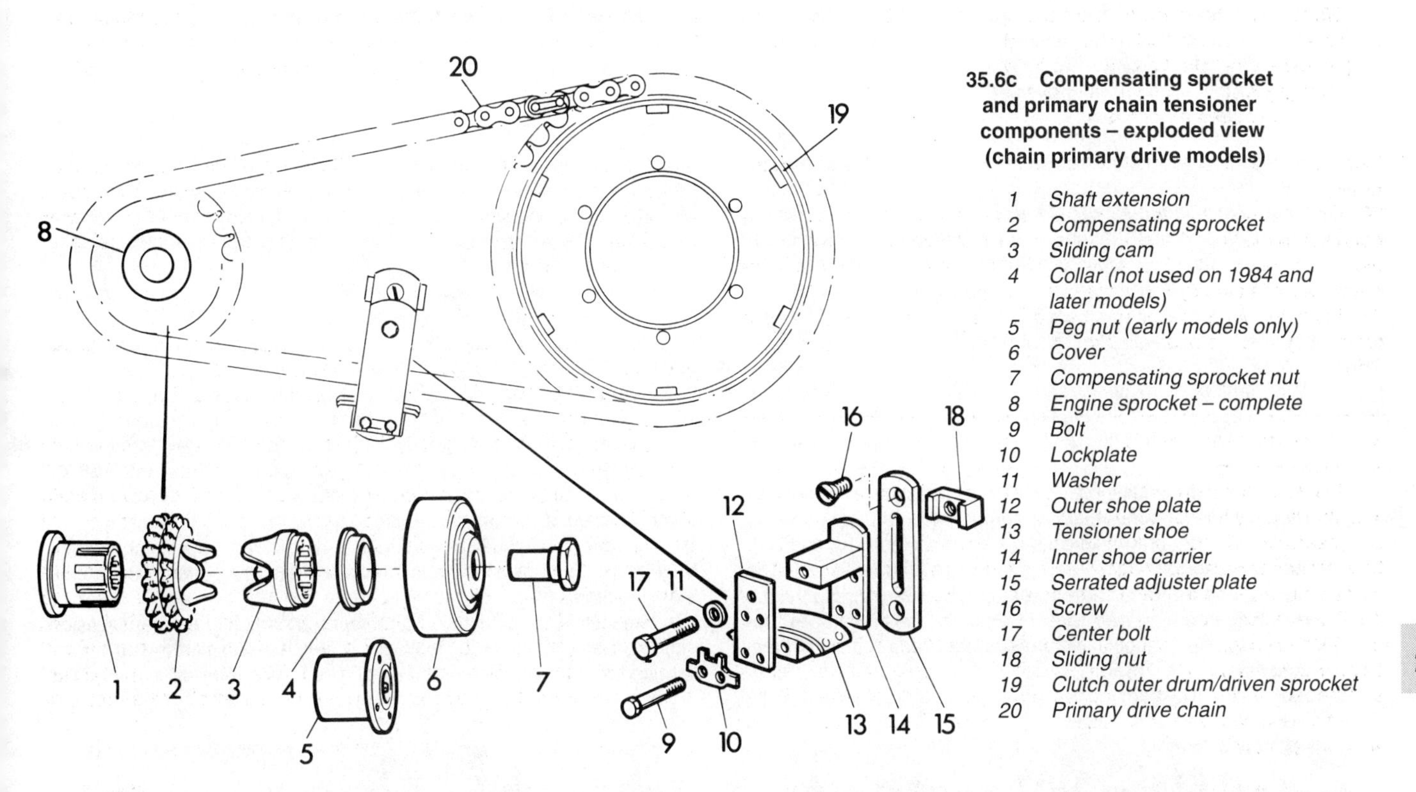

35.6c Compensating sprocket and primary chain tensioner components – exploded view (chain primary drive models)

1 *Shaft extension*
2 *Compensating sprocket*
3 *Sliding cam*
4 *Collar (not used on 1984 and later models)*
5 *Peg nut (early models only)*
6 *Cover*
7 *Compensating sprocket nut*
8 *Engine sprocket – complete*
9 *Bolt*
10 *Lockplate*
11 *Washer*
12 *Outer shoe plate*
13 *Tensioner shoe*
14 *Inner shoe carrier*
15 *Serrated adjuster plate*
16 *Screw*
17 *Center bolt*
18 *Sliding nut*
19 *Clutch outer drum/driven sprocket*
20 *Primary drive chain*

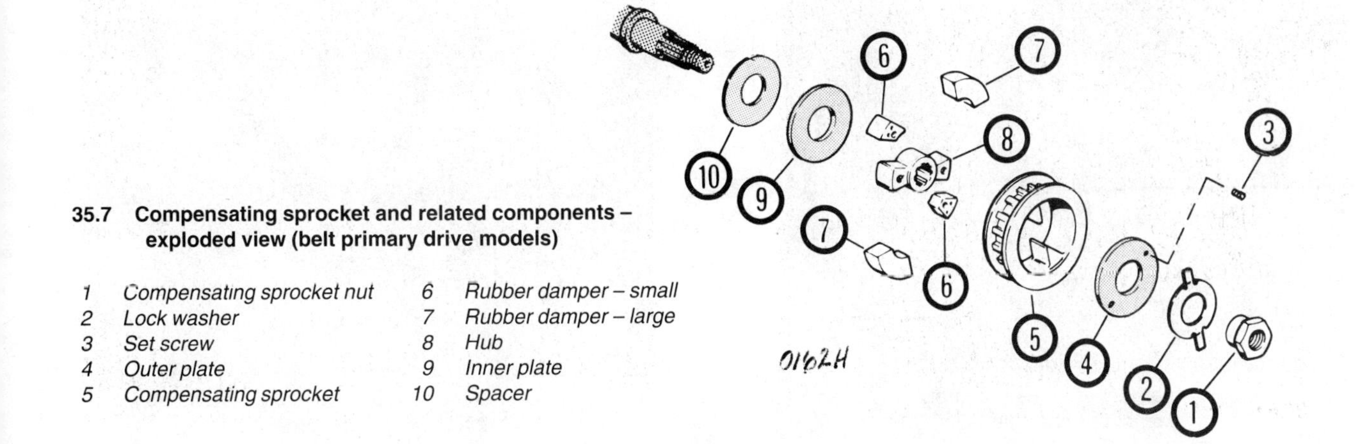

35.7 Compensating sprocket and related components – exploded view (belt primary drive models)

1 *Compensating sprocket nut*
2 *Lock washer*
3 *Set screw*
4 *Outer plate*
5 *Compensating sprocket*
6 *Rubber damper – small*
7 *Rubber damper – large*
8 *Hub*
9 *Inner plate*
10 *Spacer*

35.8 Unscrew the large center bolt (arrow) to free the chain tensioner assembly

35.10 Use a large washer (arrow) behind the adjusting screw locknut to compress the clutch springs

only one set screw and its main purpose is to keep the outer plate aligned with the hub in case a puller must be used to remove the sprocket assembly. After the set screw is removed, push the outer plate, the rubber dampers, the hub and the inner plate out of the sprocket.

8 On primary chain models, remove the complete chain tensioner assembly by unscrewing the large center bolt that secures it to the rear of the crankcase **(see illustration)**.

Dry clutch (through early 1984 models)

Refer to illustrations 35.10, 35.11, 35.14 and 35.17

9 The clutch spring backplate, the springs and the clutch pressure plate are removed as a unit. Because the two plates are under considerable pressure from the springs, a special procedure must be followed to ensure safe removal.

10 Remove the clutch adjusting screw locknut. Place a large washer, with an outside diameter larger than the center hole in the spring backplate, over the adjusting screw and reinstall the locknut **(see illustration)**.

11 Tighten the locknut until the springs are slightly compressed. The three spring tensioning nuts can now be removed and the plate/spring assembly detached safely by pulling it out of the clutch outer drum **(see illustration)**. The components should only be separated if one or more of them require replacement.

12 To separate them safely, insert three bolts through the original spring tensioning nut stud holes and attach a nut to each bolt. Tighten the nuts until the springs are compressed slightly, then remove the locknut and adjusting screw. Separate the parts by loosening the three nuts a little at a time. The bolts must have threads long enough to allow complete release of the spring pressure.

13 Remove the clutch plates, one at a time, noting their relative positions to simplify reassembly.

14 Grasp the clutch outer drum with one hand and the engine sprocket with the other and pull the two components off at the same time, along with the primary chain and chain tensioner **(see illustration)**.

15 On pre-1976 models, withdraw the clutch operating pushrod from the hollow transmission mainshaft. **Note:** *On belt drive models, remove the spacer(s) from behind the compensating sprocket* **(see illustration 35.7).** *They're used for alignment of the sprockets and must be installed in their original locations during reassembly.*

16 Remove the clutch hub nut after flattening the ears on the tab washer. Note that the nut has a left-hand thread and must be loosened in a clockwise direction.

17 The clutch hub is installed on a tapered portion of the transmission mainshaft and is located by a Woodruff key. As a result, the hub is very tight and a puller will be needed to remove it **(see illustration)**. A puller

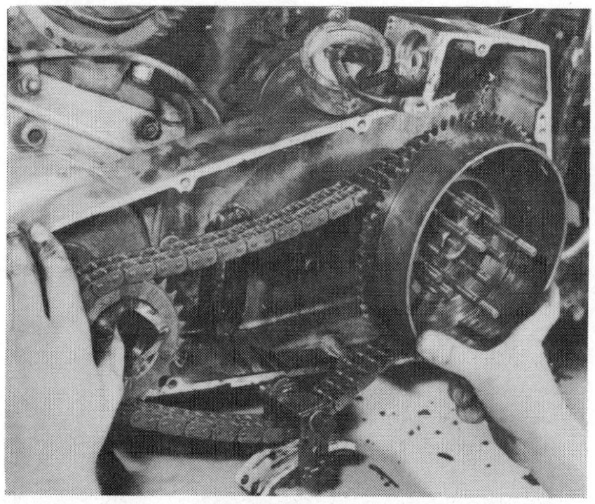

35.14 Remove the primary drive assembly (and chain tensioner) as a unit

35.17 A puller is required to remove the clutch hub from the transmission mainshaft

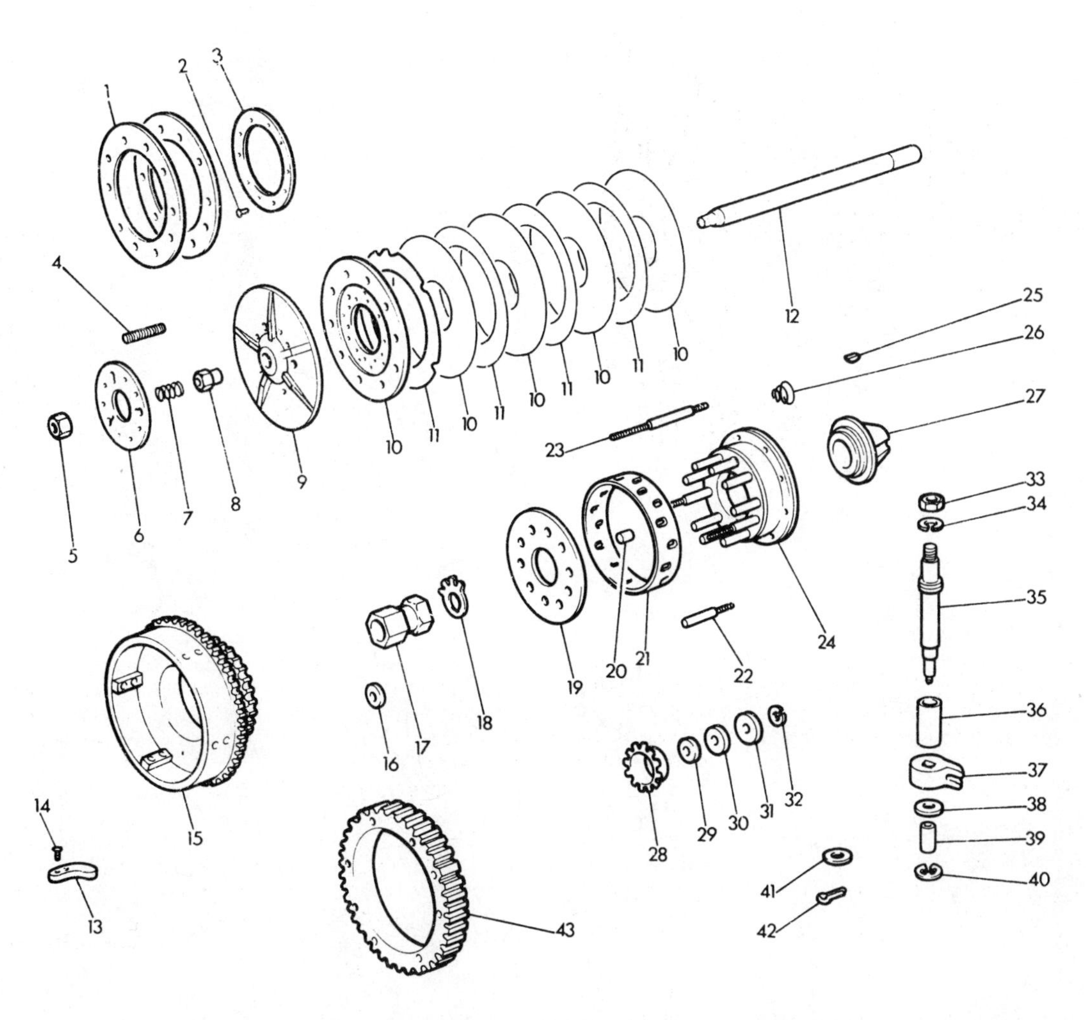

35.11 Dry clutch components – exploded view

1 *Friction disc lining*
2 *Rivet*
3 *Clutch center friction disc*
4 *Adjusting screw*
5 *Spring tensioning nut*
6 *Spring backplate*
7 *Clutch spring*
8 *Adjusting screw locknut*
9 *Pressure plate*
10 *Friction disc*
11 *Steel (plain) plate*
12 *Clutch operating pushrod*
13 *Steel plate buffer*
14 *Rivet*
15 *Clutch outer drum/driven sprocket*
16 *Pushrod oil seal*
17 *Clutch hub nut/seal housing (LH thread)*
18 *Tab washer*
19 *Bearing backing plate*
20 *Bearing roller*
21 *Bearing cage*
22 *Plate stud*
23 *Stud*
24 *Clutch hub (clutch center)*
25 *Woodruff key*
26 *Spiral spring*
27 *Clutch release bearing (1973 thru 1975)*
28 *Clutch release bearing cap (1975-on)*
29 *Thrust washer (1975-on)*
30 *Thrust bearing (1975-on)*
31 *Thrust washer (1975-on)*
32 *Circlip (1975-on)*
33 *Nut*
34 *Lock washer*
35 *Shaft*
36 *Upper bushing*
37 *Release fork*
38 *Washer*
39 *Lower bushing*
40 *Circlip (1978-on)*
41 *Washer (1970 thru 1978)*
42 *Cotter pin (1970 thru 1978)*
43 *Starter ring gear*

2

that applies force with bolts at three points should be used. It should be positioned over the three longest studs projecting from the clutch hub so it can be held in place by the three spring tensioning nuts. **Note:** *Before installing the puller, thread the hub nut onto the mainshaft approximately six turns to prevent the shaft threads from being damaged by the pressure of the bolt. The clutch pushrod oil seal installed in the end of the nut must also be protected from the end of the bolt by a washer.*

18 After removal of the clutch hub, pry the Woodruff key out of the shaft with a screwdriver.

Wet clutch (late 1984 and later models)

Refer to illustration 35.19

Caution: *The clutch installed on 1990 and later models should not be disassembled by the home mechanic – a special tool is required to compress the diaphragm spring, which can fly out with extreme force if the tool isn't used. Have the clutch disassembled by a Harley-davidson dealer service department.*

19 On late 1984 through 1989 models proceed as follows: Loosen the

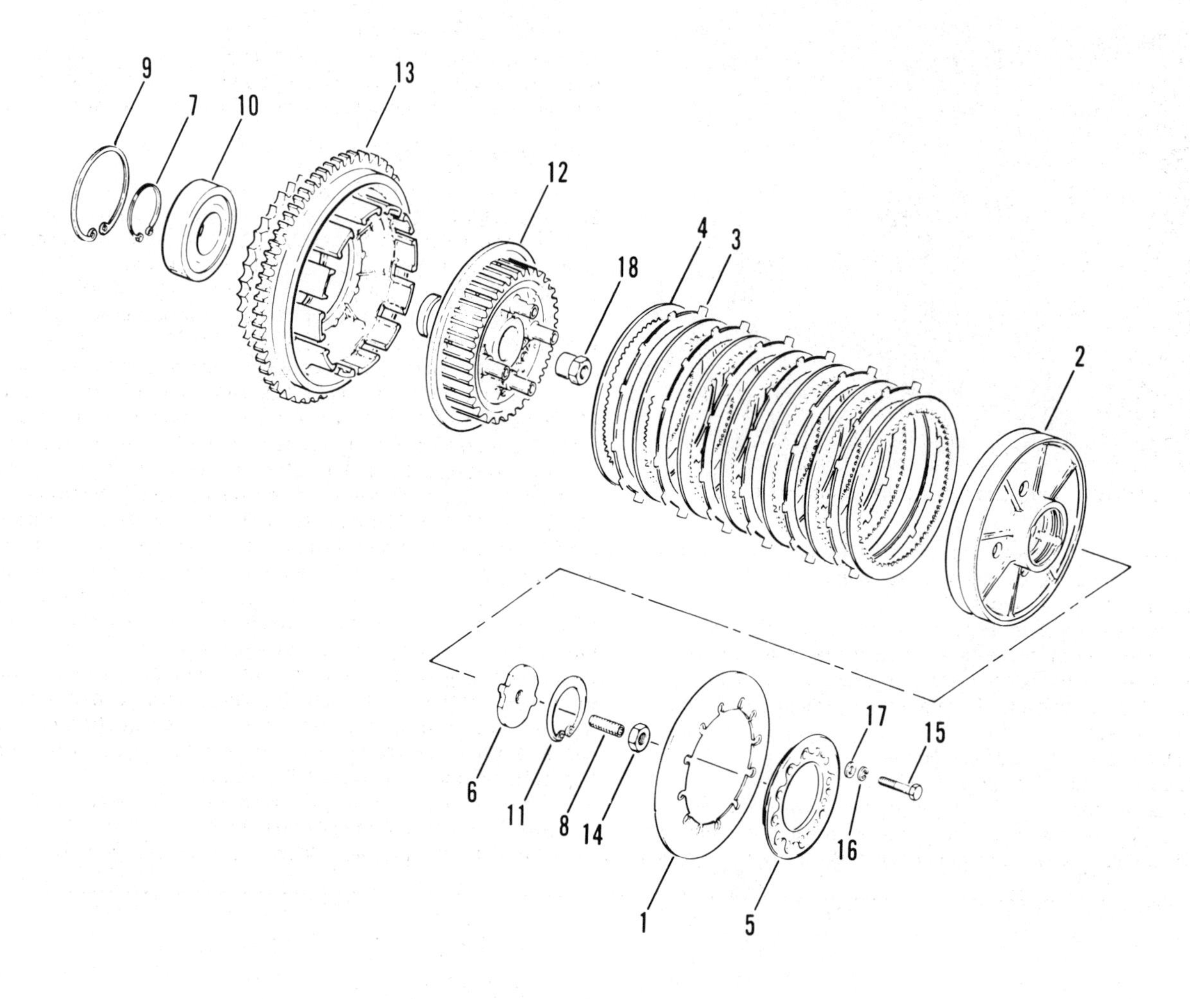

35.19 Wet clutch components – exploded view (late 1984 through 1989 models)

1 *Diaphragm spring*
2 *Pressure plate*
3 *Friction disc (6)*
4 *Steel clutch plate (7)*
5 *Adjuster plate*
6 *Release plate*
7 *Circlip – external*
8 *Adjuster screw*
9 *Circlip – internal*
10 *Pilot bearing*
11 *Retaining ring*
12 *Clutch hub (clutch inner)*
13 *Clutch drum (clutch outer)*
14 *Locknut*
15 *Bolt (4)*
16 *Lock washer (late 1984/early 1985 only)*
17 *Washer*
18 *Transmission mainshaft nut (LH thread)*

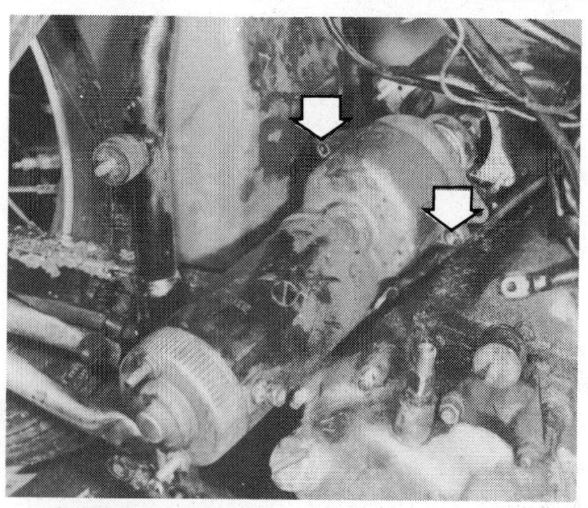

35.29 On some models, the starter motor is held in place by two nuts

35.32a The shift linkage rod is secured by a circlip

35.32b Slide the pivot shaft out of the support bracket

adjuster plate bolts in 1/4-turn increments until the spring pressure is released, then remove them and detach the adjuster plate and diaphragm spring **(see illustration)**.

20 Remove the pressure plate. The steel clutch plates and friction discs can now be changed or inspected without additional disassembly. Remove the clutch plates and friction discs and keep them in order.

21 Remove the nut from the end of the transmission mainshaft. Note that it has left-hand threads!

22 Remove the pushrod threaded end piece.

23 Attach a puller to the clutch hub.

24 Refer to Step 14 above and remove the primary drive chain, the compensating sprocket and the clutch drum/hub unit as an assembly. The clutch drum/hub unit must be removed from the transmission shaft with the puller – it won't slide off.

25 The clutch drum/hub unit shouldn't be disassembled unless the pilot bearing is defective. If the pilot bearing must be replaced, remove the circlips and press the clutch hub out of the bearing inner race.

Primary chaincase

Refer to illustrations 35.29, 35.32a, 35.32b and 35.34

26 On electric start models, the starter motor should now be detached and the starter solenoid disconnected. Carefully pry the rubber cap off the solenoid, which is mounted on the primary chaincase.

27 Disconnect the two large wires and the small wire. When loosening the main terminal nuts, make sure the terminal post is supported to prevent rotation and damage.

28 Working on the right-hand side of the machine, disconnect the single wire attached to the starter motor end cap. The wire is held by a nut.

29 Remove the bracket from the rear of the starter motor and then unscrew the two nuts that clamp the starter motor housing to the back of the primary chaincase. The starter motor can now be withdrawn, leaving the starter secondary shaft in the primary chaincase **(see illustration)**.

30 On late 1978 and later models with a four-speed transmission, remove the through bolts after the bracket is detached. Hold the starter motor at both covers to prevent it from falling apart and lift it out of position.

31 On all models, the inner primary chaincase can now be removed. Remove the gear shift lever, if not already done, and disconnect the pivot bracket chrome cover, which is retained by two screws.

32 Disconnect the shift linkage rod, at the operating arm, by removing the small circlip **(see illustration)**. Carefully ease the operating arm/pivot shaft out towards the rear of the pivot bracket **(see illustration)**. In some cases, removal of the chaincase inner front mounting bolt will be required to provide clearance for pivot shaft removal.

33 Loosen the bolts securing the transmission to the baseplate.

34 Remove the mounting bolts from the front section of the chaincase **(see illustration)**. On some models, two of the bolts are safety wired.

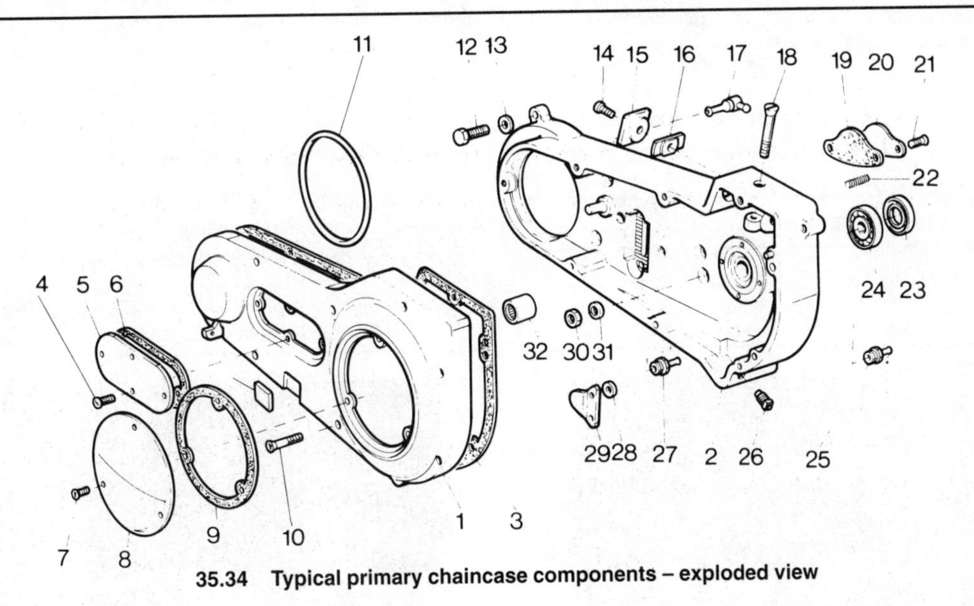

35.34 Typical primary chaincase components – exploded view

1. *Primary chaincase cover*
2. *Primary chaincase inner section*
3. *Gasket*
4. *Screw*
5. *Inspection cover*
6. *Gasket*
7. *Screw*
8. *Clutch cover*
9. *Gasket*
10. *Screw*
11. *O-ring*
12. *Bolt*
13. *Washer*
14. *Screw*
15. *Solenoid hole cover (non-electric start models)*
16. *Cover securing bar (non-electric start models)*
17. *Oil line union*
18. *Pivot screw*
19. *Gasket (non-electric start models)*
20. *Starter motor blanking plate (non-electric start models)*
21. *Screw*
22. *Stud*
23. *Oil seal*
24. *Clutch shaft bearing*
25. *Breather hose union*
26. *Drain plug*
27. *Oil return union*
28. *Washer*
29. *Feed orifice guard*
30. *Nut*
31. *Washer*
32. *Starter shaft outer bearing*

2

36.7 The clutch pushrod oil seal can be pried out of the hub nut

36.8 The bearing cage is secured by three spiral springs (arrow)

35 Remove the bolts securing the chaincase to the transmission. On five-speed models, two of the chaincase mounting bolts must be removed from the back side of the case and the other two bolts are secured with locking tabs and safety wire.
36 On FLT models, remove the bolts securing the chain boot to the primary chaincase.
37 Disconnect any oil lines or vent hoses that may be attached to the chaincase. Label them to be sure they're reinstalled in the proper place during reassembly.
38 Use a block of wood passed through from the right-hand side of the machine and a hammer to separate the chaincase from the transmission. **Note:** *On 1984 FLT/FXR models, the chaincase must be rotated clockwise around the transmission mainshaft to detach it.* The case is now free and can be lifted off.
39 If two washers fall off when the case is removed, be sure to install them on the two front transmission-to-primary case studs during reassembly.

36 Clutch components – inspection

Dry clutch

Refer to illustrations 36.7, 36.8, 36.11a and 36.11b

1 Check the condition of the clutch sprocket to make sure none of the teeth are chipped, broken off or badly worn.
2 Wash all of the clutch components, except the friction discs, and check the steel plates for warpage.
3 Visual inspection will show whether the clutch plate tangs are burred and whether corresponding indentations have formed in the slots they ride in. Burrs should be removed with a file, which can also be used to dress the slots square, provided the depth of the indentations isn't too great.
4 On pre-1981 models, the steel plates are equipped with a buffer device on the trailing edge of each slot. The buffer consists of a cage holding a spring loaded ball. If the cage is damaged or the ball no longer returns, the buffer unit must be replaced with a new one. The cage is held in position by two rivets – the heads can be ground off to allow removal with a punch.
5 Check the thickness of the friction discs. When the discs have worn thin, they must be replaced. Always replace them as a complete set, even if some have not reached the service limit. Worn friction discs promote clutch slip.
6 Check the free length of the clutch springs and compare the results to the Specifications in this Chapter. Don't stretch the springs if they have compressed. They must be replaced when the service limit has been reached.
7 Check the clutch pushrod for distortion by rolling it on a sheet of glass. Heavy action is often caused by a bent rod, which may hang up. If oil has

36.11a The modified outrigger bearing is pre-packed with grease and sealed

36.11b The original oil seal should also be installed

been leaking out along the mainshaft, check the condition of the clutch pushrod seal, which is mounted in the center of the clutch hub nut. The seal can be pried out and a new one driven in **(see illustration)**.

8 The bearing the clutch outer drum rides on is a single cage, double-row roller type and is mounted on the clutch hub boss. Check the fit of the clutch outer drum on the bearing. If radial play is evident, attention to the bearing is required **(see illustration)**.

9 If the bearing races are pitted or scored, they must be replaced, together with the main component they're attached to. To remove the bearing cage and rollers, pry the three spiral springs off the long clutch studs. Slide the bearing backing plate out of position. As the cage is removed, the rollers will fall free.

10 On some models, a journal ball bearing is installed in a housing in the rear of the primary chaincase to act as an outrigger bearing for the transmission mainshaft. This bearing is prone to lack of lubrication and failure. Check the bearing for up-and-down play and roughness when rotated. If the bearing requires replacement, the chaincase must be removed by separating it from the transmission and engine crankcase as described in the previous Section.

11 The outrigger bearing can be driven out, taking the oil seal with it. Heat the case to approximately 150-degrees F before attempting removal. A special ball bearing, prepacked with grease and incorporating an oil seal, should be used as a replacement. This will prevent future failure due to lack of lubrication **(see illustrations)**.

Wet clutch

12 Clean all of the parts with solvent and dry them with compressed air (if available).

13 Check the friction discs for worn, checked and chipped linings. Look for grooves in the steel plates.

14 Check the steel plates for warpage with feeler gauges on a perfectly flat surface. Replace any that are warped beyond the specified limit.

15 Measure the thickness of the steel plates and friction discs. If they're thinner than specified, install new ones.

16 Check the diaphragm spring for cracks and damage.

17 Check the clutch hub keyway and Woodruff key for damage and distortion. Make sure the splines and slots in the hub and drum are smooth and undamaged. Check the tangs on the friction discs for burrs and wear.

18 See if the pilot bearing turns smoothly and make sure it isn't discolored from excessive heat.

37 Primary drive components – inspection

Refer to illustration 37.6

1 Examine the teeth on the compensating (crankshaft) sprocket and the clutch drum. If any are chipped, hooked or broken, the sprocket must be replaced. It's a good idea to replace the clutch drum, the compensating sprocket and the primary chain or belt together as a matched set. A badly worn sprocket will cause the chain or belt to wear more rapidly and cause the engine to lose power.

2 Check the condition of the splines on the sliding cam and the shaft extension **(see illustration 35.6c)**. If they're worn, the components should be replaced as a matched set. Although it's unlikely the sliding cam has worn to any great degree, it should be checked where it makes contact with the compensating sprocket.

3 The component most likely to wear is the sprocket spring, which will compress after extended use. Increased cam action is a sure sign the spring should be replaced.

4 On 1980 and early 1981 models with a primary drive belt, check the belt and measure the amount of play. The belt isn't adjustable, so if the play is over one-inch with 10 pounds of force applied to the top run of the belt at the midpoint, it must be replaced with a new one.

5 Check the belt carefully for oil and dirt deposits. Slight deposits can be removed with a shop towel. DO NOT clean the belt with solvent.

6 Check the belt for hardened or cracked back surface rubber, cracked or separated plies, worn, cracked or missing teeth and cracked or worn sides **(see illustration)**.

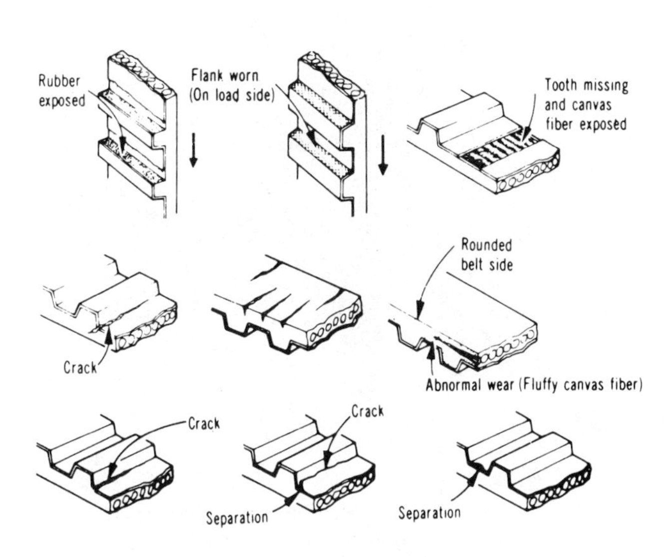

37.6 Inspect the primary drive belt for wear as shown here

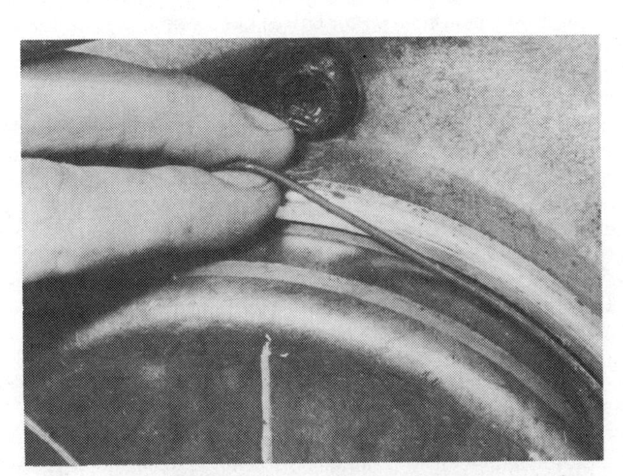

38.2 Place a new O-ring over the crankcase boss before installing the inner primary chaincase

2

38 Primary drive and clutch – installation

Refer to illustrations 38.2 and 38.5

1 If the gear shift linkage rod was disconnected, it should be reconnected before the inner primary chaincase is installed. Be sure the rod is facing the correct way or the clevis will bind on the selector box arm, causing difficult or missed shifts.

2 Place a new O-ring on the crankcase boss that mates with the inner chaincase **(see illustration)**.

3 Apply red Loctite to the outer race of the transmission mainshaft ball bearing and install it in the chaincase bore. **Note:** *On late 1985 and later FLT/FXR models, the bearing is secured with two retaining rings, so Loctite isn't necessary. Later FX/Softail models don't have a bearing. However, a new seal must be installed and the splines on the transmission mainshaft should be wrapped with electrician's tape so the seal isn't damaged when the chaincase is installed.*

4 Refer to the Caution in Step 6 and then install the primary chaincase, pushing it into position until it engages with the protruding studs or the bolt holes line up. On dry clutch models, connect the oil feed and return lines, using new hose clamps on their unions.

5 Install the bolts that secure the chaincase to the engine and tighten them to the specified torque in 1/4-turn increments. On some models, the

38.5 Safety-wire or secure the two inner bolts with thread-locking compound

two inside bolts must be locked together with safety wire **(see illustration)**. If the threads in the case and on the bolts are perfectly clean, thread-locking compound can be used instead of safety wire.

6 **Caution:** *The engine and transmission mounting nuts/bolts must be loose when the inner chaincase is installed to prevent strain on the case and to allow it to self-align.* Install the chaincase-to-transmission nuts or bolts and locking tabs (if used), together with the oil breather orifice cover plate (if equipped). Note that the cover plate is installed with two washers on the inside of the plate, next to the chaincase. Tighten the nuts/bolts to the specified torque in 1/4-turn increments. Note that some later FLT/FXR models are equipped with locking tabs and the bolts have small holes in the heads – tie the bolts and tabs together with safety wire. Also, some models have locking tabs that must be bent up against the bolt heads.

7 Tighten the engine and transmission mounting nuts/bolts, then make sure the transmission mainshaft turns freely.

8 Apply grease to the clutch operating pushrod and insert it into the hollow transmission mainshaft.

9 Place the Woodruff key in position in the tapered section of the mainshaft and slide the clutch hub into position over the mainshaft. The Woodruff key will engage with the keyway in the clutch hub.

10 Install the tab washer and the clutch hub nut. Lock the mainshaft of the transmission to keep it from turning by shifting the transmission into First gear and applying the rear brake, then tighten the nut. Remember, the nut has left-hand threads. Bend the tabs on the lock washer over the nut to secure it.

Primary chain models

11 Install the compensating sprocket spacer (if used) and the shaft extension on the end of the crankshaft.

12 Lubricate the clutch bearing with clean engine oil. Assemble the compensating sprocket and the clutch outer drum on the workbench, so they're meshed with the primary drive chain. The chain should be installed with the tensioner assembly.

13 Lift the primary drive assembly and slide the compensating sprocket and clutch drum into place on the shafts.

Primary belt models

Refer to illustration 38.15

14 Lubricate the rubber dampers in the compensating sprocket with Harley-Davidson POLY-OIL. Assemble the clutch outer drum and the compensating sprocket on the workbench so they're meshed with the primary drive belt.

15 The clutch hub and compensating sprockets must be in alignment when they're installed. The alignment is altered by using different thickness spacers between the compensating sprocket and the alternator rotor. Place the spacers, removed during disassembly, over the crankshaft. If necessary, the spacer thickness can be determined as follows:

a) Measure the distance from the alternator rotor hub outer surface to the primary chaincase gasket surface.
b) Measure the distance from the clutch hub friction surface to the primary chaincase gasket surface.
c) The difference between the two measurements (dimension step "C" in the table) must be looked up on the accompanying table **(see illustration)** to determine the correct thickness spacer(s) required to align the sprockets.

Dimension Step "C"	Size	Part No.
0.324 to 0.304 in	0.060	24054-80
0.304 to 0.284 in	0.040	24053-80
0.284 to 0.264 in	0.020	24052-80
0.264 to 0.254 in	0.010	24051-80
0.254 to 0.244 in	0.000	None

38.15 Compensating sprocket spacer thickness table (primary belt drive models only)

38.18 Note the OUT mark on the steel clutch plates and install them accordingly

16 Install the assembled components as a complete unit and slide the compensating sprocket and clutch drum into place on the shafts.

17 Install the inner plate, the hub, the rubber dampers and the outer plate in the compensating sprocket in the correct order **(see illustration 35.7).** Secure the outer plate to the compensating sprocket with the set screw.

Dry clutch models

Refer to illustrations 38.18, 38.19a and 38.19b

18 Replace the clutch plates one at a time, beginning with a steel plate, followed by a friction disc. Continue installing the plates, alternating between the two types. The steel plates are marked OUT on one side and should be installed accordingly **(see illustration).** In addition, on models so equipped, the spring loaded buffers should be staggered in the outer drum.

19 Replace the spring/pressure plate assembly and install the three nuts with the indented face of each nut facing IN **(see illustration)**. Tighten the nuts evenly until the large washer on the adjusting screw becomes loose. Remove the locknut, discard the large washer and reinstall the locknut. Tighten or loosen the three nuts until the distance between the pressure plate and the rear of the spring backplate is 1-1/32 inch (26 mm) **(see illustration)**.

Wet clutch models

20 Assemble the clutch plates and friction discs in the hub and drum assembly. Start and end with a steel plate.

38.19a Install the clutch spring tensioning nuts with the indentations facing IN

38.19b Tighten the spring tensioning nuts evenly until the correct clearance between the pressure plate and spring backplate is obtained

21 Check the Woodruff key to make sure it's parallel with the transmission mainshaft taper and that it doesn't extend more than 0.119-inch above the shaft.
22 Install the compensating sprocket, the primary chain and the clutch drum/hub assembly at the same time. Align the key in the transmission mainshaft with the keyway in the clutch hub as it's installed on the shaft. **Caution:** *Make very sure the key is installed correctly – the clutch hub could be damaged if it isn't.*
23 Apply two drops of red Loctite to the compensating sprocket nut and the clutch hub nut before installation. Install the hub nut and tighten it to the specified torque. **Caution:** *DO NOT overtighten it.*
24 Install the pressure plate, the diaphragm spring and the adjuster plate in the clutch hub with the holes aligned. Make sure the spring is installed with the convex side out.
25 Apply purple Loctite to the bolt threads, then install the bolts and tighten them to 6.5 to 8 ft-lbs (9 to 11 Nm) in a criss-cross pattern.
26 Adjust the clutch as described in Chapter 1. **Note:** *If the retaining ring (no. 11 in illustration 35.19) has been removed, make sure the bevelled edge faces out when it's reinstalled.*

All models

27 Install the compensating sprocket and related components. Tighten the nut after locking the crankshaft as described in Section 35.
28 Install the chain tensioner center bolt so it engages with the sliding nut in the rear of the serrated adjuster plate. Adjust the chain tension as described in Chapter 1.
29 Adjust the tension of the primary drive belt (if equipped) as described in Chapter 1.
30 On electric start models, the starter motor must be installed now to allow replacement of the clutch operating arm and adjustment of the clutch.
31 On late 1978 and later four-speed models, set the starter motor in position. The starter must be held at both covers during installation to prevent it from falling apart. Position the mounting bracket on the starter and insert the through-bolts.
32 On all other models, lubricate the starter secondary shaft bearing in the starter motor housing with grease. Place a new gasket over the starter motor retaining studs, then replace the oil seal holder plate with the raised edge facing the starter motor. Insert the starter driven gear into the housing attached to the front of the starter motor. Slide the starter motor into position so the splines on the secondary shaft engage with those in the driven gear. Push the starter motor until it's seated on the studs and install and tighten the two mounting nuts.
33 Attach the clutch operating arm to the shaft which runs into the transmission. Adjust the clutch as described in Chapter 1.
34 Connect the wires to the solenoid and place the rubber cap in position over them.
35 On five-speed models, coat the rear chain boots and their mating surfaces with RTV sealant. Place the boots in position and install and tighten the mounting bolts.
36 Install a new gasket on the primary chaincase, after checking the mating surfaces to make sure they're perfectly clean.
37 On models so equipped, replace the bronze washer on the end of the starter shaft.
38 On chain primary drive models, lubricate the primary drive chain with clean engine oil.
39 Carefully set the primary chaincase cover in position and install the mounting screws in their original locations. Tighten the screws a little at a time in a criss-cross pattern. Replenish the primary chaincase oil supply on wet clutch models.

Models with chain primary drive and dry clutch only

40 In order to ensure the primary chaincase doesn't leak oil and, more importantly, so the automatic oil feed and return system functions correctly, the primary chaincase must be perfectly airtight.
41 Harley-Davidson recommends testing the chaincase with a vacuum gauge to check the seal. The vacuum should be 20-inches or more of water with the engine running at 1500 rpm and the breather pipe to the oil tank pinched closed. This test is impractical and in any case shouldn't be required if care has been taken during reassembly. A loss of vacuum and therefore failure of the chaincase lubrication may be due to the following defects:

Leakage at the crankcase/chaincase O-ring
A damaged primary chaincase gasket
Damaged starter motor housing gasket or shaft O-ring
Leakage of air through the chaincase bearing or between the bearing inner race and the transmission mainshaft

42 The defects listed probably won't prevent the feed of oil to the chaincase, but will prevent proper scavenging of the lubricant and cause a rise in the level.
43 Install the shift lever and secure it with the pinch bolt.
44 If nylon rope was inserted into the combustion chamber to lock the crankshaft, remove it now and reinstall the spark plug.

39 Completing engine reassembly

1 Grease the gearshift lever shaft and insert it into the pivot bracket from the inside.
2 Reconnect the gearshift link arm to the shaft arm and secure it with the circlip.

3 Replace the pivot bracket cover and attach the shift lever to the shaft.
4 Secure the speedometer cable with the clip at the base of the cylinder barrels (if equipped).
5 Carefully reconnect the rectifier/regulator by pushing the socket into the plug in the crankcase.
6 Reconnect the wires from the starter motor, battery and starter relay to the starter solenoid and position the protective boot.
7 Hook up the primary ignition wires at the timing cover wire harness connector (for the contact breaker points or the electronic ignition sensor plate). Replace the ignition coil and connect the primary wires to the terminals. Install the chrome coil cover (if equipped).
8 Install the spark plugs and hook up the wires.
9 Reattach the carburetor to the intake manifold (see Chapter 3).
10 Reconnect the choke cable. When the choke knob is pulled completely out, the choke must be closed. Reconnect the throttle in a similar manner so there's a small amount of free play before the throttle valve begins to open.
11 Install the air cleaner assembly (see Chapter 3).
12 Reconnect all the oil lines to the oil pump. Refer to Chapter 3 if necessary. Secure the oil lines with new hose clamps.
13 If the battery carrier was removed, it should be reinstalled now.
14 Reconnect the wire(s) to the oil pressure switch. If detached, reconnect the brake light switch wires.
15 Reconnect the starter motor lead at the terminal projecting from the motor and the wires to the starter relay mounted below the carrier.
16 Replace the battery. Make sure all wires on the machine are correctly connected before attaching the battery cables. Remember, the ground wire runs to the negative (–) terminal on the battery.
17 Install the battery cover (where used) and secure the battery with the strap.
18 Reinstall the exhaust pipes. Use new gaskets at the exhaust ports.
19 Attach the forward right-hand footrest bracket to the transmission mounting studs. On FX models, note that the rear brake operating pushrod must be inserted into the master cylinder as the footrest is installed.
20 Install the footboards, if not already done. Grease the pivot bolts before inserting them and tightening the nuts.
21 Replace the seat (if removed) and the fuel tank (see Chapter 3).
22 Reconnect the fuel line to the carburetor after passing the hose through the guide clip secured by the lower carburetor mounting stud.
23 Refill the oil tank with the specified oil. Don't overfill it. Wait until the engine has been run and the level has dropped, then pour in more oil if necessary.

40 Starting and running the rebuilt engine

1 Before starting the engine after a rebuild, make sure all oil and fuel line connections are tight and all wires are connected.
2 Turn the ignition kill switch to the Off position. Check the oil level and remove the spark plugs.
3 Turn the key switch on and crank the engine over with the electric starter or kick it over (where applicable) to circulate oil through the engine. Install the spark plugs, hook up the wires and turn the ignition kill switch to the On position.
4 Make sure there's fuel in the tank, turn the fuel valve to the On position and operate the choke. Start the engine and allow it to run at a fast idle until it reaches operating temperature. Check the oil return to the oil tank.
5 The engine may tend to smoke initially due to the amount of oil used during reassembly. The excess oil should burn away gradually as the engine components seat themselves.
6 Check the exterior of the engine for oil leaks. Make sure each gear engages correctly and all controls function properly, particularly the brakes. This is an essential last check before taking the machine on the road.
7 As soon as the engine is running satisfactorily, the ignition timing should be checked and, if necessary, readjusted as described in Chapter 1.
8 When the engine is first started, the valves may be noisy. The noise should disappear as soon as the hydraulic tappets have built up pressure and self-adjusted.
9 Refer to the following Section for the proper break-in procedure.
10 After a road test and after the engine has cooled down completely, retorque the cylinder heads (pre-Evolution engine only).

41 Recommended break-in procedure

1 Any rebuilt engine needs time to break-in, even if parts have been installed in their original locations. For this reason, treat the machine gently for the first few miles to make sure oil has circulated throughout the engine and any new parts installed have started to seat.
2 Even greater care is necessary if the engine has been rebored or a new crankshaft has been installed. In the case of a rebore, the engine will have to be broken in as if the machine were new. This means greater use of the transmission and a restraining hand on the throttle until at least 500 miles have been covered. There's no point in keeping to any set speed limit – the main idea is to keep from lugging the engine and to gradually increase performance until the 500 mile mark is reached. These recommendations can be lessened to an extent when only a new crankshaft is installed. Experience is the best guide, since it's easy to tell when an engine is running freely.
3 If a lubrication failure is suspected, stop the engine immediately and try to find the cause. If an engine is run without oil, even for a short period of time, irreparable damage will occur.

42 Transmission – general information

The configuration of the engine and transmissions used on these machines is called a "pre-unit" type, where the transmission is a remote assembly contained within its own aluminum case and mounted behind the engine. Power to the transmission is transmitted via a multi-plate clutch and roller chain or cogged drive belt enclosed in a detachable unit on the left side of the machine (known as the primary drive).

The selection of gears in the transmission is carried out by shifting forks. The forks slide the movable gears into and out of mesh with the various other gears. On Four-speed models, the gear shifter unit is mounted on the outside of the transmission. Five-speed models have the shifter cam assembly bolted to the case, under the transmission top cover.

43 Transmission repair operations possible with the transmission in the frame

1 Attention to the kickstart mechanism, such as replacement of the ratchet and gear, can be carried out with the transmission in place, after removal of the outer cover. Replacement of the kickstart return spring can be carried out merely by detaching the kickstart lever and the spring cover plate.
2 Adjustment of the gear selector mechanism can be done without removing the transmission, but it requires removal of the clutch, primary drive and primary chaincase. The procedure for shifter adjustment is described in Section 49 of this Chapter.

44 Transmission – removal and installation

Removal

Four-speed models

Refer to illustrations 44.7 and 44.12

1 The transmission can be removed from the frame with the engine still in place, but only after removal of the primary drive components and clutch, the starter motor (if equipped) and the battery and battery carrier. Refer to Section 35. With these major components removed to allow access to the transmission, continue as follows.
2 Remove the oil tank drain plug and allow all the oil to drain into a container.

44.7 The transmission sprocket nut has left-hand threads

44.12 The speedometer cable on early models is retained by a gland nut

3 Disconnect the oil tank breather hose, the return hose at the top of the tank and the oil feed line from the union at the oil pump. The hose clamps should be pried apart with a screwdriver and discarded.
4 Remove the oil tank filler cap and dipstick.
5 Remove the two nuts securing the oil tank to the rubber mounted support studs. Maneuver the oil tank off the studs and remove it from the machine to the left-hand side.
6 Remove the front right-hand footrest and mounting bracket. On some models, the footrest shaft also serves as the brake pedal pivot, and the brake operating pushrod must be withdrawn from the master cylinder as the assembly is removed.
7 Before disconnecting the final drive chain, flatten the tab washer ears securing the mainshaft sprocket nut and remove the nut. Note that the nut has a left-hand thread and must be removed in a clockwise direction **(see illustration)**. Remove the chain by removing the master link, then pull the transmission sprocket off the splines. On belt drive models, loosen the rear wheel adjusting nuts and remove the drive belt.
8 Disconnect the clutch cable from the release arm.
9 Remove the exhaust system components that are in the way of the transmission. On some models the entire exhaust system will have to be removed.
10 Separate the shift rod from the transmission. The shift rod is secured by either a cotter pin and clevis pin or by a bolt and nut.
11 Disconnect the wire for the Neutral indicator switch.
12 Remove the speedometer drive cable and housing from the transmission (not all models) **(see illustration)**.
13 Remove the bolt securing the transmission to the support bracket on the right side. Remove the bolts and screws attaching the transmission mounting plate to the frame.
14 Remove the transmission, with the mounting plate attached, out the left side of the frame.

Five-speed models

15 Drain the transmission oil (see Chapter 1). On 1993 and later FLT and 1991 and later Dyna models also drain the engine oil from the oil pan (see Chapter 1). On 1991 and later Dyna models go on to remove the oil pan (see Chapter 3).
16 Remove the cover from the right side of the transmission. Note that the clutch cable can remain attached to its release arm.
17 Withdraw the pushrod end piece (complete with oil slinger on later models) from the mainshaft end. A snap-ring retains the thrust bearing assembly to the pushrod end piece.
18 Disconnect the wires from the neutral switch. Detach the vent hose and on later models, engine oil hoses, from the transmission case.
19 Remove the bolts securing the top cover to the transmission and detach the top cover and gasket. Lift the shifter cam assembly out the transmission after unscrewing the four mounting bolts; take care not to loose the guide pins from the support blocks.
20 Remove the clutch and primary drive components as described in Section 35 (if not already done).
21 Mesh the gears in two speeds at a time to lock the transmission, then remove the locknuts and spacers.
22 Remove the mounting hardware from the side door of the transmission. Pull the side door, countershaft and mainshaft out of the transmission case as an assembly. Withdraw the main clutch pushrod from the mainshaft bore if necessary.
23 Label the oil hoses attached to the top of the transmission case and remove them. Plug the ends of the hoses.
24 Remove the oil filter from the bottom of the case (models through 1991).
25 On models with an enclosed drive chain, the rear chain boots must be disconnected from the primary chaincase. It may be necessary to use a Harley-Davidson special tool (part no. 97101-81) to accomplish this.
26 On chain drive models disconnect the master link and remove the chain from the sprocket. On belt drive models move the rear wheel forward in the swingarm to provide enough slack to maneuver the belt off the sprocket.
27 Lift the motorcycle and support it securely on blocks. Place a block of wood below the engine to support it.
28 Remove the nut and spacer from the swingarm pivot shaft on the right side of the frame.
29 On all models, remove the footrests from both sides of the motorcycle. On FXR models, the pivot shaft mounting brackets must also be removed.
30 Tap the pivot shaft out of the swingarm from the right side of the motorcycle, using a large punch and a hammer. It isn't necessary to remove the nut and washer from the left side of the pivot shaft.
31 Separate the gear shifter rod from the shifter arm located at the top of the transmission.
32 Remove the two bolts and the washers and nuts that secure the transmission to the engine.
33 Flatten back the tabs of their lockwashers, and remove the three or four bolts (as applicable) which retain the transmission case to the inner primary chaincase. On certain models the lower bolt is accessed from the rear side of the primary chaincase and also retains the engine ground strap.
34 Lift the transmission case out of the frame.
35 The oil pan on 1993 and later FLT models may be detached from the transmission as described in Chapter 3, Section 13.

Installation

36 Installation is basically the reverse of removal.

45 Transmission – disassembly

Four-speed models

Refer to illustrations 45.2, 45.5, 45.6, 45.7, 45.10a, 45.10b, 45.11, 45.13 and 45.19

1 If the oil wasn't drained, do so now. Remove the drain plug from the bottom of the transmission until as much oil as possible has drained, then clean the drain plug and screw it back into the case.

45.2 Note the special breather screw in the top cover (four-speed)

45.5 Pull the clutch release bearing off (four-speed)

45.6 Unscrew the end nut from the mainshaft (four-speed)

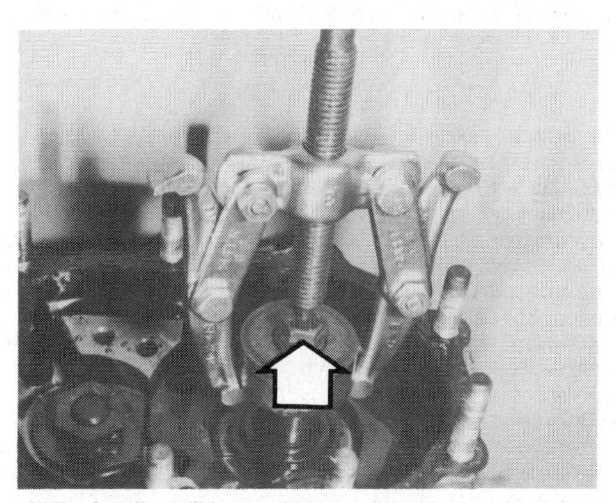

45.7 A puller will be needed to remove the bearing boss; note the bolt (arrow) inserted in the end of the mainshaft (four-speed)

45.10a The selector fork rod is secured by a screw (four-speed)

45.10b Slide the rod out and lift the forks out of the case (four-speed)

2 Remove the screws retaining the top cover to the transmission. A special breather screw is also installed in the top cover. Note the location of the screw, then remove it **(see illustration)**. It must be installed in the same location during reassembly.

3 The top cover can now be lifted off, complete with the gear shifter assembly, and stored out of the way.

4 Remove the nuts securing the end cover to the transmission. On some models you'll also have to remove the battery carrier bracket. The cover is then free to be removed, complete with the kickstart shaft and spring (on models so equipped), and the clutch release shaft.

45.13 Flatten the tab washer, then remove the nut and the washer end plate (four-speed)

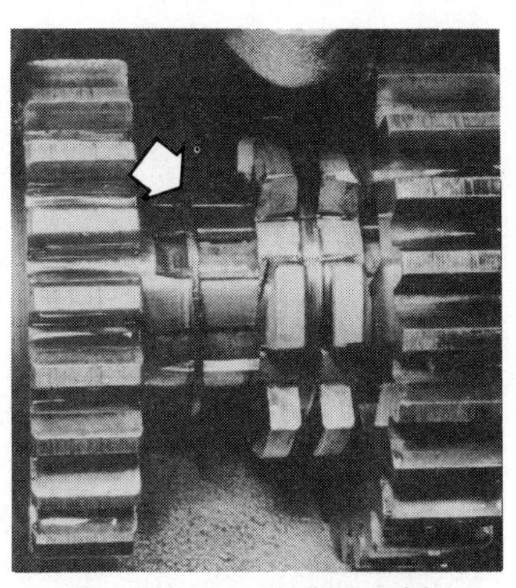

45.19 Move the circlip out of the groove toward the right (four-speed)

45.25 Sprocket locking plate is secured by two screws – 1992 and later models

5 Pull the clutch release bearing off the end of the mainshaft **(see illustration)**. Position the transmission so the drive side end of the mainshaft is held securely in the jaws of a vise. Use wood blocks in the vise jaws to protect the mainshaft from damage.

6 Bend the ear of the tab washer, securing the end nut on the mainshaft, down and remove the nut and washer **(see illustration)**.

7 Remove the boss the clutch release bearing is attached to and to which the kickstart ratchet is attached (on kickstart models) from the tapered shaft end. The boss is a tight fit and will require a two or three-jaw puller to be used. When the puller is being attached, place a short bolt in the hollow end of the mainshaft to take the thrust of the puller screw **(see illustration)**.

8 After the boss has been removed, the kickstart ratchet gear engagement spring and bushing can be pulled off the shaft (kickstart models).

9 Pry the Woodruff key(s) from the end of the mainshaft with a small screwdriver. Store them in a safe place to prevent loss.

10 The shift fork shaft is a push fit in the case and is retained by a set screw that passes through the top mating surface of the case. Remove the set screw and carefully pull the shaft out **(see illustrations)**. Note the O-ring in the groove at the left end of the shaft. Don't lose the shift fork rollers, which are a loose fit on the fork pins.

11 Remove the four screws or Allen bolts holding the mainshaft bearing retainer plate to the inside wall of the transmission's outer compartment **(see illustration on next page)**. The screws are very tight and may require an impact driver to loosen them.

12 Lift the retainer plate and the oil deflector (if equipped) out of the case.

13 Remove the countershaft end nut after flattening the tab washer on pre-1980 models. Remove the tab washer and the backing washer **(see illustration)**.

14 Remove the single screw passing through the speedometer driveshaft housing flange and pull the complete assembly out (if equipped).

15 Carefully drive out the countershaft with a brass punch.

16 Slide the needle roller bearing spacer washer out from between the countershaft First gear and the right-hand transmission wall. The complete countershaft gear cluster assembly can be lifted out as a unit. Note: On early models, an uncaged needle roller bearing is installed in each end of the countershaft double gear. Be sure the needles don't fall out or, if they do, that the needles from one bearing don't get mixed up with those from the other. If loss or interchange of the 22 rollers in each bearing does occur, the two bearings must be replaced with new ones.

17 After lifting the countershaft clusters out, remove the needle rollers and store them separately.

18 Using a soft-face hammer or a block of wood and a hammer (to protect the shaft end), drive the mainshaft out from the left side until the ball-bearing just clears the large opening in the right transmission wall.

19 Using a scribe or awl, pry the circlip out of the groove in the mainshaft between the main drive gear and the mainshaft Third gear **(see illustration)**.

20 Withdraw the mainshaft, complete with the bearing and mainshaft double gear, while at the same time moving the circlip along the shaft.

21 Lift the mainshaft Third gear and the sliding dog clutch out of the transmission. Note that the dog clutch is installed with the face stamped HIGH facing the main drive gear.

22 Push the main drive gear into the case and remove it from the top. On early models, the main drive gear runs on an uncaged needle roller bearing with 44 rollers. Remove the thrust washer from the bearing (if not already done), remove the 44 rollers and keep them in a safe place. Note the L-shaped key which engages with the drive gear splines and the keyway in the oil seal spacer.

5-speed models

Refer to illustration 45.24

23 Mount the transmission case securely in a vise equipped with soft jaws. If the sprocket nut has not already been loosened, place the drive chain/belt in position over the mainshaft (final drive) sprocket and secure the ends of the chain/belt to prevent the sprocket from turning. In the case of belt drives, be very careful not to bend the belt any more than necessary to hold the sprocket – the manufacturer advises not to form the belt into a loop smaller than 3 inches diameter otherwise it may weaken.

24 On all models through 1991, flatten the ears on the tab washer securing the sprocket nut **(see illustration on next page)**. The nut has left-hand threads so turn it clockwise to loosen it. Remove the retaining nut, tab washer, sprocket and sprocket spacer.

25 On 1992 and later models, remove the two lockscrews and slide off the locking plate **(see illustration)**. Unscrew and remove the sprocket nut, noting that it has a left-hand thread. Withdraw the sprocket and spacer.

26 In order to remove the main drive gear, a press is needed. Take the transmission case to a dealer service department or an automotive machine shop to have the gear removed.

46 Shifter assembly – overhaul

Four-speed transmission (1970 through early 1979)

Refer to illustrations 46.1, 46.3a, 46.3b, 46.6, 46.15, 46.17, 46.18a, 46.18b and 46.22

Disassembly

1 Remove the three countersunk screws securing the shift arm to the

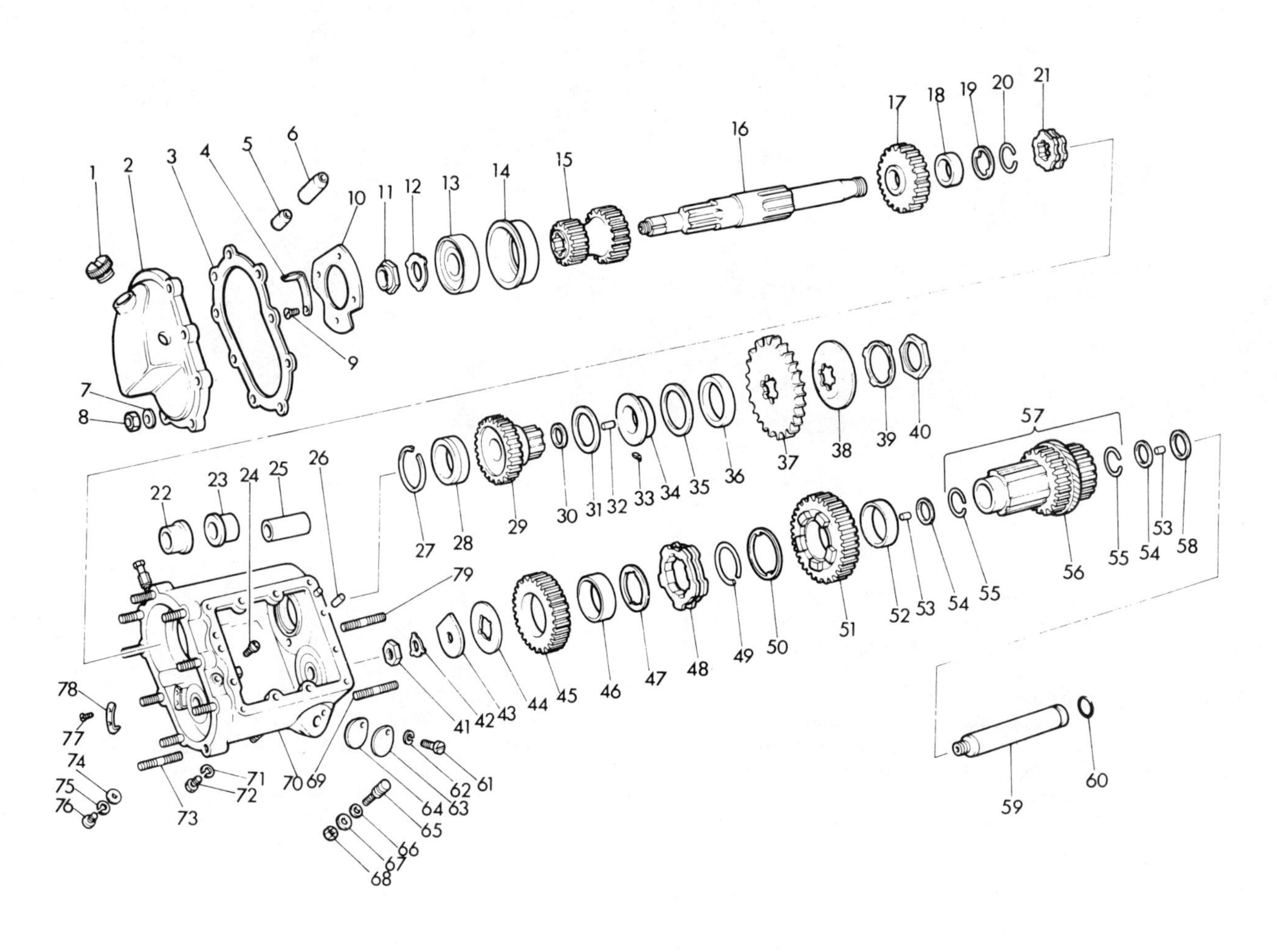

45.11 Four-speed transmission components – exploded view

1 *Oil filler cap*
2 *End cover*
3 *Gasket*
4 *Oil deflector*
5 *Clutch shaft lower bushing*
6 *Clutch shaft upper bushing*
7 *Plain washer*
8 *Nut*
9 *Screw*
10 *Bearing retainer plate*
11 *Sleeve nut*
12 *Tab washer*
13 *Ball bearing*
14 *Bearing sleeve*
15 *Mainshaft 1st and 2nd gear*
16 *Mainshaft*
17 *Mainshaft 3rd gear*
18 *Bushing*
19 *Spacer*
20 *Circlip*
21 *Shift (dog) clutch*
22 *Countershaft bushing*
23 *Countershaft bushing*
24 *Breather screw*
25 *Main drive gear bushing*
26 *Dowel*
27 *Circlip*
28 *Main drive gear bearing outer race*
29 *Mainshaft main drive gear*
30 *O-ring*
31 *Roller backing plate*
32 *Roller*
33 *Drive L-key*
34 *Oil seal collar*
35 *Cork seal*
36 *Oil seal*
37 *Final drive sprocket*
38 *Oil deflector plate*
39 *Tab washer*
40 *Nut*
41 *Nut*
42 *Tab washer*
43 *Countershaft retaining plate*
44 *Roller backing washer*
45 *Countershaft 1st gear*
46 *Bushing*
47 *Spacer*
48 *Shift (dog) clutch*
49 *Circlip*
50 *Spacer*
51 *Countershaft 2nd gear*
52 *Bushing*
53 *Roller*
54 *Roller bearing washer*
55 *Circlip*
56 *Speedometer drive gear*
57 *Countershaft gear*
58 *Roller backing washer*
59 *Countershaft*
60 *O-ring*
61 *Screw (FX only)*
62 *Lock washer (FX only)*
63 *Speedometer drive blanking plate (FX only)*
64 *Gasket (FX only)*
65 *Stud*
66 *Plain washer*
67 *Lock washer*
68 *Nut*
69 *Stud*
70 *Transmission case*
71 *Lock washer*
72 *Bolt*
73 *Stud*
74 *Plain washer*
75 *Lock washer*
76 *Bolt*
77 *Screw*
78 *Kickstarter crank stop*

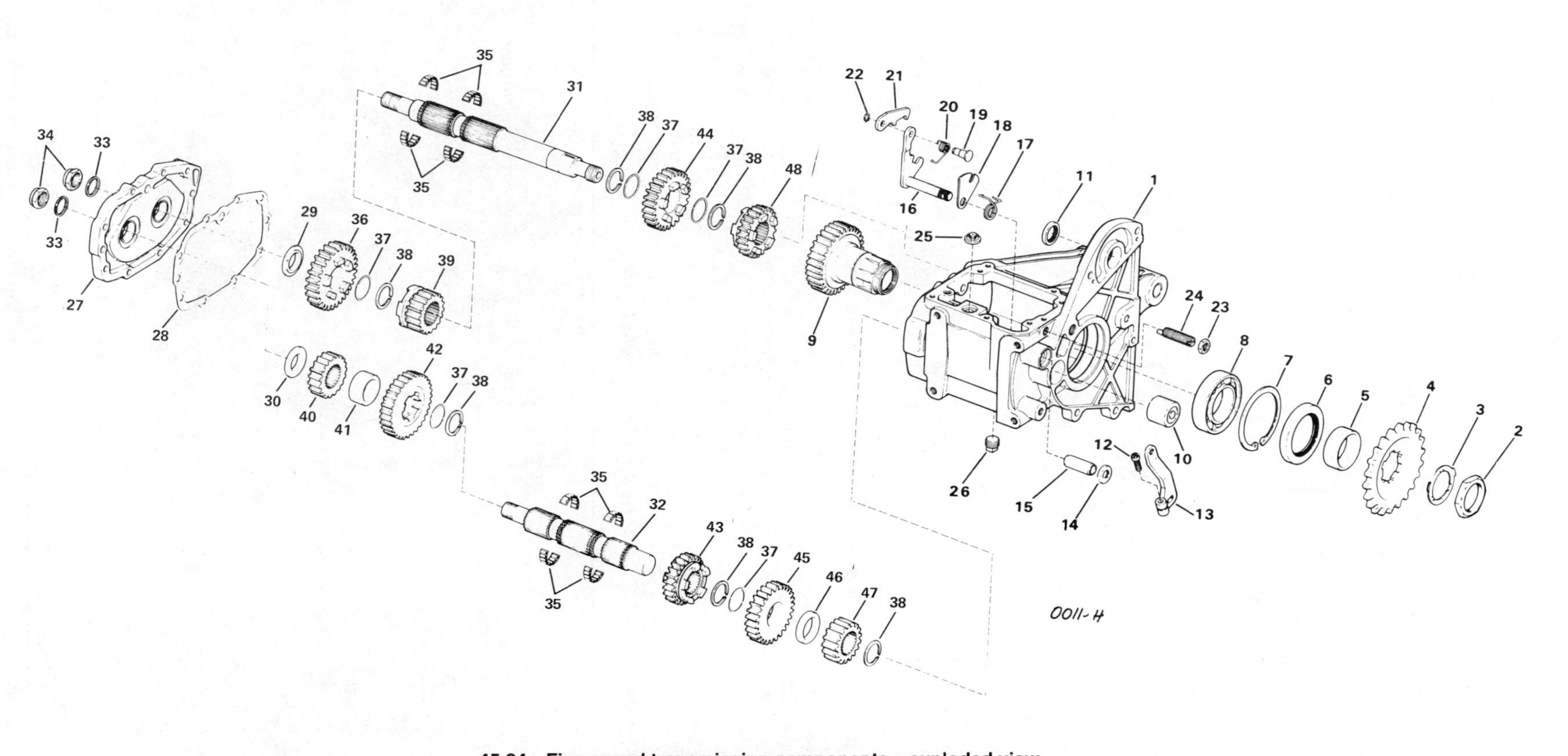

45.24 Five-speed transmission components – exploded view

1 Transmission case
2 Nut
3 Lock washer
4 Sprocket
5 Spacer
6 Oil seal
7 Retaining ring
8 Bearing
9 Main drive gear
10 Bearing
11 Oil seal
12 Screw
13 Shift lever
14 Oil seal
15 Spacer
16 Shift arm
17 Spring
18 Centering plate
19 Pin
20 Spring
21 Pawl
22 Retaining ring
23 Locknut
24 Adjusting screw
25 Relief valve
26 Drain plug
27 Side door
28 Gasket
29 Mainshaft spacer
30 Countershaft spacer
31 Mainshaft
32 Countershaft
33 Spacer
34 Nut
35 Bearings
36 Mainshaft 4th gear
37 Thrust washer
38 Retaining ring
39 Mainshaft 1st gear
40 Countershaft 4th gear
41 Spacer
42 Countershaft 1st gear
43 Countershaft 3rd gear
44 Mainshaft 3rd gear
45 Countershaft 2nd gear
46 Spacer
47 Countershaft 5th gear
48 Mainshaft 2nd gear

2

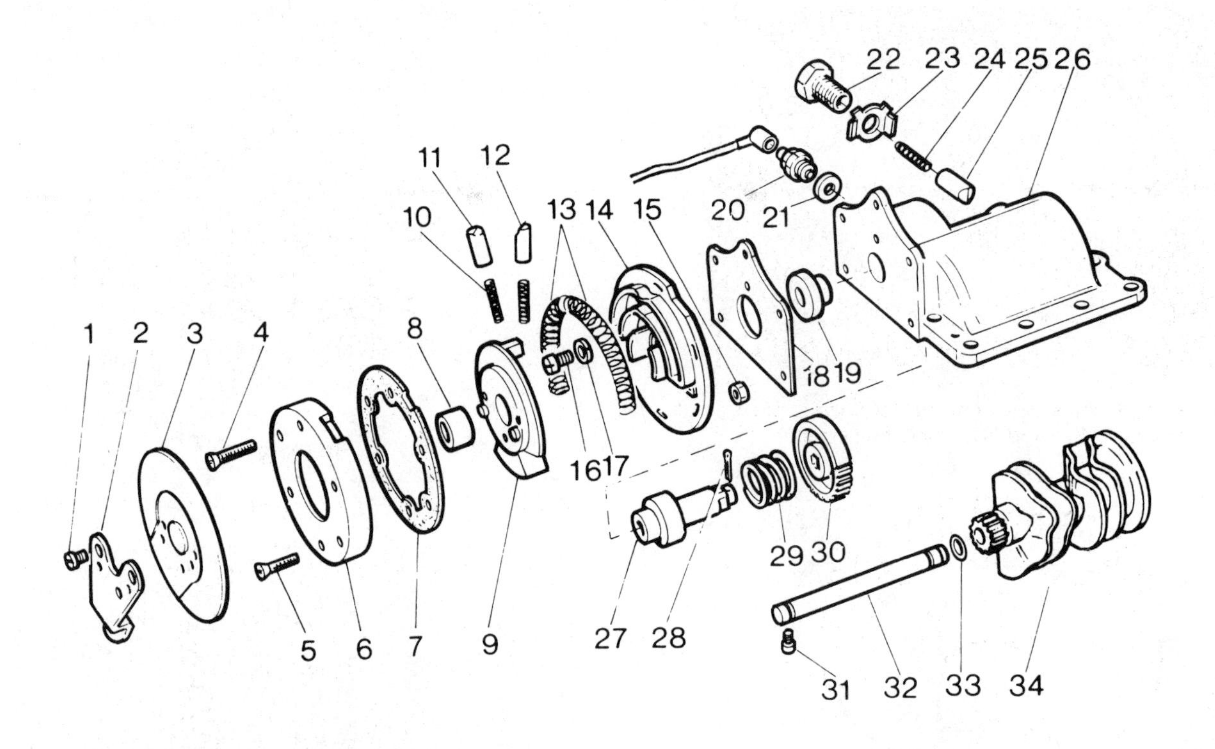

46.1 Four-speed transmission shifter components – exploded view

1 Countersunk screw
2 Shift arm
3 Dust cover
4 Screw – long
5 Screw – short
6 Shifter box body
7 Gasket
8 Bushing
9 Pawl carrier
10 Detent spring
11 Left-hand pawl
12 Right-hand pawl
13 Centralizer (pawl carrier) spring
14 Adapter plate
15 Nut
16 Lock screw
17 Lock washer
18 Gasket
19 Bushing
20 Neutral indicator switch
21 Sealing washer
22 Detent housing
23 Tab washer
24 Detent spring
25 Cam follower
26 Transmission top cover
27 Shift shaft
28 Cotter pin
29 Coil spring
30 Shifter gear
31 Screw
32 Camshaft
33 O-ring
34 Shift cam

46.3a Remove the center screw and . . .

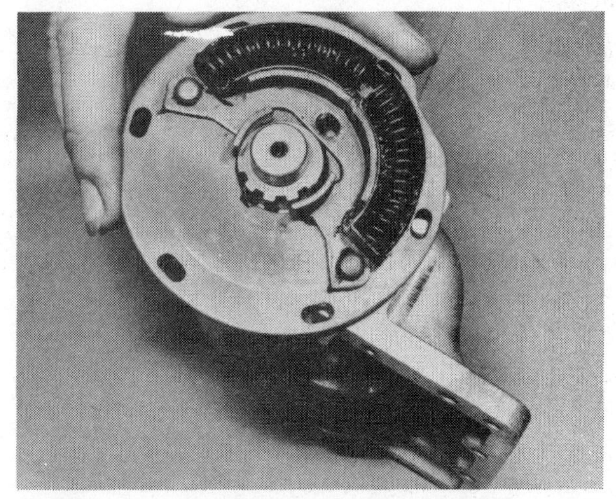

46.3b . . . the adapter plate from the cover (four-speed)

dust cover. Remove the dust cover and the shift arm **(see illustration)**.

2 Remove the five long screws and the single short screws from the pawl carrier cover (they're retained by nuts behind the adapter plate). The pawl carrier cover, the carrier and the gasket are then free to be removed. Note: The pawls in the carrier are spring-loaded. When the cover is removed the plungers may be released. Be sure to keep the two pawls and the springs separated, as well as the pawl carrier centralizer springs.

3 The adapter plate and gasket can be separated from the cover after removing the single retaining screw **(see illustrations)**.

4 Disconnect the wire and unscrew the Neutral indicator switch from the

46.6 Remove the screw from the top cover to allow removal of the shift cam shaft (four-speed)

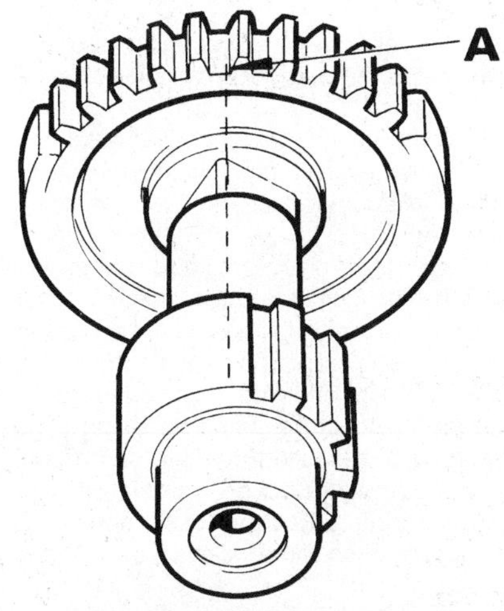

46.17 Aligning the shift gear and the shift shaft (the gear must be installed so the marked tooth gap aligns as shown)
A = Timing mark

46.15 Check the tips of the pawls for wear (four-speed)

46.18a The relieved tooth is for timing purposes (four-speed)

cover.

5 The detent housing is secured by a tab washer. Bend the ears of the tab washer down, then unscrew the housing. Remove the housing, tab washer, spring and cam follower.

6 The shaft the shifter cam rotates on is secured by a single set screw **(see illustration)**. Remove the set screw from the left side of the cover and carefully tap the shaft out with a punch. When the shaft is removed, the shifter cam can be lifted out of the cover.

7 Removal of the shifter shaft isn't required unless the shaft bushing is worn. To remove the shaft, remove the cotter pin securing the shifter gear and pull the shaft out, leaving the gear and spring in the cover. These two remaining components will lift out.

Inspection

8 Clean all of the components, except the Neutral indicator switch and gaskets, with solvent. Dry all of the parts with a clean, lint-free rag or compressed air (if available).

9 Work in clean, well lit surroundings so defects don't go undetected. Failure to notice damage or signs of advanced wear may necessitate another teardown at a later date, due to the premature failure of the part concerned.

10 Except in extreme cases, little wear will develop in the shifter components within the transmission itself, other than those described in Section 47. If the machine has covered a lot of miles, the shifter gear teeth and the teeth with which they mesh on the shifter cam may wear, requiring replacement of both components.

11 The bushing the shifter shaft rides on and the bushing the pawl carrier rotates on may wear after extended periods of use. The pawl carrier bushing can be driven out of position with a tubular drift. The shouldered bushing in the cover must be drawn out with a puller because the shape of the casting obstructs access to it. A puller can be fabricated from a length of pipe, slightly wider than the bushing, a long threaded bolt and nut, and a pair of washers, one of which must be slotted and have a diameter slightly less than the smaller external diameter of the bushing.

12 Insert the bolt, head first, through the bushing. Slip the slotted washer over the bolt, from inside the case, to secure the bolt head.

13 Place the pipe over the shank of the bolt, followed by the second washer and the nut. Heat the case in an oven to about 300-degrees F before the bushing is removed. As the nut is tightened, the bushing will be drawn out of the case.

14 The new bushing can be installed in the case by reversing the bolt and pulling the bushing into position by tightening the nut inside the case.

15 Inspect the tips of the two pawls and the cam follower **(see illustration)**. In order for these components to function properly, the tips must be fairly sharp. Worn components must be replaced with new ones.

16 Check the pawl return springs and the pawl carrier centralizer springs for distortion and loss of strength. Replace any suspect components as a matched set.

Reassembly

17 Lubricate the bushings in the cover with clean oil, then turn the cover over so the shifter gear and spring can be installed. Insert the shifter shaft so it engages with the shifter gear and spring. The shifter shaft can only be installed in one position. Align the scribed line on the shifter gear with the portion of the shaft boss adjacent to the extreme left-hand spline **(see illustration)**. Secure the shifter gear and spring to the shaft by inserting a new cotter pin through the radial hole in the shaft.

18 The shifter cam must also be installed in the correct relationship to the shifter gear. When the shifter cam is installed, the slightly relieved (ground

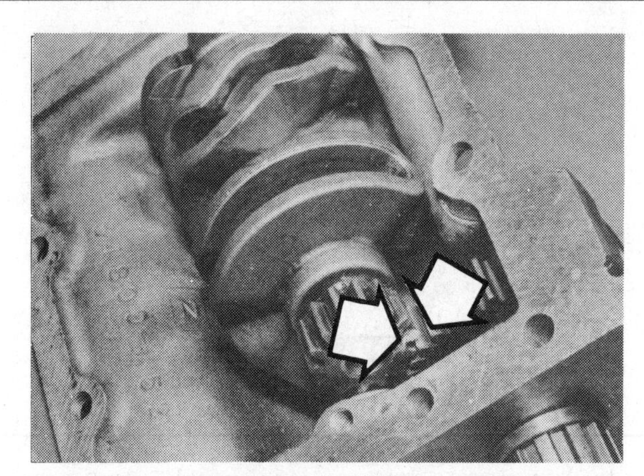

46.18b The shift gear and the shift cam gear must be timed (four-speed)

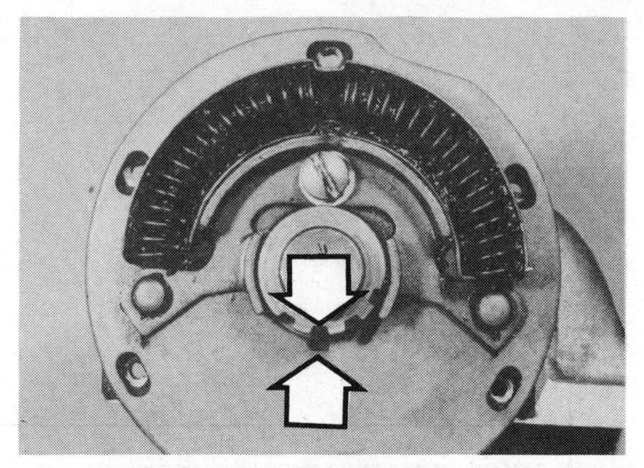

46.22 The adapter plate must be timed with the shifter shaft (four-speed)

back) tooth must align with the scribed line on the shifter gear **(see illustrations)**.

19 Insert the shaft through the transmission cover and the shifter cam. Remember to install a new O-ring in the widest of the two grooves at the right end of the shaft before installing the shaft. Check the timing of the cam and gear and adjust it if necessary before securing the shaft with the set screw inserted into the left-hand groove.

20 Position the adapter plate, and a new gasket, over the outer end of the shifter shaft. Loosely install the center screw to support the plate.

21 Install the cam follower, the spring, the tab washer and housing. Rotate the shifter cam so the cam follower engages any of the notches in the cam except the Neutral notch. First gear is when the shifter cam is in the extreme counterclockwise position, looking at the selector end.

22 Rock the shifter cam slightly to be sure the cam follower is perfectly seated. Rotate the adapter plate so the punched timing mark on the plate aligns exactly with the shifter shaft spline, which is second from the left **(see illustration)**.

23 Tighten the center screw securing the adapter plate and carefully check the alignment. **Note:** *Before continuing, be sure that all five positions of the shifter cam can be reached.* Installing the pawl carrier and the pawls and springs requires patience. The carrier is installed against the pressure of the pawl springs, together with the reluctance of the carrier centralizer spring to stay in place in the guides.

24 Install the carrier centralizer springs in the guide trough and lubricate them with grease.

25 Insert the two pairs of springs and pawls into the pawl carrier. Position the two pawls so the stepped portions face each other.

26 Lubricate the bronze bushing in the carrier as well as the pawls. Carefully place the pawl carrier, slightly offset, in position over the shifter shaft boss.

27 Push the carrier towards the center position so the bronze bushing fits over the boss simultaneously with the centralizer projection (on the carrier) entering the gap between the two centralizer springs. The centralizer springs will probably try to spring out of position and the pawls will attempt to rotate in the pawl carrier. If this happens, take the pawl carrier off, reposition the pawls and springs and attempt the procedure again.

28 Position the pawl carrier cover on the adapter plate so the recessed portion is adjacent to the similarly shaped cutout in the adapter plate. Insert the shifter cover screws and attach and tighten the securing nuts.

29 Place the shifter arm/dust cover unit in position and install and tighten the three screws or Allen bolts.

30 The transmission cover is now ready to be attached to the main transmission case.

Four-speed transmission (late 1979-on)

Disassembly

31 Remove the five bolts securing the top cover to the transmission. The bolt directly above the final drive sprocket can only be loosened when removing the cover. To remove this bolt, the shift linkage must be disassembled.

32 Separate the cover from the transmission and remove the gasket.

33 Remove the two bolts securing the shifter shaft cover to the end of the transmission top cover.

34 Unscrew the bolt securing the shift lever to the transmission cover. Remove the shift linkage assembly from the transmission top cover.

35 With the shift linkage removed, the only remaining cover mounting bolt can be removed.

36 The shift linkage can be disassembled by removing the nut and washer that connects the linkage to the shifter shaft cover. On FX models, the two retaining rings on the shift linkage must also be removed.

37 Drill a 1/4-inch hole in the plug in the top of the transmission cover and pry the plug from position. The plug must be replaced with a new one.

38 Reach into the top cover through the opening in the top and remove the snap-ring and washer securing the shifter cam and the pawl assembly.

39 Turn the cover over and remove the cam follower and spring from the corner of the cover. Remove the cam follower body from the transmission cover after bending the tabs of the lockplate out of the way and removing the retaining bolts.

40 Unscrew the four Allen bolts holding the pawl stops in position. Detach the pawl stops and spring from the cover.

41 Remove the neutral indicator switch from the cover.

Inspection

42 Clean all of the components, except the neutral indicator switch and gaskets, with solvent. Dry all of the parts with a lint-free rag or compressed air (if available). Work in clean, well lit surroundings so defects don't go undetected. Failure to notice damage or signs of advanced wear may necessitate another stripdown at a later date, due to premature failure of the part concerned.

43 Inspect the tips of the cam follower and shifter pawls. In order for these components to function properly, the tips must be fairly sharp. Worn tips indicate the need for replacement with a new component.

44 Check the cam follower return spring and the pawl carrier spring for distortion and apparent weakening. Replace any defective components as a matched set.

45 Check the neutral indicator switch by depressing the plunger and releasing it. If the plunger doesn't spring back, the switch must be replaced with a new one.

Reassembly

46 Reassembly is the reverse of the disassembly procedure. Be sure to coat the pawl springs with multi-purpose grease before installation.

47 After the cam follower body is attached to the cover, bend the tabs of the lockplate up against the flats of the mounting bolts to lock them in place.

48 Coat the threads of the neutral indicator switch with thread sealant before installing it.

49 Be sure that the pawls engage with the teeth of the shifter cam gear when the pawl assembly is installed. Also be sure that the tab on the pawl carrier locates between the two springs before the washer and snap-ring are installed.

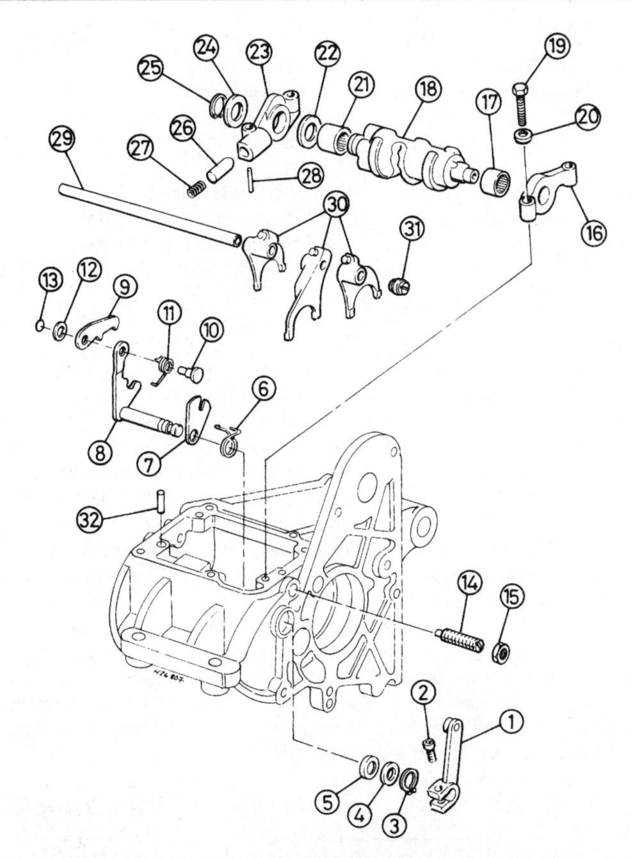

46.53 Five-speed transmission shifter components – exploded view

1	*Shifter arm*	18	*Shift cam*
2	*Screw*	19	*Bolts*
3	*Snap-ring*	20	*Washers*
4	*Washer*	21	*Right shift cam bearing*
5	*Oil seal*	22	*Inner thrust washer (early models)*
6	*Return spring*		
7	*Plate*	23	*Right support block*
8	*Shifter rod*	24	*Outer thrust washer*
9	*Shifter pawl*	25	*Snap-ring*
10	*Spring post*	26	*Neutral switch plunger*
11	*Return spring*	27	*Spring*
12	*Washer*	28	*Roll pin*
13	*Retaining ring*	29	*Shift fork shaft*
14	*Gear engagement adjuster*	30	*Shift forks*
15	*Locknut*	31	*Set screw*
16	*Left support block*	32	*Support block guide pins*
17	*Left shift cam bearing*		

50 Coat the edges of the plug with gasket sealer and set the plug in place on top of the cover with the domed side of the plug facing up. Seat the plug by hitting the center with a ball-pein hammer.

51 Be sure to insert the transmission cover bolt into the left rear mounting hole before attaching the shifter linkage components.

Five-speed transmission

Disassembly

52 Remove the top cover from the transmission and the shifter cam, as described in Section 44 (Steps 15 through 17).

53 Slide the support block off of the left side of the cam.

54 Remove the snap-ring from the right side of the shift cam with a snap-ring pliers.

55 With the snap-ring removed, the outer thrust washer and the right support block can be slid off the shifter cam. Mark the outer thrust washer in some way so it's not mixed up with the inner thrust washer (where fitted).

Inspection

56 Clean all of the shifter components, except the bearings and the neutral indicator switch, with solvent. Dry all of the parts with a clean lint-free rag or, if available, with compressed air.

57 Work in clean, well lit surroundings so defects don't go undetected. Failure to notice damage or signs of advanced wear may necessitate another stripdown at a later date, due to the premature failure of the part concerned.

58 Check that the plunger in the neutral indicator switch will spring back without binding after depressing it. If it binds it must be replaced with a new one.

59 Clean the bearings with a clean lint-free rag and examine them closely. If the bearings are pitted, grooved or worn, they will have to be replaced with new ones. The bearings must be pressed in and out of the support block. Be sure to install the bearings with the letters stamped in the bearings facing the outside of the support block when it is installed on the shifter cam.

60 Check the ends of the shifter cam. If they are pitted or grooved, the shifter cam and the bearings in the support blocks must be replaced with new ones. Make sure that the shaft inside the shifter cam is not loose. Check the shifter cam for wear and cracks and replace it with a new one if necessary.

Reassembly

61 Slide the inner thrust washer (not fitted from mid-1991) into position on the large end of the shifter cam. Install the right support block with the cam follower on the cam lobes.

62 Place the outer thrust washer on the end of the cam and secure the components with the snap-ring. Be sure the snap-ring is firmly seated in the groove and that the thrust washer(s) can spin after the snap-ring is installed.

63 Attach the left support block to the small end of the shifter cam and place the assembly in position in the transmission. Be sure to have the numbers on the support blocks facing down.

64 As the shifter cam is being lowered into position, engage the shifter forks in the slots. There are guide pins in the transmission case that the right support block (and from mid-1991, also left support block) must line up with.

65 Install the washers and bolts securing the support blocks and make sure the left support block is straight and does not bind on the bearing. This is especially important on early models, where the block is not located by guide pins. Tighten the mounting bolts to 7 to 9 ft-lbs (9 to 12 Nm). **Caution:** *Be careful not to overtighten the bolt nearest the neutral switch plunger – distortion may result.*

66 Rotate the shifter cam and make sure all of the gears engage. Check the shifter cam alignment and shifter cam end play as described below.

67 If the neutral indicator switch was removed or replaced, apply thread sealing compound to its threads and tighten the switch to 3 to 5 ft-lbs (4 to 7 Nm). Check the gear engagement as described below.

68 Install the top cover on the transmission with a new gasket and tighten the cover bolts.

Shifter cam alignment

69 On all models through mid-1991 check the shifter cam alignment, and if necessary adjust by fitting a thicker or thinner inner thrust washer to the right support block. Refer to the appropriate paragraph below. Note that this check is not necessary on models from mid 1991 and later (identified by the omission of the inner thrust washer and 'pressed-in' neutral pin in shifter cam as opposed to cast ramp).

70 To align the shifter cam on models through 1983, the right side door and gasket must be removed. Refer to Section 44 for the removal procedure. Measure the distance the center of the groove is from the outer surface of the transmission case while the transmission is in 3rd gear. It should be 3.043 inch (77.29 mm). If the distance is wrong, the inner thrust washer must be replaced with a thicker or thinner one until the exact distance is obtained.

71 On 1984 to mid-1991 models, position the drum in neutral and take up any play against the right support block. Measure the distance from the outer machined surface of the right support block to the center cam groove's nearest edge. It should be 1.992 to 2.002 inch (50.59 to

50.85 mm); if not, replace the inner thrust washer with one of different thickness until the measurement falls between that specified.

Shifter cam end play

72 Attach a dial indicator to the transmission case and measure the end play of the shifter cam. If it isn't within the 0.001 to 0.004 inch (0.025 to 0.102 mm) range, the outer thrust washer must be replaced with one of different thickness.

Gear engagement check

73 Shift the transmission into 3rd gear and check that the upper two pins on the shifter cam are perfectly centered in the slot of the shifter pawl. There should be 0.010 inch (0.254 mm) of clearance between the edges of the shifter pawl slot and the cam pin nearest that edge.

74 Adjust by loosening the locknut on the end of the transmission case and turning the adjusting screw until the desired clearance is obtained. Tighten the locknut to 20 to 24 ft-lbs (27 to 33 Nm) and recheck the clearance.

47 Transmission components – inspection

1 Give the transmission components a close visual examination for signs of wear and damage such as chipped or broken teeth, worn dogs or splines and bent selector arms. If the machine has had a tendency to jump out of gear, look carefully for worn dogs on the backs of the gears and similar wear on the projections on the dog clutches they engage. Check also for wear in the selector tracks in the shifter cam and the cam plate with which the detent pawl locates. In the former case, wear will be evident in the form of rounded corners or even a wedge-shaped profile. The corners of the cam plate tracks will wear first; all such wear is characterized by brightly polished surfaces.

2 The shifter arms usually wear across the fork that engages with a gear, causing a certain amount of sloppiness in the gear change. A bent shifter will immediately be obvious, especially if overheating has blued the surface.

3 All defective transmission components should be replaced. There is no satisfactory method of repairing them.

4 Don't forget to check the transmission mainshaft and countershaft. If it's suspected they're bent, the gear cluster must be stripped down and the shafts checked with a dial indicator. Neither one should have more than 0.003-inch (0.076 mm) runout. If this reading is exceeded, the shaft must be replaced.

Four-speed transmission

5 After a thorough cleaning with solvent, check the mainshaft journal ball bearing for radial play and roughness when rotated. If either one is noted, install a new bearing. Removal of the bearing from the mainshaft can take place after the retaining nut has been unscrewed.

6 The two countershaft and single mainshaft needle roller bearings require a small amount of running clearance, as shown in the Specifications. If additional play is evident, the bearings must be replaced. With the exception of the mainshaft needle roller bearing whose outer race is a press fit in the transmission case, the needle rollers on pre-1979 models run directly on the components they support and rotate on. If bearing wear is very advanced, damage may have occurred to the roller tracks on the components, requiring replacement. The sleeve gear bearing outer race may be driven or pressed out of the case after removing the oil seal collar and circlip. The case MUST be heated in an oven to approximately 300-degrees F before the race is driven out. Installation of the race should be carried out in a similar manner.

7 The clearance between the gears, bushings and shafts should be checked against the Specifications at the beginning of the Chapter. Re-

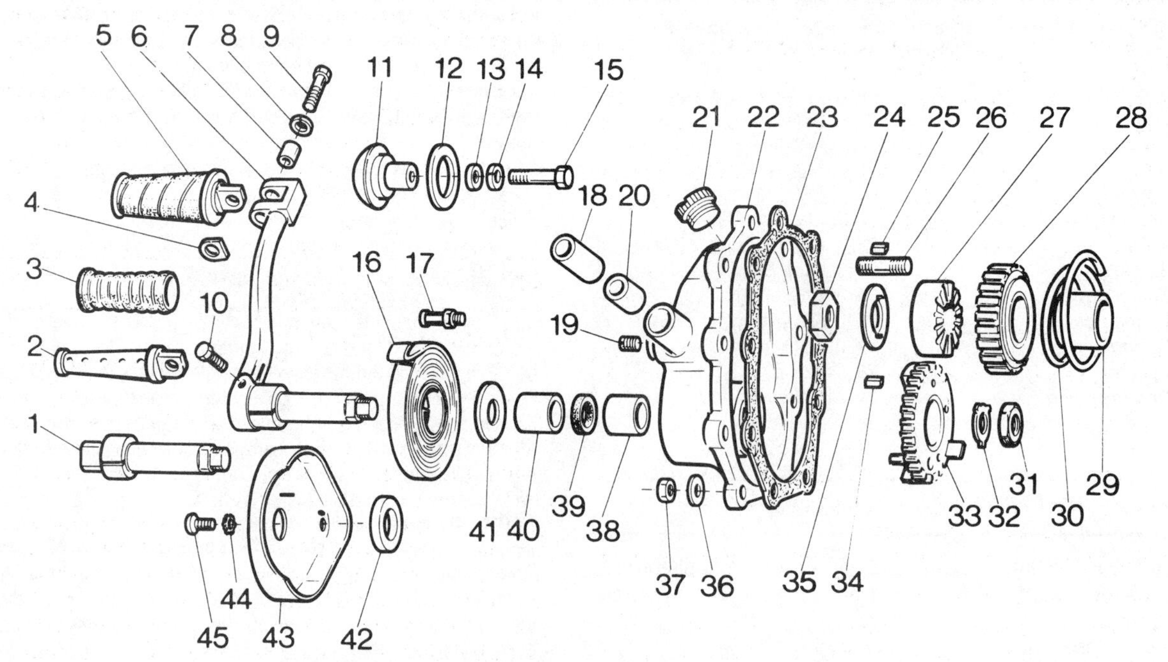

48.1 Kickstart components (four-speed) – exploded view

1	*Kickstart shaft*	10	*Pinch bolt*	19	*Screw*	27	*Kickstart ratchet*	36	*Washer*
2	*Kickstart pedal*	11	*Blanking cap*	20	*Clutch shaft lower bushing*	28	*Kickstart mainshaft gear*	37	*Nut*
3	*Rubber*	12	*Seal*	21	*Filler cap*	29	*Bushing*	38	*Bushing*
4	*Shim*	13	*Plain washer*	22	*Transmission end cover*	30	*Tension spring*	39	*Oil seal*
5	*Kickstart pedal – complete*	14	*Lock washer*	23	*Gasket*	31	*Nut*	40	*Bushing*
6	*Kickstart crank*	15	*Bolt*	24	*Nut*	32	*Washer*	41	*Thrust washer*
7	*Bushing*	16	*Return spring*	25	*Key*	33	*Kickstart gear*	42	*Spring cover insert*
8	*Washer*	17	*Spring anchor*	26	*Stud*	34	*Key*	43	*Spring cover*
9	*Bolt*	18	*Clutch shaft upper bushing*			35	*Tab washer*	44	*Washer*
								45	*Screw*

49.1 Secure the needle rollers in place with heavy grease (four-speed)

49.6a Be sure the circlip is correctly seated in the groove in the mainshaft (four-speed)

49.6b Note the HIGH mark on the shifter clutch (four-speed)

place any bushings that are outside the maximum clearance.

8 Oversize rollers are available to enable a selective fit between the components. Bushings and bearings should be replaced by a Harley-Davidson dealer service department with the necessary expertise and instruments for measurement of bearing play and bearing selection.

Five-speed transmission

9 Check the shifter fork shaft to be sure it's straight. If it's bent or damaged, it must be replaced with a new one. Using a small carpenter's square, check the shift forks to see if they're perfectly square on the shaft. If not, replace the shift forks.

10 Compare the shift forks to a new one. If the old forks have worn more than 0.020-inch, they must be replaced with new ones.

11 Make sure the bearings in the side door feel smooth when rotated and check to see if they're pitted or otherwise damaged. Replace the bearings with new ones if necessary. If the mainshaft or countershaft bearings in the side door must be replaced, they have to be pressed out of position and new ones pressed in. This should be done by a Harley-Davidson dealer service department or an automotive machine shop.

12 Check all of the bearings and oil seals in the transmission case for wear and damage. Replace any components that appear to be even slightly worn. When replacing oil seals, coat the lips of the seal with oil or grease and apply sealant or Locktite to the outside edge.

13 Check the condition of the springs on the shifter arm assembly. If the spring fails to hold the pawl on the cam pins, it must be replaced with a new one.

48 Kickstart components and clutch release shaft (four-speed models) – inspection

Refer to illustration 48.1

1 After removal of the transmission end cover, no further disassembly is required to determine the condition of the components. Apart from sudden failure of the kickstart return spring, the main area of wear occurs in the teeth of the kickstart gear and ratchet pinion and the teeth of the ratchet components **(see illustration)**. The latter type of wear will cause slipping. Worn teeth can't be reprofiled; the only remedy is replacement of the parts. The ratchet pinion is installed with a plain bushing, which can be driven out if excessive play dictates replacement.

2 To remove the kickstart spindle, unscrew the nut after bending down the tab washer. Using a soft-face hammer, drive the spindle out, driving the kickstart gear off the shaft. As the spindle is withdrawn, the return spring will disengage automatically from the spindle. The spring can be removed after detaching the chrome cover, which is held in place by a single screw. Check the fit of the spindle in the two plain bushings. If wear is evident, the bushings can be driven out and new ones installed. An oil seal is installed between the two bushings – it must be replaced if the bushings are removed, or if oil leakage has occurred along the spindle.

3 Reassemble the kickstart mechanism as follows. Place the return spring in position with the thrust washer between the spring and case. The chamfered side of the thrust washer should face the spring. Lubricate the spindle and insert it through the bushings, attaching the inner end of the spring in the slotted spindle boss. Place the spindle end in the jaws of a vise and rotate the cover approximately one-turn in a clockwise direction to tension the spring (make sure the vise jaws are lined with some type of soft metal to avoid damage to the spindle). Connect the kickstart gear so the spindle is unable to unwind. Install the tab washer and nut, tighten the nut and bend the tab washer up to secure it.

4 In time, the clutch shaft bushings will wear, causing excessive movement of the shaft. To remove the shaft, dislodge the circlip from inside the case and pull the shaft out, leaving the release lever and thrust washer in the case. Remove the lever and thrust washer. The old bushings can be driven out and new ones installed.

49 Transmission – reassembly

Four-speed models

Refer to illustrations 49.1, 49.6a, 49.6b, 49.12a, 49.12b, 49.13, 49.14, 49.27 and 49.39

1 On pre-1978 models, insert the 44 needle rollers the main drive gear rides on into the transmission case. The needle rollers will be easier to keep in place if you apply heavy grease to the bearing race **(see illustration)**. Later models have caged bearings.

2 Set the thrust washer in place over the bearings on the inside of the transmission case. Late 1978 and later models don't have a thrust washer between the inside of the transmission case and the main drive gear.

3 Install a new oil seal on the end of the main drive gear. Carefully slide the main drive gear into position, from the inside of the case. Make sure the uncaged bearings, on the earlier models, aren't disturbed or knocked out of place.

4 Slide the First and Second gears onto the mainshaft, followed by the bearing housing, the mainshaft bearing, the tab washer and the retaining nut.

5 Tighten the ball bearing retaining nut to the specified torque and lock it in position by bending over the ears of the tab washer.

6 Insert the partially assembled mainshaft into the transmission case until the Third gear, the bushing, the retaining washer, the circlip and the shifter (dog) clutch can be installed on the mainshaft. Slide the circlip along the shaft until it can be seated in the groove **(see illustration)**. **Note:** *The shifter clutch must be installed with the side marked HIGH facing the main drive gear* **(see illustration)**.

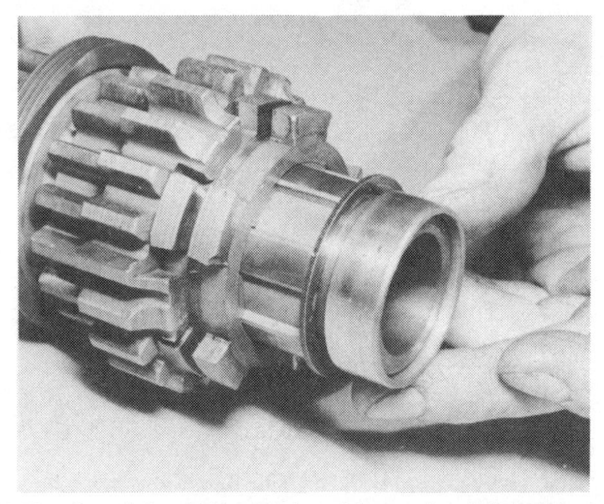

49.12a Install the shifter clutch and the washer, followed by . . .

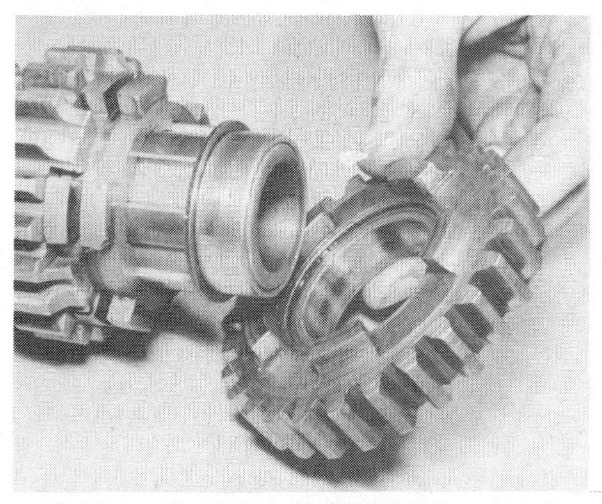

49.12b . . . the First gear and bushing (four-speed)

49.13 Make sure the needle rollers, in both ends of the countershaft gear, are in order (four-speed)

49.14 The thrust washer on the right end can be installed either before or after the countershaft assembly (four-speed)

7 Using a soft-face hammer, drive the mainshaft into the case until the shoulder of the bearing race is seated against the case.

8 On pre-1977 models with uncaged needle rollers on the countershaft, apply heavy grease to the countershaft gear and install the needle rollers. The lock ring and the bearing retaining washer must be installed on the countershaft gear before the bearings are placed in position.

9 On early models, attach the thrust washer on the left side of the countershaft gear in the recess. On late 1978 and later models, attach the caged needle bearings to the countershaft gear.

10 Coat the bearings with grease, then slide the bushing, the countershaft Second gear and the retaining washer onto the countershaft gear.

11 Secure the components on the countershaft gear with a new retaining ring.

12 Place the shifter clutch, the thrust washer, the First gear bushing and First gear in position on the countershaft gear **(see illustrations)**.

13 Make sure the uncaged needle rollers on early models are still in place **(see illustration)**.

14 Apply some grease to the countershaft end washers and place them in position on the inside of the transmission case. Carefully set the countershaft gear assembly in place in the case **(see illustration)**. With a new oil seal attached to the countershaft, insert the countershaft through the case and the countershaft gear assembly from the sprocket side of the case. **Note:** *The oil seal should be on the sprocket side of the case.*

15 Check the countershaft end play with a dial indicator. If the end play isn't within the specified limits, the end washer on the First gear end of the assembly can be exchanged for a washer of a different thickness to obtain the desired end play.

16 On pre-1980 models, position the mainshaft retaining plate so the V-shaped cutout is toward the right side of the shaft.

17 Place the oil deflector plate in position, on models so equipped, and secure the retaining plate with the four screws.

18 Position the countershaft retaining plate with the flat side against the mainshaft retaining plate. Slide the tab washer over the countershaft and thread the retaining nut onto the shaft.

19 Tighten the retaining nut to the specified torque and bend the ears of the tab washer over the flats on the nut to secure it in position.

20 On 1980 and later models, position the retaining plate on the mainshaft so the extended corner of the plate engages the groove in the end of the countershaft.

21 Install and tighten the retaining plate mounting screws.

22 If the oil seal near the final drive sprocket was removed from the case, the new one should now be installed. Use a small file to clean up the case where the old oil seal was staked in place. On pre-1978 models, position a new cork seal over the main drive gear sleeve.

23 Lubricate the lips of the new oil seal and insert it into the case so the spring side of the seal is facing the main drive gear spacer. Carefully drive the oil seal into place. Be sure the seal enters the bore squarely.

24 When the seal is in place, use a cold chisel to stake it in two places, 180-degrees apart.

25 On pre-1978 models, install the L-shaped key so the longest portion

49.27 Engage the shift forks with their respective shifter clutches (four-speed)

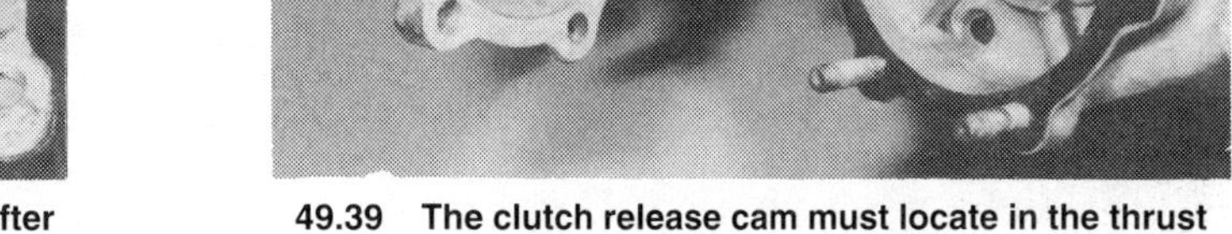

49.39 The clutch release cam must locate in the thrust bearing recess (four-speed)

locates in the main drive gear and the short end locates in the recess in the oil seal collar.

26 If new shift forks have been installed, or if the shift forks have been disassembled, they must be reassembled and adjusted so they're in the correct relationship to each other. This procedure requires shims 0.007 and 0.014-inch (0.18 and 0.36 mm) thick and a special transmission setting tool. The adjustment can't be done without the special tool, so it should be left to a Harley-Davidson dealer service department. The procedure is simple, but it's very important for the proper operation of the transmission.

27 Place the two fully assembled shift forks in the transmission so each one engages with the channel on its respective shift (dog) clutch. **Note:** *The forks are not interchangeable. The fork with the larger diameter must engage the countershaft shifter clutch* **(see illustration)**.

28 Check the condition of the O-ring attached to the groove in the left end of the shift fork shaft. Replace it with a new one if necessary.

29 Slide the shaft into position, engaging it with the shift forks, then install the set screw used to secure the shaft. Be sure the narrow end of the set screw engages the groove in the shaft.

30 Rotate the shift cam in the transmission cover so it's in the neutral position. Move the shift forks independently until the gears are also in neutral.

31 Thoroughly lubricate the gears in the transmission with oil and apply some lubricant to the channels in the shift cam.

32 Position a new transmission cover gasket on the case, over the two dowel pins.

33 Attach the shift fork roller to each fork pin. Lower the cover into place and make sure the shift fork rollers enter the channels in the shift cam.

34 Install and tighten the screws used to secure the cover to the transmission. Install the special breather bolt that fits in the right-hand side of the cover on some models.

35 Operate the shifter arm to see if each gear can be engaged – problems are usually caused by errors in one of the adjustment procedures.

36 On kickstart models, install the kickstarter gear and the thrust backing spring.

37 Place the Woodruff key(s) in position in the tapered end of the mainshaft. Attach the kickstart ratchet. On electric start models, a plain boss is installed in place of the ratchet.

38 Install the tab washer and the mainshaft nut. Lock the transmission and tighten the mainshaft nut to the specified torque. Bend the ears of the tab washer over the flats of the nut to lock it in place.

39 Attach the clutch thrust bearings and, on kickstart models, install the starter mechanism **(see illustration)**.

40 Place a new gasket over the end cover studs and install the cover.

41 Insert the speedometer drive assembly, if equipped, into the side of the case, using a new gasket between the flange and the case. Tighten the retaining screw securely.

42 The transmission is now ready to install in the frame. Reverse the removal procedure, referring to Section 44.

43 Fill the transmission with the specified amount of oil or until the oil level is at the bottom edge of the plug hole.

Five-speed models

44 Position the sprocket spacer over the main drive gear, then slide the mainshaft (final drive) sprocket into position.

45 Place the tab washer (models through 1991 only) on the main drive gear. Install the nut, noting that it has left-hand threads and must be tightened counterclockwise to the specified torque. **Note:** *You'll have to hold the countershaft sprocket as described in Section 45 while the nut is tightened.* Lock the nut by bending over the ears of the tab washer (models through 1991) or by installing the lockplate and its retaining screws (models from 1992 and later).

46 The remainder of transmission reassembly must be done with the transmission case installed in the frame. Reverse the disassembly procedure in Section 44 for reassembly.

47 If the transmission won't fit in the swingarm, a special tool (HD-33805) is available to spread the swingarm cleveblocs.

48 When installing the swingarm pivot shaft through the swingarm and transmission from the left side, place the dowel pin holes in the plastic washers at the top. The chamfer on the washer for the right side must face out. Coat the pivot shaft with anti-seize compound before installation.

49 When installing the pivot shaft or the footrest mounting brackets, be sure the roll pins engage the locating hole in the rubber mount. Also make sure the flat on the pivot shaft registers with the flat on the rubber mount on the right.

50 Position the footrests so they'll fold up at a 45-degree angle to the rear before tightening them.

51 On FLT models, the rear chain boots and the mating surfaces on the inner primary chaincase and transmission must be coated with RTV sealant. Tighten the mounting screws to four ft-lbs.

52 See if all of the gears can be engaged. If not, adjust the shifter as described in Section 46.

53 Lubricate the seal and screw the oil filter into place on the bottom of the transmission. Install the oil drain plug and fill the transmission to the recommended level.

54 When installing the final drive chain, make sure the closed end of the master link faces the direction of chain travel.

Chapter 3 Fuel and lubrication systems

Contents

Specifications

Carburetor type

1970	Tillotson
1971 through 1975	Bendix
1976 through 1989	Keihin
1990-on	Keihin CV

Idle speed

	RPM
Tillotson	900 to 1000
Bendix	700 to 900
Keihin	
1976 through early 1978	900
Late 1978 through 1983	800 to 900
1984 through 1989	
FLT and FXR	
1984 through 1988	900 to 950
1989	1000 to 1050
FX and Softail	1000 to 1050
1990-on	
FLT and FXR	1000
FX/Softail and Dynas	1000 to 1050

Low speed mixture screw setting *(late 1977 through 1979 models only)*

Model (carburetor)	Turns open
FLH – 80 (27466 – 78)	3/4
FLH – 1200 (27467 – 78A)	1-1/8
FX – 1200 (27468 – 78A)	1-1/8

Fast idle speed *(late 1978 through 1983 models only)* . . 1500 rpm

Main jet sizes

Tillotson	0.049, 0.051, 0.053, 0.055, 0.057, 0.059, 0.061 and 0.063 in
Bendix	90, 95, 100, 105, 110, 115, 120 and 125
Keihin	
1976 through early 1978	1.60, 1.65, 1.70, 1.75, 1.80 and 1.85 mm
Late 1978 through 1983	150, 155, 160 and 165
1984 through 1989	
FLT	
Early 1984	165
Late 1984 through 1986	175
1987	170
FXR	
1984 and 1985	160
1986	170
1987	165
FLT and FXR (1988 and 1989)	
California only	140
All others	165
All other models	
1986 FXWG and FXST/C only	170
1988 and 1989 California only	140
All others	165
1990 and 1991	
California only	165
All others	185
1992-on	
FLT and Dyna	
California only	160
All others	175
FXR and Softail	
California only	160
All others	165

Slow jet sizes *(1984 and later models only)*

1984 through 1987 (all)	50
1988 and 1989	
California only	42
All others	52
1990 and 1991	
California only	42
All others	45
1992-on	40

Torque specifications

	Ft-lbs (unless otherwise indicated)	Nm
Tillotson carburetor		
Inlet needle valve seat	40 to 45 in-lbs	4.5 to 5.0
Diaphragm cover plug	23 to 28 in-lbs	2.5 to 3.0
Keihin carburetor (1976 through 1983)		
Carburetor mounting nuts or bolts	10 to 14	14 to 19
Air cleaner bracket-to-rocker arm cover	13 to 20	18 to 27
Air cleaner bracket-to-backplate	10 to 15	14 to 20
Air cleaner backplate-to-carburetor	75 to 80 in-lbs	8.5 to 9.0
Keihin carburetor (1984-on)		
Carburetor mounting nuts/bolts	15 to 17	20 to 23
Rubber compliance fitting		
Mounting bolts	40 to 60 in-lbs	4.5 to 7.0
Hose clamps	15 to 20 in-lbs	1.7 to 2.2
Intake manifold mounting bolts/nuts (1990 only)	6 to 10	8 to 14
Air cleaner mounting nuts/bolts		
1984 and 1985 (FLT)		
Bracket-to-cylinder heads	13 to 17	18 to 23
Bracket-to-backplate	10 to 15	14 to 20
Backplate-to-carburetor	75 to 80 in-lbs	8.5 to 9.0
Cover screws	12 to 17	16 to 23

Torque specifications (continued)

Keihin carburetor (1984-on)	Ft-lbs (unless otherwise indicated)	Nm
1984 and 1985 (all others)		
Bracket-to-cylinder heads	13 to 17	18 to 23
Backplate-to-bracket and carburetor	7 to 10	9 to 14
Backplate bottom bolt	13 to 17	18 to 23
Cover screws	12 to 17	16 to 23
1986-on (FLT and FXR)		
Backplate		
To cylinder heads and bracket	10 to 12	14 to 16
To carburetor	3 to 5	4 to 7
Cover screws	3 to 5	4 to 7
1986-on (all others)		
Head bolt and backplate bottom bolt	10 to 12	14 to 16
Captive carburetor bolt	3 to 5	4 to 7
Cover screw	3 to 5	4 to 7

1 Fuel system – general information

The fuel system consists of the gas tank, the control valve, the fuel line and the carburetor.

Gas is fed by gravity to the carburetor through the control valve, which contains a built-in filter. The control valve has three positions; On, Off and Reserve. The reserve position provides a small amount of fuel after the main supply has run out, so the engine will still run for a short time.

Three different makes of carburetors were installed on the motorcycles covered by this manual. Tillotson carburetors were used on 1970 models, Bendix carburetors were used on 1971 through 1975 models and Keihin carburetors were used on 1976 and later models. Beginning in 1990, a constant velocity (CV) carburetor was standard on all models. All carburetors have a butterfly throttle (CV carburetors also have a slide, but it's vacuum operated and responds to throttle butterfly movement) and incorporate an accelerator pump. The Bendix and Keihin carburetors have an integral float chamber. Each carburetor has a manual choke to facilitate easy starting in low outside temperatures.

A large capacity air cleaner is attached to the carburetor intake on all models. Refer to Chapter 1 for filter maintenance instructions.

All 1985 and later California models are equipped with an Evaporative emission control system to reduce air pollution that stems from evaporation of gasoline in the fuel tank when the motorcycle is parked.

Several fuel system routine maintenance procedures are included in Chapter 1. They include *Fuel system – check, Idle speed – adjustment, Fuel filter – cleaning and replacement* and *Throttle – check and lubrication.*

2 Fuel tank – removal and installation

Removal (all models)

Refer to illustration 2.3

Warning: *Gasoline is extremely flammable and highly explosive under certain conditions – safety precautions must be followed when working on any part of the fuel system! Don't smoke or allow open flames or unshielded light bulbs in or near the work area. Don't do this procedure in a garage with a natural gas appliance (such as a water heater or clothes dryer). Also, before starting work, disconnect the negative battery cable from the battery.*

Note: *Where applicable, do not turn the instrument panel upside-down after it's removed – the damping oil used in the fuel gauge will leak out and stain the gauge face.*

1 Turn the fuel control valve to the Off position and disconnect the fuel line from the carburetor. Some hose clamps installed at the factory must be cut or pried off and can't be reused.

2 Insert the end of the fuel line into a clean gasoline container and turn the fuel valve to the On position to drain the tank. Use a funnel to direct the gasoline into the container.

2.3 Loosen or remove the hose clamp (arrow) and detach the crossover hose from one side of the tank

3 On two-piece fuel tanks and some one-piece tanks, the crossover hose, that connects the two tanks or the lower portions of the one-piece tank, must be disconnected from one of the fittings. Release the clamp and slide it down the hose **(see illustration)**. Some clamps have to be cut off with wire cutters and can't be reused – install a worm-drive clamp when the hose is reinstalled.

Removal – one-piece fuel tanks

1970 through 1979

4 The front of the tank is secured by a single bolt running through the frame at the head stock and passing through the two flanges welded to the front of the fuel tank.

5 The rear of the tank is secured by a coil spring which hooks onto tabs attached to the tank. After removal of the bolt and the spring, the tank can be lifted off.

1980-on

6 Remove the seat (most are held in place by two bolts at the rear [underside] and a slip-fit bracket at the front). If the seat is hinged, simply open it up.

FLT models (1980 through 1988)

7 Remove the seat bracket and plastic frame cover (if equipped). Unplug the fuel gauge wire harness connector (you may have to cut one or more plastic wire ties as well, if they're used to secure the harness to the frame).

8 Remove the screws at the rear of the tank and the mounting bolt at the front to detach the tank. If the tank is being replaced, remove the trim and sending unit for installation on the new one.

FLT models (1989 and later)

9 Follow the instructions in Step 7 above.

10 Remove the three console mounting screws (one at the rear and two at the front, under the locked cover).
11 Remove the gas cap and gently lift up on the console to separate it from the tank. Mark the hoses and fittings if necessary, then disconnect the rubber overflow and emissions hoses from the console.
12 Follow the instructions in Step 8 above. **Caution:** *DO NOT remove the nut attaching the yellow wire to the sending unit. The float will drop into the tank if you do.*

FXR models (except FXLR/FXRS – 1988 and later)

13 Remove the three tank center panel screws and the gas cap.
14 Carefully lift up on the center panel and detach all wires. On some models, the speedometer cable will have to be disconnected as well.
15 Detach the fuel gauge ground wire clipped to the underside of the tank. Remove the front tank mounting bolt and detach the center panel.
16 Remove the rear mounting bolts and carefully lift up on the tank to remove it. Don't damage the wire harness between the tank and frame.
17 Transfer the sending unit to the new tank if the old one is being replaced.

FXLR models

18 Follow the procedure for FXR models (Steps 13 through 17), but ignore any references to the center panel and fuel gauge.

FXRS models

19 Remove the two instrument panel screws, then carefully lift it up and disconnect the speedometer cable and wires.
20 Remove the front tank mounting bolt.
21 Follow the instructions in Steps 16 and 17 above.

Dyna models

22 Remove the instrument panel from the tank (two screws or single domed nut, according to model) and disconnect the drive cable(s) and wiring.
23 Disconnect the crossover hose and emission/vent hose from the tank.
24 Remove the single front and rear mounting bolts and lift the tank free.
25 The fuel gauge is situated in the false left-hand gas cap. If the tank is being replaced, remove the gauge and transfer it and the sending unit to the new tank. Pry the gauge gently from the tank and disconnect its wiring. The sending unit is secured by five screws.

Removal – two-piece fuel tanks

Refer to illustration 2.22

26 If a choke control knob is attached to the instrument panel, unscrew the knob and the locknut **(see illustration)**.
27 Remove the odometer knob and screw from the right side of the instrument panel (if equipped).
28 Remove the mounting screws and detach the panel.
29 The tanks are held independently at the rear by two bolts. On early models the front mountings consist of two long bolts which pass through frame brackets and into nuts on the other side. Later models use shorter bolts which screw into the frame brackets, and thus allow one tank to be removed independently from the other.
30 Remove the pin securing the front of the seat to the seat post. Hinge the seat all the way up and support it in this position.
31 Remove the two front bolts/nuts first, then remove the rear bolt/nut holding each tank half. **Note:** *Spacers and washers are used at all mounting bolt locations. Keep track of them and be sure to install them in the same locations during reassembly.*
32 Disconnect the crossover hose from one of the tanks and disconnect any emission hoses (where fitted).
33 Lift the tanks off the frame.

Installation (all models)

34 When replacing the fuel tank(s), reverse the above procedure. Be sure to install any spacers or washers in their original locations. Replace cracked or deteriorated hoses and clamps with new ones.
35 If a new tank is installed on 1985 and later FLT and FXR models equipped with a California EVAP system, you'll have to drill or punch a 0.03 to 0.06-inch (0.762 to 1.524 mm) hole in the fitting on the new tank's filler neck. Use a sharp awl and a hammer or a 1/16-inch drill bit to make the hole. If a drill is used, be sure to clean any metal chips produced out of the tank before installing it.

3 Fuel control valve – removal and installation

Refer to illustrations 3.4, 3.5a and 3.5b

Warning: *Gasoline is extremely flammable and highly explosive under certain conditions – safety precautions must be followed when working on any part of the fuel system! Don't smoke or allow open flames or unshielded light bulbs in or near the work area. Don't do this procedure in a garage with a natural gas appliance (such as a water heater or clothes dryer). Also, before starting work, disconnect the negative battery cable from the battery.*

1 If the control valve is leaking or the filter must be cleaned, the valve must be detached from the bottom of the tank. Drain the fuel into a clean gasoline container first (see Section 2), then detach the fuel line from the valve. Most hose clamps installed at the factory must be pried or cut off and can't be reused.
2 Unscrew the large nut and detach the valve from the bottom of the fuel tank.
3 Refer to Chapter 1 for instructions on cleaning and replacing the fuel filter.
4 If the valve leaks badly or doesn't work correctly, it must be replaced with a new one or repaired. Pre-1975 models have a sealed fuel valve which can't be repaired. The control valve on 1975 and later models can be disassembled by removing the two screws from each side of the lever **(see illustration)**.

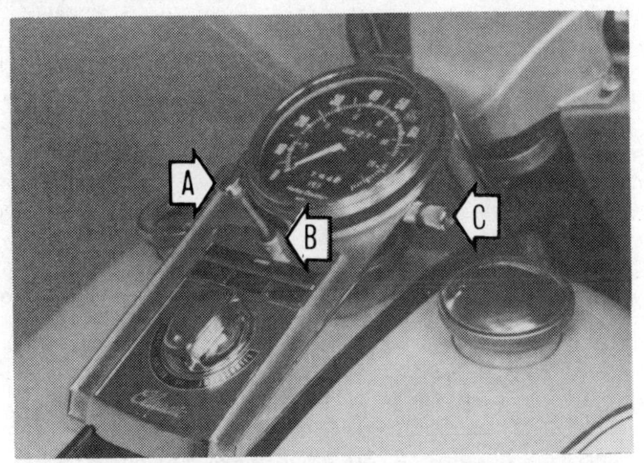

2.22 Remove the choke knob (A), the locknut (B) and the odometer knob and screw (C) to detach the instrument panel (two-piece fuel tank)

3.4 Remove the screws (arrows) to disassemble the fuel control valve

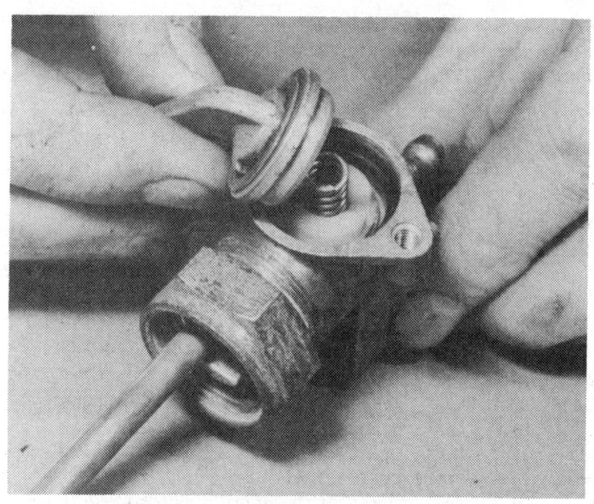

3.5a Remove the lever, spring and . . .

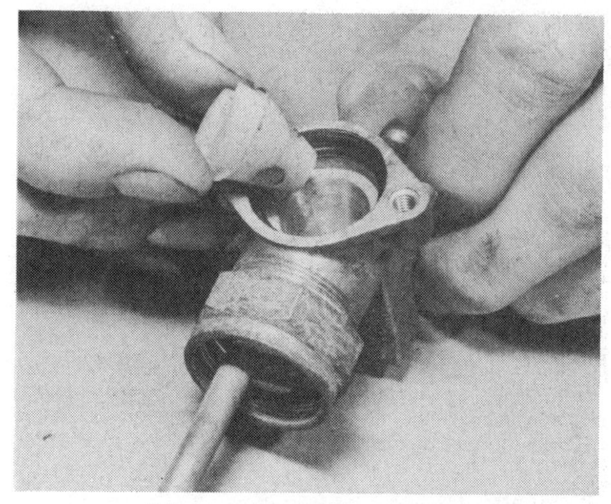

3.5b . . . nylon valve assembly

4.2a Turn the control valve to the Off position and . . .

4.2b . . . detach the clamp securing the fuel line to the carburetor fitting (original equipment clamps must be cut off and discarded)

5 The lever, spring and nylon valve can be removed after the screws are removed **(see illustrations)**.

6 Before installing the valve in the fuel tank, apply teflon tape or sealant that's resistant to gasoline to the tank threads. On 1975 and later models, the valve has a left-hand thread and the fitting on the bottom of the fuel tank has a right-hand thread. As the large nut is tightened, the fuel valve and the tank are drawn together.

7 When the fuel valve is securely fastened to the tank, connect the fuel line to it. Install a new hose clamp if necessary.

4 Carburetor – removal and installation

Refer to illustrations 4.2a, 4.2b and 4.4

Warning: *Gasoline is extremely flammable and highly explosive under certain conditions – safety precautions must be followed when working on any part of the fuel system! Don't smoke or allow open flames or unshielded light bulbs in or near the work area. Don't do this procedure in a garage with a natural gas appliance (such as a water heater or clothes dryer). Also, before starting work, disconnect the negative battery cable from the battery.*

Note: *Although it isn't absolutely necessary on every model, you probably will find it easier to remove the carburetor if the fuel tank is removed first (see Section 2).*

1 Refer to Section 8 and remove the air cleaner assembly.

2 Turn the fuel control valve to the Off position and disconnect the fuel line from the carburetor. Most hose clamps installed at the factory, must be pried or cut off and can't be reused **(see illustrations)**.

3 Disconnect the throttle cable(s) from the carburetor. Tillotson and Bendix carburetors use set screws to secure the throttle cable. To disconnect the cable(s) from a Keihin carburetor, turn the throttle valve open by hand and pull the cable ferrule out of the hole in the throttle lever. On 1981 and later models, there are two cables attached to the throttle lever.

4 Disconnect the choke cable from the carburetor – it's attached with a set screw on all carburetors **(see illustration)**. **Note:** *The CV carburetor used on 1990 and later models doesn't have a choke. It has an enrichener valve that's cable-operated just like the choke. To detach it, unscrew the fitting and pull the valve out of the left (rear) side of the carburetor. Be careful not to damage the end of the valve while it's exposed.*

5 Disconnect the vacuum and EVAP (emission) system hose(s) from the carburetor (some later California models only).

6 On 1990 and later FLTC and FLHTC Ultra models, detach the cruise-control servo cable from the carburetor bracket and the throttle lever. It's held in place with C-clips and a washer.

7 On all but 1990 and later models (CV carburetor), remove the nuts/bolts that secure the carburetor to the intake manifold. Carefully separate the carburetor from the intake manifold. Late 1976 through early 1978 Keihin carburetors have an O-ring seal between the carburetor and intake manifold. All other models have a gasket.

4.4 Loosen the set screw (arrow) securing the choke cable to the carburetor lever

8 On 1990 and later models, simply pull the carburetor out of the sealing ring in the intake manifold.

9 Installation is the reverse of the removal procedure, but be sure to install a new gasket, O-ring or sealing ring between the intake manifold and carburetor. Coat the CV carburetor sealing ring with a small amount of liquid dish soap before installation.

10 Replace the gasket between the air cleaner baseplate and the carburetor with a new one.

5 Carburetor overhaul – general information

1 Poor engine performance, hesitation and little or no engine response to idle fuel/air mixture adjustments are all signs that major carburetor maintenance is required.

2 Keep in mind that many so-called carburetor problems are really not carburetor problems at all, but mechanical problems in the engine or ignition system faults. Establish for certain that the carburetor needs maintenance before assuming an overhaul is necessary.

3 For example, fuel starvation is often mistaken for a carburetor problem. Make sure the fuel filter, the fuel line and the gas tank cap vent hole are not plugged before blaming the carburetor for this relatively common malfunction.

4 Most carburetor problems are caused by dirt particles, varnish and other deposits which build up in and block the fuel and air passages. Also, in time, gaskets and O-rings shrink and cause fuel and air leaks which lead to poor performance.

5 When the carburetor is overhauled, it's generally disassembled completely and the metal components are soaked in carburetor cleaner (which dissolves gasoline deposits, varnish, dirt and sludge). **Caution:** *Don't soak any rubber parts (especially the vacuum piston diaphragm on the CV carburetor) in carburetor cleaning solvents. They will be damaged if you do.* The parts are then rinsed thoroughly with solvent and dried with compressed air. The fuel and air passages are also blown out with compressed air to force out any dirt that may have been loosened but not removed by the carburetor cleaner. Once the cleaning process is complete, the carburetor is reassembled using new gaskets, O-rings, diaphragms and, generally, a new inlet needle and seat.

6 Before taking the carburetor apart, make sure you have a rebuild kit (which will include all necessary O-rings and other parts), some carburetor cleaner, solvent, a supply of rags, some means of blowing out the carburetor passages and a clean place to work.

7 Some of the carburetor settings, such as the sizes of the jets and the internal passageways are predetermined by the manufacturer after extensive tests. Under normal circumstances, they won't have to be changed or modified. If a change appears necessary, it can often be attributed to a developing engine problem.

6 Carburetor – disassembly, inspection and reassembly

Warning: *Gasoline is extremely flammable and highly explosive under certain conditions – safety precautions must be followed when working on any part of the fuel system! Don't smoke or allow open flames or unshielded light bulbs in or near the work area. Don't do this procedure in a garage with a natural gas appliance (such as a water heater or clothes dryer).*

1 Before disassembling the carburetor, clean the outside with solvent and lay it on a clean sheet of paper or a shop towel.

2 After it's been completely disassembled, submerge the metal components in carburetor cleaner and allow them to soak for approximately 30 minutes. Do not place any plastic or rubber parts in it – they'll be damaged or dissolved. Also, don't allow excessive amounts of carburetor cleaner to get on your skin.

3 After the carburetor has soaked long enough for the cleaner to loosen and dissolve the varnish and other deposits, rinse it thoroughly with solvent and blow it dry with compressed air. Also, blow out all the fuel and air passages in the carburetor body. **Note:** *Never clean the jets or passages with a piece of wire or drill bit – they could be enlarged, causing the fuel and air metering rates to be upset.*

Tillotson carburetor

Refer to illustrations 6.4 and 6.19

4 Carefully turn the idle mixture adjustment screw in until it bottoms, while counting the number of turns, then remove it, along with the spring **(see illustration on next page)**. Record the number of turns – you'll need to refer to it later. Remove the intermediate mixture adjusting screw in the same manner. By counting the number of turns until they bottom, the adjustment screws can be returned to their original positions and adjustments will be kept to a minimum.

5 Note the position of the throttle valve before removing it to ensure it's reinstalled in the same position. Remove the two screws securing the throttle valve to the shaft and detach the valve.

6 Remove the screw securing the throttle shaft and accelerator pump. Pull the throttle shaft out of the carburetor body along with the throttle shaft spring and washers, then remove the dust seals from both sides of the carburetor body.

7 Invert the carburetor and remove the screws securing the diaphragm cover. Carefully lift off the cover, then remove the diaphragm and gasket. Separate the diaphragm from the gasket by peeling them apart.

8 Take out the screw that secures the accelerator pump plunger, then withdraw the plunger.

9 Remove the plug screw from the diaphragm cover.

10 Remove the inlet control lever screw. This will permit the control lever pin, the control lever and the inlet needle to be removed. These parts are very small and easily lost if you aren't careful. Remove the control lever tension spring from below the assembly.

11 Remove the inlet needle valve seat and gasket with a thin-wall 3/8-inch socket. Note the position of the seat insert with the smooth side toward the inside of the cage. The gasket can be lifted out with a scribe.

12 Unscrew the main jet plug, then remove the main jet and gasket.

13 Drill a 1/8-inch hole in the center of the main nozzle welch plug. Be careful not to drill beyond the welch plug, since damage to the main nozzle can result. Insert a small punch through the hole and carefully pry the plug out of the casting.

14 Remove the idle port welch plug as described in the previous Step.

15 Using a small punch, remove the welch plug over the economizer check ball and let the check ball roll out.

16 Remove the screws securing the choke valve and lift out the bottom part of the valve.

17 Slide the choke shaft assembly out of the carburetor (when this is done, the upper part of the choke valve will be released and can be removed). Remove the choke spring and the choke shaft friction ball and spring. Pry the choke shaft dust seal out of the carburetor body.

18 Clean and inspect the parts as described in Steps 2 and 3 of this Section. Inspect the carburetor body for cracks and make sure the throttle shaft turns freely without excessive play. If it's sloppy, a new carburetor will

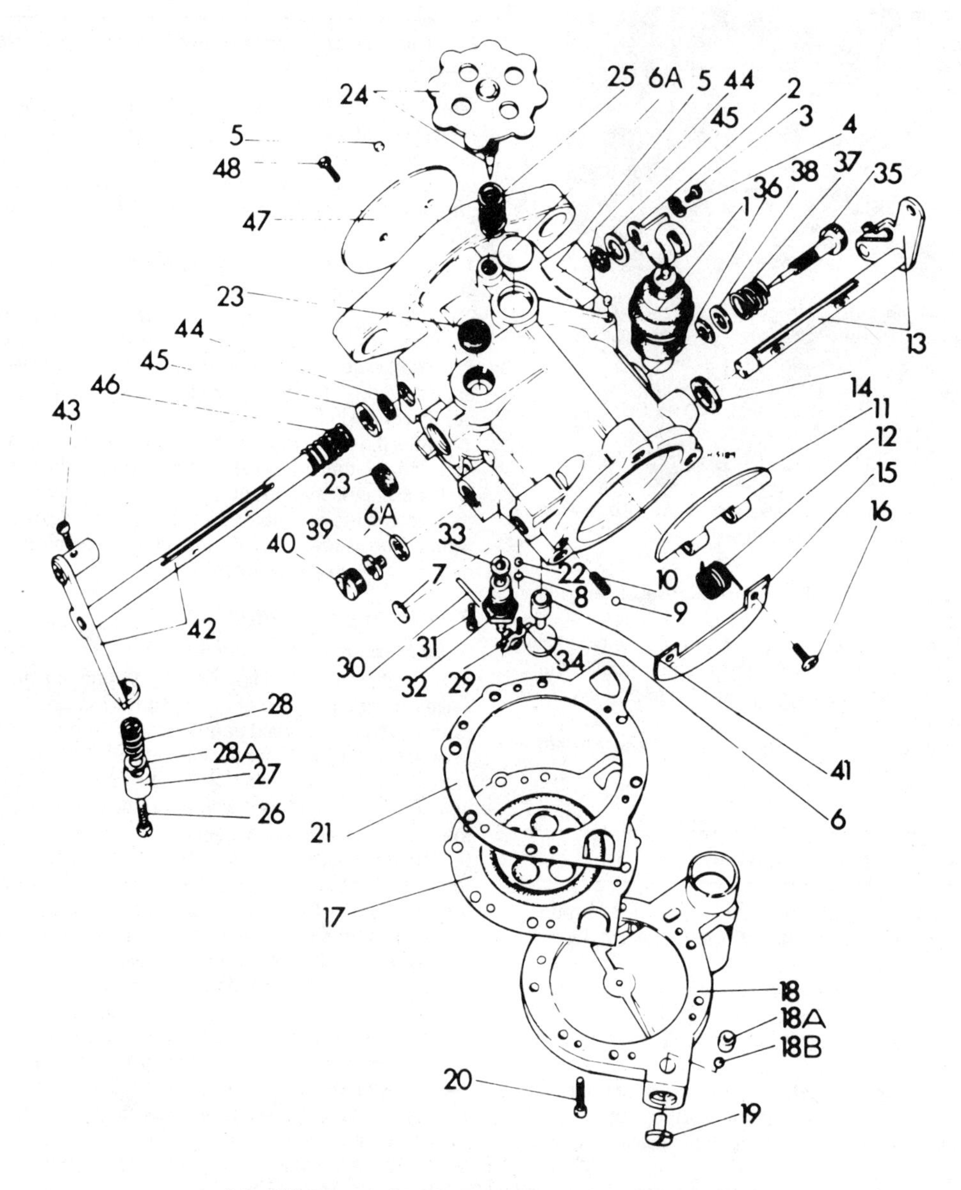

6.4 Tillotson carburetor components – exploded view

1 *Accelerator pump*
2 *Accelerator pump lever*
3 *Accelerator pump screw*
4 *Lock washer*
5 *Channel plug*
6 *Main nozzle welch plug*
6A *Welch plug*
7 *Idle port welch plug*
8 *Welch plug*
9 *Choke shaft friction ball*
10 *Choke shaft friction spring*
11 *Choke valve (top)*
12 *Choke valve spring*
13 *Choke shaft*
14 *Choke shaft dust seal*
15 *Choke valve (bottom)*
16 *Screws*
17 *Diaphragm*
18 *Diaphragm cover*
18A *Accelerator pump check ball retainer*
18B *Accelerator pump check ball*
19 *Diaphragm cover plug*
20 *Diaphragm cover screws*
21 *Diaphragm cover gasket*
22 *Economizer check ball*
23 *Fuel filter screen*
24 *Idle mixture adjustment screw*
25 *Idle mixture adjustment screw spring*
26 *Throttle stop screw*
27 *Throttle stop screw cup*
28 *Throttle stop screw spring*
28A *Lock washer*
29 *Inlet control lever*
30 *Inlet control lever pin*
31 *Inlet control lever screw*
32 *Inlet needle valve and seat*
33 *Gasket*
34 *Inlet control lever tension spring*
35 *Intermediate mixture adjusting screw*
36 *Intermediate mixture adjusting screw gasket*
37 *Washer*
38 *Washer*
39 *Main jet*
40 *Main jet plug*
41 *Main nozzle check valve*
42 *Throttle shaft*
43 *Throttle lever cable screw*
44 *Dust seal*
45 *Washer*
46 *Throttle shaft spring*
47 *Throttle valve*
48 *Screws*

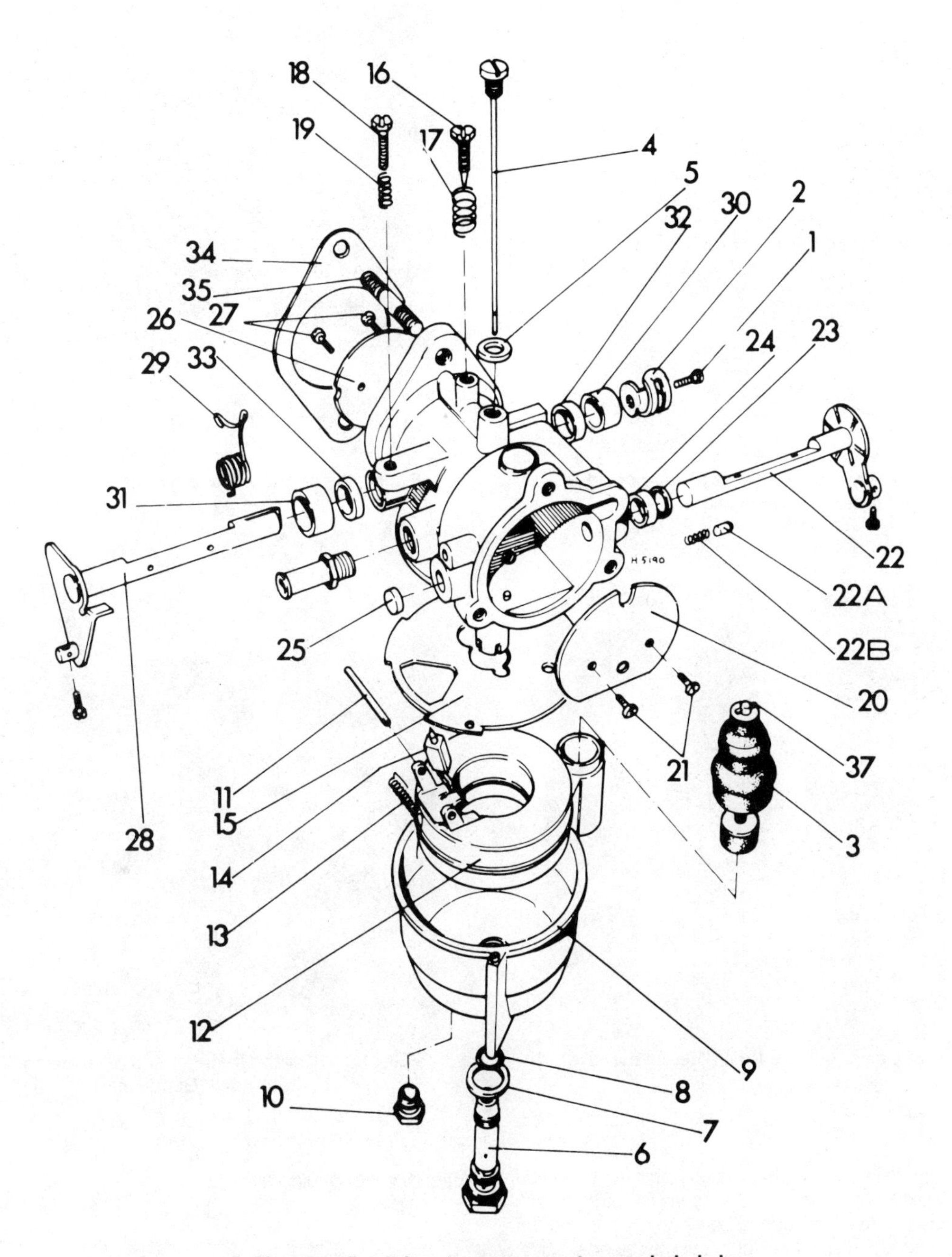

6.25a Bendix carburetor components – exploded view

1	*Screw*	14	*Inlet needle valve*	24	*Choke shaft seal*
2	*Accelerator pump lever*	15	*Gasket*	25	*Choke shaft cup plug*
3	*Accelerator pump*	16	*Idle mixture adjusting screw*	26	*Throttle valve*
4	*Idle tube*	17	*Spring*	27	*Screw*
5	*Idle tube gasket*	18	*Throttle stop screw*	28	*Throttle shaft and lever*
6	*Main jet and tube assembly*	19	*Spring*	29	*Throttle shaft spring*
7	*Fiber washer*	20	*Choke valve*	30	*Throttle shaft seal retainer*
8	*O-ring*	21	*Screw*	31	*Throttle shaft seal retainer*
9	*Float bowl*	22	*Choke shaft and lever*	32	*Throttle shaft seal*
10	*Float bowl drain plug*	22A	*Plunger*	33	*Throttle shaft seal*
11	*Float pivot pin*	22B	*Spring*	34	*Manifold gasket*
12	*Float assembly*	23	*Choke shaft seal retainer*	35	*Stud*
13	*Float spring*			36	*Accelerator pump shaft pin*

3

6.19 Check the inlet needle valve for a groove or ridge in the tapered area (arrow)

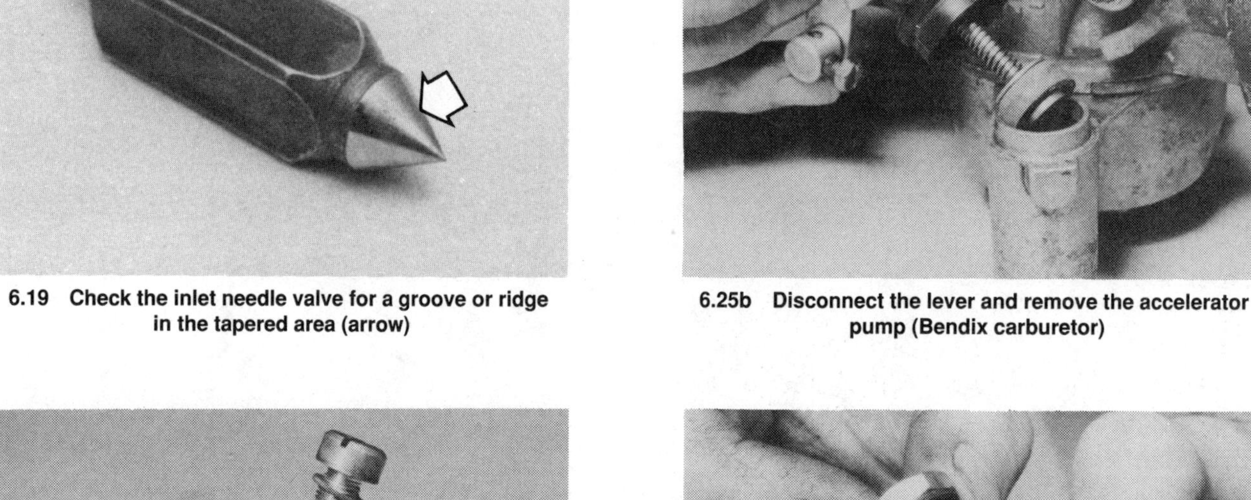

6.25b Disconnect the lever and remove the accelerator pump (Bendix carburetor)

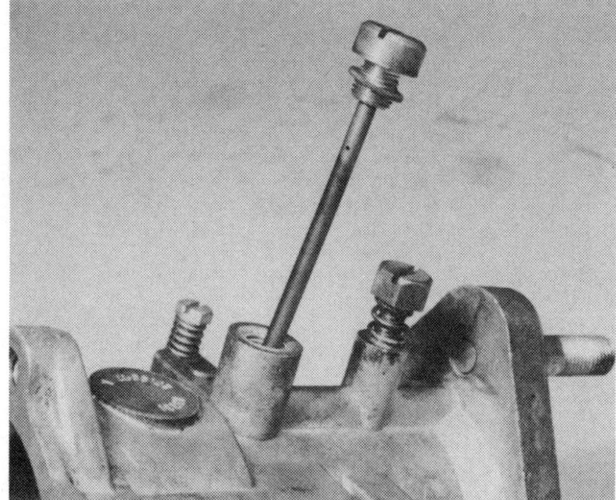

6.27 Unscrew and withdraw the idle tube (Bendix carburetor)

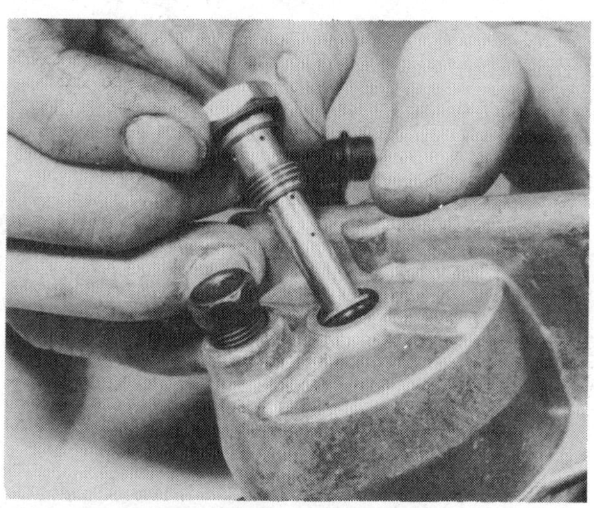

6.28 Unscrew the main jet and tube assembly from the float bowl (Bendix carburetor)

be needed (although sometimes the throttle shaft bores can be reamed out and bushings installed – check with a dealer service department).

19 Make sure the inlet control lever rotates freely on the pin and the forked end of the lever engages with the slot in the inlet needle valve. Check the end of the control lever for wear and burrs. Check the spring to be sure it isn't stretched or distorted. Check the inlet needle valve and seat for nicks and a pronounced groove or ridge on the tapered end of the valve **(see illustration)**. If there is one, a new needle and seat should be used when the carburetor is reassembled.

20 Examine the rest of the parts for wear and damage.

21 Reassemble the carburetor by reversing the disassembly sequence. Use a new diaphragm as well as new gaskets and seals and don't overtighten any of the small fasteners or they may break off.

22 Seat the new welch plugs by striking them with a punch slightly smaller than the plug itself. When the plug is seated, it should be flat, not concave. This will ensure a tight fit around the edge of the casting opening.

23 The inlet control lever tension spring should be installed in the carburetor body. Be sure it attaches to the protrusion on the inlet control lever. Bend the diaphragm end of the control lever so when the lever is installed, it's flush with the floor of the metering chamber.

24 Be sure to tighten the inlet needle valve seat and the diaphragm cover plug to the torque figures listed in this Chapter's Specifications.

Bendix carburetor

Refer to illustrations 6.25a, 6.25b, 6.27, 6.28, 6.29a, 6.29b, 6.29c, 6.40 and 6.42

25 Remove the screw securing the accelerator pump lever to the throttle shaft **(see illustration on previous page)**. Disconnect the accelerator pump boot from the float bowl. Remove the accelerator pump and lever **(see illustration)**.

26 Compress the spring on the accelerator pump shaft, rotate the pump lever 1/4-turn and disengage the pin at the top of the shaft from the lever.

27 On 1972 through 1974 models, unscrew the idle tube from the top of the carburetor body and detach the gasket at the same time **(see illustration)**.

28 Unscrew the main jet and tube assembly from the bottom of the float bowl **(see illustration)**. This will release the float bowl. Remove the O-ring and fiber washer from the main jet assembly.

29 Note how the float spring is positioned **(see illustration)**, then push the float pivot pin out of the throttle body. You may have to use a small punch to push the pin out. Remove the float assembly along with the inlet needle valve and the float spring **(see illustrations)**. Remove the float bowl gasket.

30 Carefully screw the idle mixture adjusting screw in until it bottoms,

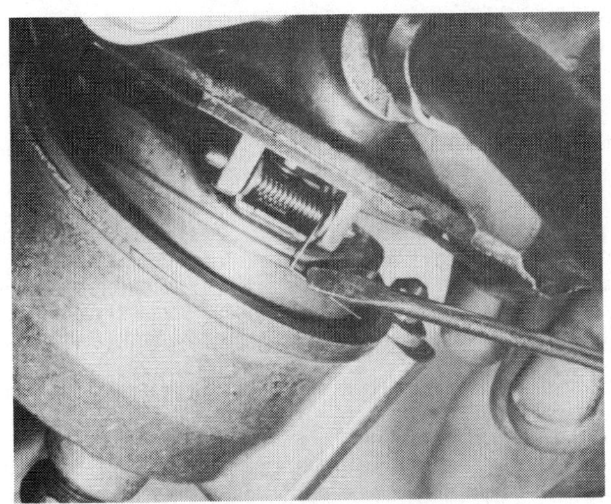

6.29a Note the position of the float spring before removing the pivot pin (Bendix carburetor)

6.29b You may have to drive the float pivot pin (arrow) out with a small punch (Bendix carburetor)

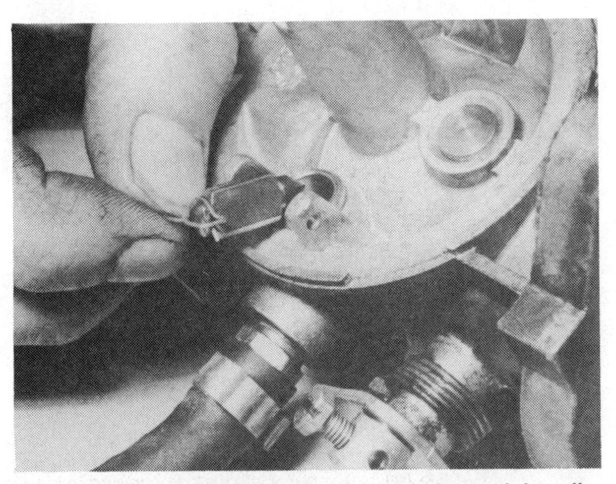

6.29c Lift out the inlet needle valve with the retaining clip attached (Bendix carburetor)

6.40 Be sure the inlet needle valve is correctly seated during reassembly – the retaining clip fits over the tab on the float (arrow) (Bendix carburetor)

while counting the number of turns, then remove it along with the spring. Remove the throttle stop screw and spring in the same manner. Counting and recording the number of turns required to bottom the screws will enable you to return them to their original positions and minimize the amount of adjustment required after reassembly.

31 Close the choke and remove the screws securing the choke valve to the shaft. Pull the choke shaft and lever out, releasing the spring and the plunger.

32 Pry the choke shaft seal and retainer out of the carburetor body. Remove the choke shaft cup plug only if it's damaged and must be replaced.

33 Close the throttle valve and remove the two screws securing it to the shaft. Remove the valve, then slide the shaft out of the carburetor body. Release the spring from the throttle shaft.

34 Pry the throttle shaft retainers and seals out of both sides of the carburetor body.

35 Clean and inspect the parts as described in Steps 2 and 3 of this Section. Make sure the throttle shaft turns freely without excessive play. If it's sloppy, a new carburetor will be needed (although sometimes the throttle shaft bores can be reamed out and bushings installed – check with a dealer service department). Check the inlet needle valve and seat for nicks and a pronounced groove or ridge on the tapered end of the valve **(see illustration 6.19).** If there is one, a new needle and seat should be used when the carburetor is reassembled. Check the float pivot pin and its bores for wear – If the pin is a sloppy fit in the bores, excessive amounts of fuel will enter the float bowl and flooding will occur. Shake the float to see if there's gasoline in it. If there is, install a new one.

36 Reassembly is the reverse of disassembly. Use new gaskets and seals. Whenever an O-ring or seal is installed, lubricate it with grease or oil. Don't overtighten any of the small fasteners or they may break off.

37 Position the throttle shaft so the flat section is facing out and install the throttle valve (leave the screws slightly loose). Open and close the throttle a few times to center the valve on the shaft. Hold the valve firmly in place while tightening the screws.

38 Insert the choke shaft seal and seal retainer into the choke shaft hole and stake the retainer in place with a small punch.

39 Connect the choke valve to the choke shaft in the same manner as the throttle valve.

40 When the inlet needle valve assembly is installed, be sure the clip is attached to the float tab **(see illustration)**. Check and adjust the float level as described in Section 7.

41 Turn the idle mixture adjusting screw and the throttle stop screw in until they bottom and back each one out the number of turns required to restore them to their original positions.

42 Invert the carburetor and rotate the long end of the float spring up, against the float. Position the float bowl carefully over the throttle body, releasing the float spring so the long end of the spring is pressed against the

6.42 The long end of the float spring (arrow) must be positioned as shown (Bendix carburetor)

side of the float bowl **(see illustration)**.

43 Install the main jet and tube assembly through the bottom of the float bowl and into the throttle body. Tighten it securely.

Keihin carburetor (except CV type)

Refer to illustrations 6.44a, 6.44b, 6.45, 6.47, 6.48, 6.49, 6.50a, 6.50b, 6.51 and 6.52

44 Remove the screws securing the float bowl to the bottom of the throttle body and detach the float bowl **(see illustrations)**.

45 Loosen the float retaining screw and slide the pivot pin out **(see illustration)**, then carefully separate the float assembly from the carburetor.

46 Remove the inlet needle valve and retaining clip from the float assembly.

47 Separate the rubber boot from the float bowl, then disengage the accelerator pump rod from the rocker arm **(see illustration)**.

48 Remove the plug from the bottom of the throttle body to gain access to the low speed jet **(see illustration)**. On some models, the low speed jet is accessible only after removing the main jet and main nozzle as described below.

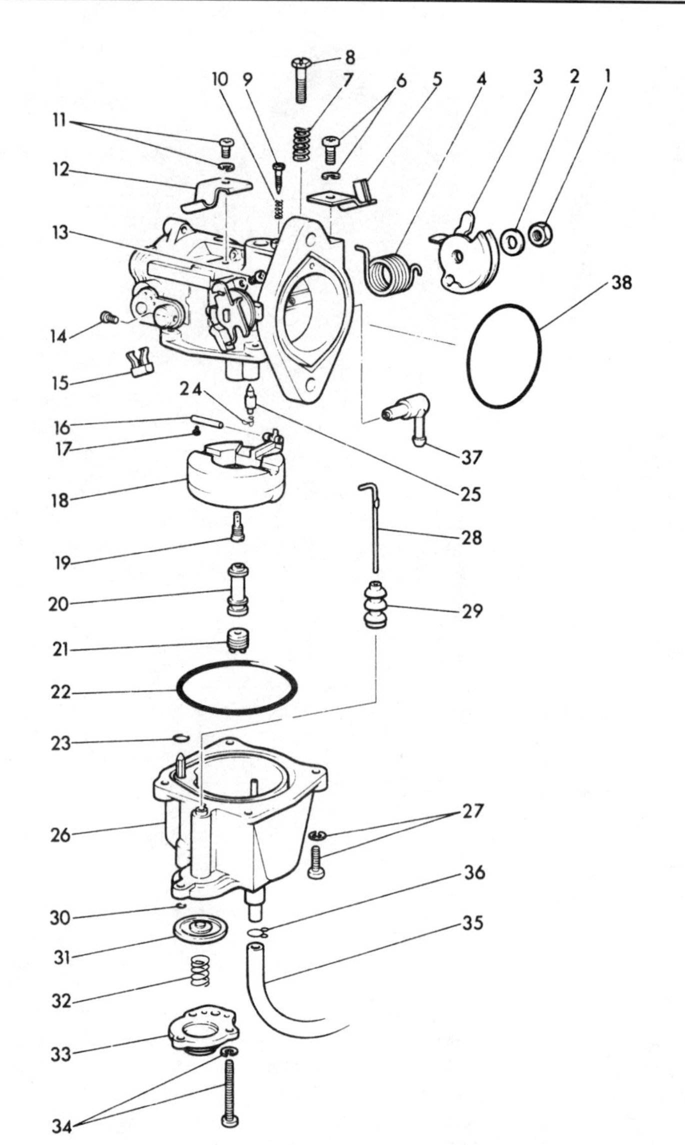

6.44a Typical non-CV type Keihin carburetor components – exploded view

1 Nut
2 Plain washer
3 Throttle lever cable pulley
4 Return spring
5 Throttle cable bracket
6 Screw/washer
7 Spring
8 Throttle stop screw
9 Low speed mixture adjusting screw
10 Spring
11 Screw/washer
12 Choke cable bracket
13 Throttle pump stroke adjusting screw
14 Screw
15 Spacer clip
16 Float pivot pin
17 Float retaining screw
18 Float assembly
19 Low speed jet
20 Main nozzle (not used on all models)
21 Main jet
22 O-ring
23 O-ring
24 Retaining clip
25 Inlet needle valve
26 Float bowl
27 Screw/washer
28 Accelerator pump rod
29 Rubber boot
30 O-ring
31 Diaphragm
32 Spring
33 Accelerator pump cover
34 Screw/washer
35 Overflow tube
36 Clip
37 Union
38 O-ring

49 Unscrew the main jet **(see illustration)**. On 1975 through early 1978 models, tip the throttle body to remove the main nozzle, then unscrew the low speed jet.

50 Unscrew the nut securing the throttle cable pulley or the fast idle cam assembly in place **(see illustration)**. Remove the throttle cable pulley and the return spring from the throttle shaft **(see illustration)**.

6.44b Remove the screws securing the float bowl (Keihin carburetor)

6.45 Loosen the retaining screw to release the float pin (Keihin carburetor)

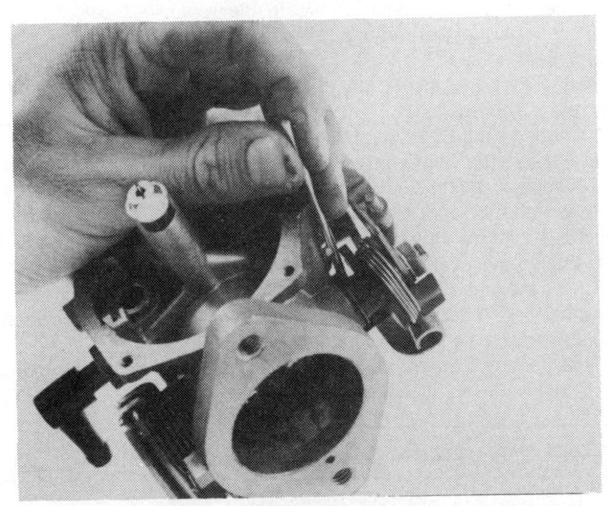

6.47 Disconnect the accelerator pump rod from the rocker arm (Keihin carburetor)

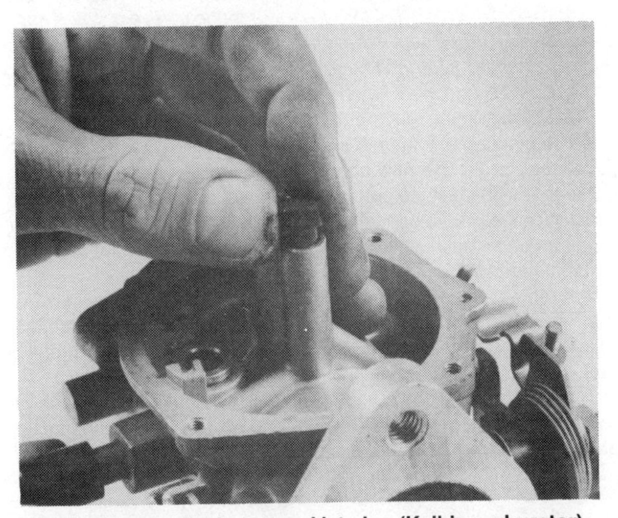

6.48 Remove the low speed jet plug (Keihin carburetor)

6.49 Remove the main jet (Keihin carburetor)

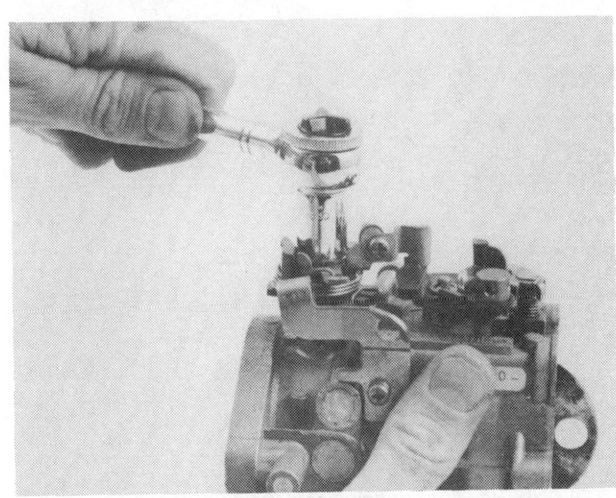

6.50a Remove the throttle cable pulley mounting nut (Keihin carburetor)

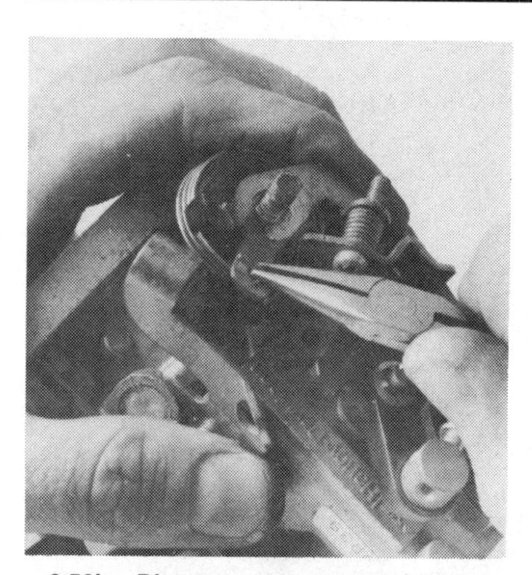

6.50b Disconnect the throttle return spring from the throttle shaft (Keihin carburetor)

6.51 Locations of the throttle cable (1) and choke cable (2) brackets (Keihin carburetor)

6.52 Remove the throttle stop screw (Keihin carburetor)

51 Remove the screws and detach the choke and throttle cable brackets **(see illustration)**.

52 Remove the low speed mixture adjusting screw (1975 through 1979 models only) and the throttle stop screw as described in Step 30 **(see illustration)**. **Note:** *Late 1977 through 1979 models have a limiter cap on the low speed mixture screw that must be pried off to remove the screw. Models built after 1979 don't have a low speed mixture screw – the idle mixture is preset and can't be changed.*

53 Don't disassemble the throttle valve – it's matched to the carburetor and isn't replaceable. If there's a problem with the valve, the carburetor must be replaced with a new one.

54 Clean and inspect the parts as described in Step 35.

55 Check the accelerator pump boot for cracks, the rod for distortion and the diaphragm for holes, cracks and other defects. If any wear or damage is evident, replace the pump components with new ones.

56 Reassembly is the reverse of disassembly. Use new gaskets and seals. Whenever an O-ring or seal is installed, lubricate it with grease or oil. Don't overtighten any of the small fasteners or they may break off. Check the float level (Section 7) before installing the float bowl.

57 Turn the low speed mixture adjusting screw (if used) and the throttle stop screw in until they bottom and back each one out the number of turns required to restore them to their original positions.

Keihin CV carburetor

Refer to illustration 6.64

58 Remove the screws and detach the vacuum chamber cover and spring.

6.64 The enrichener valve seal (arrow) must be in good condition or fuel may leak past it and cause an overly rich mixture (Keihin carburetor)

59 Lift out the vacuum piston and rubber diaphragm assembly with the jet needle attached. Be careful not to bend or nick the jet needle – store the vacuum piston assembly in a safe place.

60 Refer to Step 44 – the rest of the disassembly procedure is the same as the one for the early Keihin carburetor. After removing the float assembly, unscrew the main jet and invert the carburetor to remove the needle jet holder and needle jet.

61 In addition to the inspection procedures outlined for the early Keihin carburetor, check the following items: Hold the vacuum piston diaphragm up to a strong light. Look for pinholes and small tears.

62 Make sure the vacuum passage in the bottom of the piston is clear. The piston itself must be smooth and perfectly clean.

63 Check the jet needle for wear, nicks, scratches and distortion. It should be straight and the surface of the taper should be smooth and even. Clean the needle jet and jet holder.

64 Check the end of the enrichener valve **(see illustration)**. Make sure the rubber pad is in good condition. If it's cracked, distorted or deeply grooved, replace the valve with a new one.

65 Reassembly is the reverse of disassembly (see Steps 56 and 57).

7 Carburetor – adjustments

Tillotson carburetor

1 Carburetor adjustments should be made with the engine at normal operating temperature. Also, be sure the air filter element is clean and the

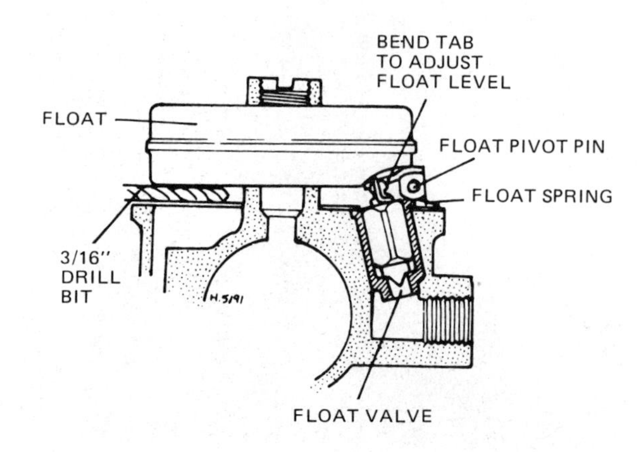

7.5 Checking the float level with a 3/16-inch drill bit as a gauge (Bendix carburetor)

7.6 **Locations of the throttle stop screw (1) and the idle mixture adjusting screw (2) (Bendix carburetor)**

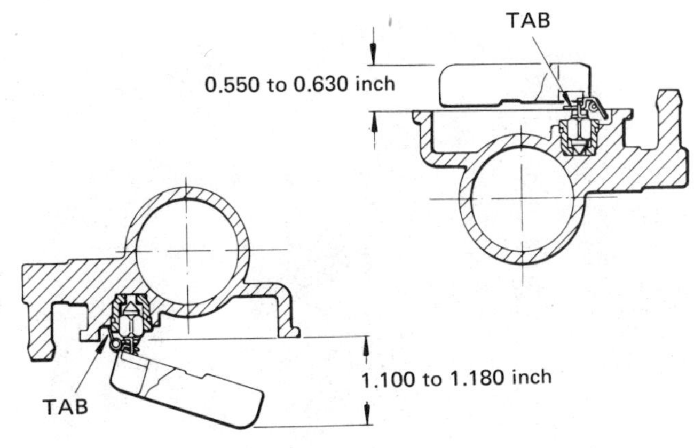

7.9 **Checking the float level on 1976 through 1978 Keihin carburetor**

air cleaner assembly is installed securely. Adjustments cannot be done accurately unless the air cleaner assembly is in place.

2 Make sure the twist grip closes the throttle lever on the carburetor completely. Turn the idle mixture screw in until it seats lightly, then back it out 7/8-turn. Adjust the intermediate mixture screw in a similar fashion.

3 Start the engine and adjust the throttle stop screw until the engine is running at approximately 2000 rpm. Turn the intermediate mixture screw in both directions until the highest engine speed is obtained without any misfiring or surging. Turn the screw an additional 1/8-turn counterclockwise.

4 Turn the throttle stop screw to adjust the idle to the recommended speed.

Bendix carburetor

Refer to illustrations 7.5 and 7.6

5 The float level must be adjusted before anything else is done to the carburetor. The carburetor must be removed and the float bowl detached to adjust the float level. Refer to Section 4 for carburetor removal and Section 6 for removal of the float bowl. Invert the carburetor and measure the distance between the lower edge of the float, opposite the pivot pin, and the gasket mating surface. The distance should be 3/16-inch (insert a 3/16-inch drill bit between the gasket surface and the float as a gauge) **(see illustration)**. If adjustment is necessary, bend the tab that contacts the inlet needle valve with needle-nose pliers. Reassemble and install the carburetor.

6 Run the engine until it reaches normal operating temperature, then shut it off. Carefully turn the idle mixture adjusting screw in until it seats lightly **(see illustration)**. Back it out 1-1/2 turns. Adjust the engine speed by turning the throttle stop screw until it runs at 700 to 900 rpm with the twist grip closed.

7 Adjust the idle mixture screw until the engine will run smoothly at idle speed and accelerate crisply. If necessary, adjust the throttle stop screw until the engine idles at 700 to 900 rpm.

8 Remember, all adjustments must be made with the engine at normal operating temperature and the air cleaner securely in place.

Keihin carburetor (except CV type)

Float level (1976 through 1978 models)

Refer to illustration 7.9

9 These models must have the float level measured in both the open and closed positions. Hold the carburetor upside-down and measure the distance between the gasket surface and the upper edge of the float. The distance should be 0.550 to 0.630-inch (14 to 16 mm) **(see illustration)**.

10 Hold the carburetor right side up and measure the distance between the gasket surface and the lower edge of the float while it's suspended. The distance should be 1.100 to 1.180-inches (28 to 30 mm). Bend the tabs on the float, as necessary, to obtain the proper float levels.

Float level (1979 through 1989 models)

Refer to illustrations 7.11a and 7.11b

11 Hold the carburetor vertically with the float pivot pin at the top. Measure the distance between the outside edge of the float and the gasket surface **(see illustration)**. The distance should be 0.630 to 0.670-inch (16 to 17 mm). If necessary, bend the tab on the float until the desired float level is attained **(see illustration)**.

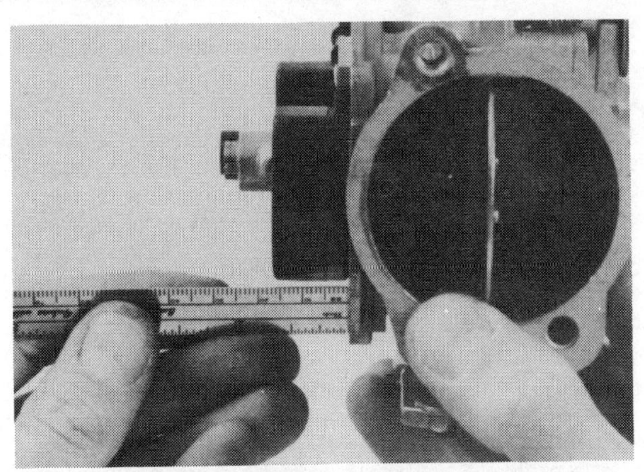

7.11a **Measure the float level with the carburetor held vertically (1979 through 1989 Keihin carburetor)**

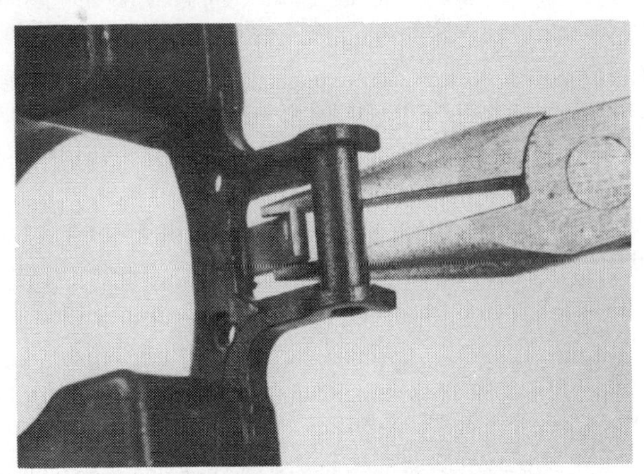

7.11b **Bend the tab to adjust the float level (Keihin carburetor)**

7.21 Fast idle adjusting screw (arrow) (1980 through 1989 Keihin carburetor)

Low speed mixture adjustment (1976 through early 1977 models)

12 Adjustments should be made with the engine at normal operating temperature and the air cleaner securely mounted.

13 Turn the low speed mixture adjusting screw **(see illustration 6.44a)** in (clockwise) carefully until it seats lightly, then back it out 7/8-turn.

14 Adjust the throttle stop screw until the engine runs at 700 to 900 rpm. Turn the low speed mixture screw until the engine runs smoothly or until it reaches its highest rpm. Readjust the throttle stop screw, if necessary, to obtain the specified idle speed.

Low speed mixture adjustment (late 1977 through 1979 models)

15 There's a limiter cap installed over the low speed mixture adjusting screw on these models. Normally the limiter cap shouldn't be removed. The low speed mixture screw should be turned only within the limits of the cap. If necessary, the limiter cap can be removed and the mixture can be altered.

16 Turn the mixture screw in until it seats lightly, then back it out the number of turns listed in this Chapter's Specifications. Reinstall the limiter cap in the center position on the adjusting screw.

17 Adjust the idle speed with the throttle stop screw to 900 rpm. Turn the limiter cap to the leanest setting that still permits the engine to run smoothly. **Note:** *Turn the limiter cap counterclockwise for a richer mixture and clockwise for a leaner mixture.*

18 Readjust the idle speed, if necessary, to the specified setting.

Low speed mixture adjustment (1980 through 1989 models)

Refer to illustration 7.21)

19 The low speed mixture for 1980 through 1989 models is set at the factory and sealed – it's not possible to adjust it.

20 Adjust the idle speed with the choke completely open. Turn the throttle stop screw until the engine is idling at the speed listed in this Chapter's Specifications.

21 Pull the choke out to the second position and turn the fast idle adjusting screw **(see illustration)** until the engine is idling at the specified speed.

Keihin CV carburetor

Float level (1990 and early 1991 models with 3-sided needle valve)

22 The carburetor must be removed and the float bowl detached when checking the float level. Make sure the floats are aligned with each other – bend them carefully to realign them if necessary.

23 Invert the carburetor and measure the distance from the float bowl mounting surface to the very bottom (curved) side of the float(s) with a dial or vernier caliper. Don't push down on the float(s) as this is done. It should be 0.725 to 0.730-inch (18.4 to 18.5 mm).

24 If the level is incorrect, carefully bend the tab on the float that contacts the inlet needle valve until it is.

25 Reinstall the float bowl and the carburetor.

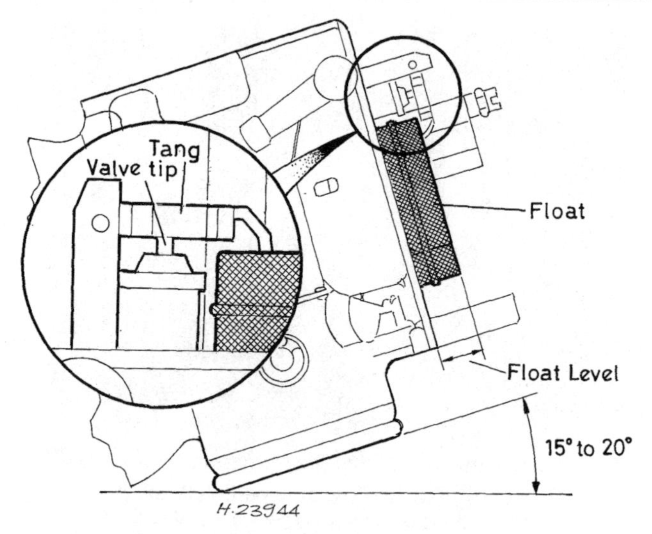

7.27 Checking the float level on late 1991 and later Keihin CV carburetor

Float level (late 1991 and later models with 4-sided needle valve)

26 Remove the carburetor and detach the float bowl. Place the carburetor body face down on a flat surface, on its engine manifold side.

27 Tilt the body at an angle of 15 to 20 degrees until the float tang is seen to just contact the needle valve tip, but not compress it **(see illustration)**. Do not tilt any more or less than the specified angle or the reading will be inaccurate.

28 Measure the distance from the top of the float to the bowl mounting surface; it should be within 0.725 to 0.730 inches (18.4 to 18.5 mm) on 1991 models, and 0.413 to 0.453 inches (10.5 to 11.5 mm) on 1992 and later models.

29 If the level requires adjustment carefully bend the tab on the float that contacts the inlet needle valve, then recheck the setting.

30 Reinstall the float bowl and carburetor.

Enrichener valve cable

31 The enrichener valve control knob should open, remain open and close without binding. The knurled plastic nut behind the knob controls the amount of resistance the cable offers.

32 If adjustment is required, loosen the locknut at the back of the cable bracket, then detach the cable from the bracket.

33 Grip the flats on the cable housing with an adjustable wrench, then turn the knurled plastic nut until the enrichener operates as described in Step 26.

34 Reattach the cable to the bracket and tighten the locknut. Do not lubricate the cable or housing – it must have a certain amount of resistance to work properly.

35 When the knob is closed, make sure the valve at the carburetor is closed completely.

Slow idle

36 With the engine at normal operating temperature and the enrichener valve closed, turn the throttle stop screw until the idle speed is correct (see this Chapter's Specifications). If the motorcycle doesn't have a tachometer, a hand-held instrument will be needed to measure the engine rpm.

8 Air cleaner – removal and installation

Refer to illustration 8.3

Note: *Although the air cleaners used on the models and years covered by this manual differ slightly in some details, they are all basically the same* ***(see illustration 4.1b in Chapter 1).*** *They consist of a backplate, fastened to the carburetor with screws and a gasket, mounting bracket(s), a filter element and a cover. The following procedure is typical of what must be done to completely remove the air cleaner assembly – take notes, label*

8.3 **Remove the screws and nuts securing the air cleaner baseplate to the carburetor (arrows)**

8.4 **Remove flexible connectors to reveal hollow breather passage bolts on 1993 and later models**

parts and make a simple sketch of the mounting bracket(s) if the air cleaner on the machine you have appears different from the one described in the text. Make sure the gasket between the backplate and carburetor is in place and in good condition and hook up all hoses during installation. Use thread locking compound on the backplate mounting screws/bolts and bend up the tabs on the lockplate (if used) to prevent the screws from backing out during engine operation.

1 Remove the screws or bolts and detach the air cleaner cover.

2 Remove the baffle plate (unless riveted in place) and filter element (see Chapter 1). If hoses are attached to the air cleaner assembly, detach them from their fittings.

3 Flatten the tabs on the lockplate (if used), then remove the screws/bolts securing the baseplate to the carburetor **(see illustration)**. Some models also have nuts/bolts attaching the baseplate to brackets or directly to the cylinder head. The mounting bolts may be safety-wired together to prevent them from loosening.

4 Most later models have a vent hose attached to the rear of the baseplate, which must be disconnected to remove the plate. Many later models also have a crankcase breather hose, attached to the bottom of the baseplate, that must be removed. From 1993 the baseplate-to-cylinder head bolts are hollow; they allow crankcase vapor, routed up through the pushrod tubes and through an umbrella valve in the middle rocker cover, to pass into the intake system. Short flexible connectors fit in the hollow bolt heads and the rear of the filter element **(see illustration)**.

5 Installation is the reverse of removal. Be sure to install a new gasket between the baseplate and carburetor. On models with safety-wired bolts, the bolts must be rewired or have a thread locking compound applied to the threads before they're reinstalled. If the bolts somehow loosen and fall out of position, the bolts or washers could possibly be drawn through the carburetor and into the engine, causing serious internal damage. **Note:** *Do not overtighten the hollow bleed bolts on 1993 and later models; the recommended torque is 10 to 12 ft-lbs (14 to 16 Nm).*

9 Evaporative emission control system – general information

Refer to illustration 9.3

This system, installed on all 1985 and later California models only, is virtually maintenance-free and shouldn't be tampered with unless a new canister, hoses or valves are required because of leaks or deterioration. An occasional check to make sure the hoses are routed properly, secured to the fittings and not kinked or blocked should be sufficient. Make sure the hoses don't come too close to or touch the exhaust system components. Also, check all canister and valve mounting fasteners to see if they're tight. On 1988 through 1991 models, check the reed valves in the air cleaner assembly backplate to make sure they aren't cracked or broken off.

The system is designed to prevent fuel vapor from escaping from the tank into the atmosphere when the engine is off. The vapor is directed from the tank through a hose and valve to a charcoal-filled canister (mounted on a frame front tube, under the seat or behind the gearbox (depending on the model), where it's absorbed by the charcoal. When the engine is started, the vapor is drawn from the canister into the carburetor and then into the engine, where it's burned in the cylinders. A large diameter hose also purges the canister with fresh air from the air cleaner during engine operation. The vapor valve prevents raw fuel from entering the vent hose when the motorcycle is at extreme angles.

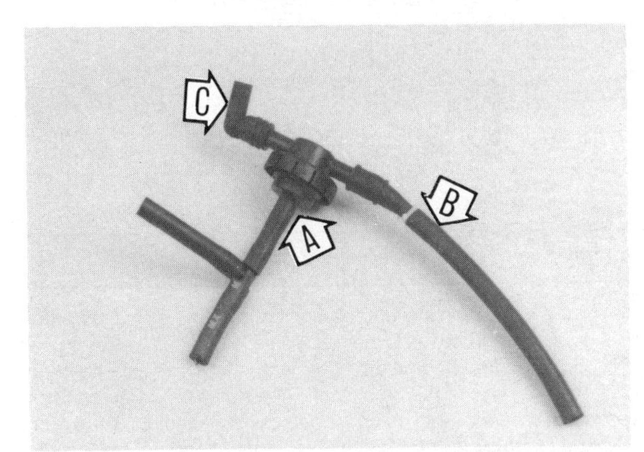

9.3 **Vacuum operated valve port locations (1988 through 1991 California models) – see text**

1988 through 1991 models also have a Vacuum Operated Electric Switch (VOES) and a vacuum operated valve. The VOES directs vacuum to the vacuum operated valve, which seals off the carburetor float bowl vent when the engine's off (it also vents it to the atmosphere when the engine's running). If the diaphragm in the vacuum valve starts to leak, the fuel/air mixture would be upset (leaned out) at high speeds. To check the valve, apply a small vacuum (1 to 2 in-Hg only) to port A **(see illustration)**. The vacuum should remain steady and the valve should be open (you should be able to blow through port B or C – air should pass through). Release the vacuum and make sure the valve is closed (no air should pass through when blowing into port B or C). If the valve doesn't function as described, install a new one.

1992 and later models retain the VOES described above, but venting of the carburetor vent hose is controlled electrically by a solenoid-operated butterfly valve, clamped to the rear of the air cleaner baseplate. The butterfly valve itself resides in a housing at the bottom of the air cleaner and if functioning correctly, should remain closed when the engine is stopped, yet open when the starter circuit operates and stay open until the engine is stopped. If failure is suspected, check first the mechanical linkage from the valve pivot to the solenoid plunger. If this is in order, check the solenoid operation as described in Chapter 8.

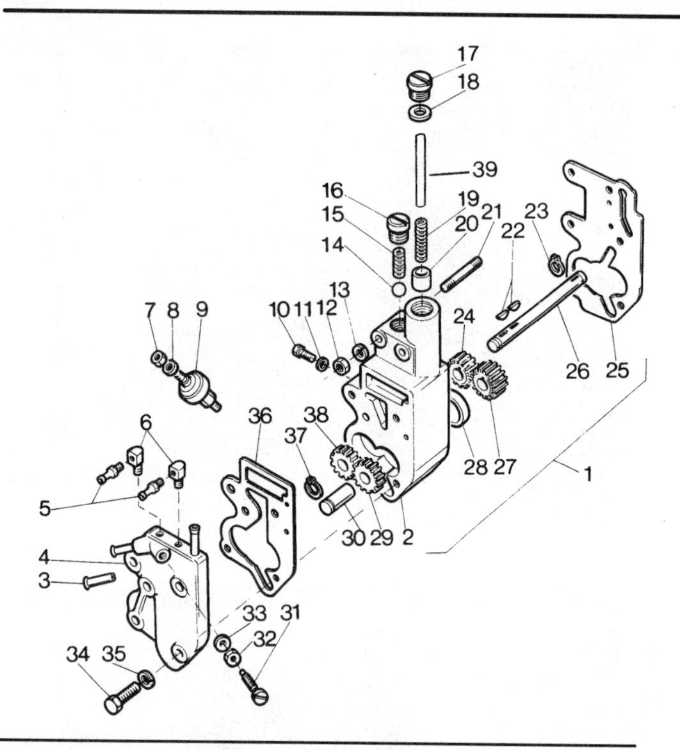

11.2 Oil pump (early model components shown, later models similar)

1	*Oil pump*	21	*Stud*
2	*Oil pump body*	22	*Woodruff keys*
3	*Union*	23	*Circlip*
4	*Pump cover*	24	*Scavenge drive gear*
5	*Union*	25	*Gasket*
6	*Union elbow*	26	*Oil pump driveshaft*
7	*Nut*	27	*Scavenge driven gear*
8	*Washer*	28	*Oil seal*
9	*Pump-mounted oil pressure switch*	29	*Feed driven gear*
		30	*Idler shaft*
10	*Plug*	31	*Rear chain oiler adjusting screw (if equipped)*
11	*Washer*		
12	*Nut*	32	*Locknut*
13	*Lock washer*	33	*Brass washer*
14	*Anti-siphon ball valve*	34	*Bolt*
15	*Spring*	35	*Lock washer*
16	*Plug*	36	*Gasket*
17	*Plug*	37	*Circlip*
18	*Sealing washer*	38	*Main feed drive gear*
19	*Spring*	39	*Dowel pin (1981 and early 1982 only)*
20	*Pressure relief valve plunger*		

10 Lubrication system – general information

The engine lubrication system is a dry sump type – oil is drawn from a remote oil tank and force fed to the major engine components by a gear-type oil pump attached behind the right-hand crankcase gearcase and driven off the right end of the crankshaft. All surplus oil drains back into the crankcase, where it's picked up by the scavenge section of the pump and returned to the oil tank for reuse.

Oil is fed under high pressure to the big-end bearings by an internal oil passage and to the rocker arm assemblies and valves by external lines (pre-Evolution engine only). A low pressure or vapor feed is supplied to all other working parts of the engine, either internally or by small diameter hoses. The high pressure feed also supplies the four hydraulic tappets. An unusual feature of the oil system is the lubrication of the primary drive chain assembly. Engine oil is supplied directly to the chaincase by a hose connected to the oil pump. Oil return is controlled by crankcase vacuum during certain phases of crankshaft rotation.

11 Oil pump – removal, inspection and installation

Refer to illustrations 11.2, 11.5, 11.8a, 11.8b, 11.13a, 11.13b, 11.14a, 11.14b and 11.15

Removal

1 Unless it's absolutely necessary, the oil pump should not be disassembled. The pump components are thoroughly lubricated, so wear is progressive and occurs very slowly over a long period of time. Damage to the pump is usually caused by metal or carbon particles passing through the gears, chipping or scoring them. In extreme cases, the gears may jam, causing the Woodruff key on the pump driveshaft to shear off.

2 If the engine is out of the frame, the oil pump can be removed as described in Section 10 of Chapter 2. Removal of the pump while the engine is in the frame can be done by removing the individual components, leaving the driveshaft in the case. Whichever method is used, the gears must be marked as they're removed so they will be reinstalled in their original positions to maintain the original clearances **(see illustration)**.

3 With the engine in the frame, you'll have to remove the right front footrest and mounting bracket to gain access to the oil pump. On some models, additional working space can be gained if the exhaust system is removed.

4 Drain the oil tank or oil pan (as applicable) as described in Chapter 1.

11.5 Oil pump line connections (arrows)

11.8a The oil pump gears seldom wear, but they must be replaced as sets if damage occurs

5 Disconnect all oil lines from the pump, labeling them first to ensure correct reconnection. The hose clamps installed at the factory are crimped aluminum, and can be removed with the blade of a screwdriver or by cutting them off with wire cutters **(see illustration)**. On models with a front-

11.8b Each pump gear is located on the shaft with a Woodruff key

11.13a Unscrew the large plug to gain access to . . .

11.13b . . . the pressure relief valve plunger

11.14a The smaller threaded plug retains . . .

11.14b . . . the anti-siphon ball valve

mounted oil filter (1992 and later) it will be necessary to disconnect the steel oil line union from the pump lower adapter.

6 On early models with a pump-mounted oil pressure switch, disconnect the wire(s) from the switch terminals.

7 Remove the bolts that pass through the pump body and, on some models, the two nuts. They must be loosened a little at a time (in 1/4-turn increments) to prevent distortion of the cover. Detach the cover and gasket.

8 The exposed gears provide the oil pressure **(see illustration)**. Remove the circlip from the end of the main feed drive gear and withdraw the gear and Woodruff key **(see illustration)**. Store the key and gear as a matched set.

9 Slide the feed driven gear off the idler shaft. **Note:** *Don't allow the pump driveshaft to shift into the gearcase – it could cause dislocation of the drive gear Woodruff key, necessitating gearcase cover removal.*

10 Remove the remaining bolts or nuts securing the pump body to the engine. Pull the pump body off the driveshaft and the scavenge drive gear.

11 Remove the scavenge drive gear and the Woodruff key, keeping them together, then withdraw the scavenge driven gear.

12 Generally, you don't have to remove the pump driveshaft unless the Woodruff key in the gearcase has been sheared. Check the condition of the key by rotating the crankshaft. If the driveshaft doesn't rotate consistently, the key is damaged and the driveshaft must be removed.

13 Unscrew the large plug at the top of the pump body and remove the relief valve assembly. Note how the parts are installed and which way they're facing during removal, to be sure they're installed properly during reassembly **(see illustrations)**.

14 Remove the smaller plug, followed by the anti-siphon valve spring and steel ball **(see illustrations)**.

15 Unscrew the plug from the rear of the pump body **(see illustration)**.

16 On models so equipped, the automatic drive chain oiler must be removed next. Turn the adjusting screw in as far as possible, counting the number of turns. Write the number down to be sure the adjusting screw will be in the correct position when it's reinstalled. Remove the adjusting screw.

11.15 Remove the blanking plug to simplify cleaning of the oil pump passages

Inspection

17 Clean all of the oil pump components with solvent and allow them to dry. Use compressed air to blow out all of the oil passages in the pump body and cover to be sure all oil, solvent and obstructions are removed. Do not mix up the gears or the tolerances in the pump will be affected.
18 Inspect the pump gears for scoring and fractured teeth. If damage or wear is evident, the gears should be replaced with new ones. Intermeshing gears must always be replaced as a matched set.
19 Check the condition of the Woodruff keys. Be sure they fit in the gears and in the shaft keyways. Although slight play in the keys is permissible, it may indicate wear. When in doubt, replace them with new ones and install a new shaft also.
20 Check the relief and anti-siphon valves and seats for pitting and other damage. Damaged seats cannot be easily repaired, so if damage is evident, a new pump body should be installed.
21 It's not unusual to have oil leakage at the cover-to-pump body joint and, more rarely, at the pump-to-gearcase joint. This condition is usually due to overtightened or unevenly tightened bolts, causing distortion. Due to the small area of the mating surfaces, only a small amount of distortion will cause a leak. If the cover is distorted, a small amount of material may be removed by rubbing the cover on a piece of very fine (400 to 600 grit) emery paper spread over a perfectly flat surface. A surface plate or plate glass (not window glass) is ideal for the lapping operation. This method of resurfacing cannot be used for the pump body. If the mating surfaces on the pump body are altered, the manufacturer's tolerances will be changed, possibly causing the gears to lock.
22 Leakage at the unions on the oil pump body can be stopped by removing them and applying thread sealant or teflon tape to the threads. Do not apply too much sealant because there is danger of blocking an oil passage.

Installation

23 Reassemble the pump by reversing the disassembly procedure using new gaskets, a new oil seal and circlip.
24 Inspect each gasket to be sure it doesn't block off any of the oil holes or obstruct the free movement of the oil pump gears. DO NOT use any gasket cement at any of the joints.
25 During reassembly, tighten the pump mounting bolts evenly, in 1/4-turn increments, and make sure the components rotate freely after each tightening sequence.
26 Jamming of the pump during reassembly can often be traced to the new gaskets interfering with the gears. If this happens, use a razor knife to trim the gasket until clearance is obtained.
27 Don't overtighten the pump bolts and nuts. The plastic gasket material will squeeze out under excessive pressure and prevent correct sealing.
28 Connect the oil lines to the pump in their original positions and secure them with new hose clamps.
29 Connect the wire(s) to the oil pressure switch (pump-mounted switch).
30 Fill the oil tank/oil pan with the recommended amount and type of oil (see Chapter 1).

12 Oil filter mount – removal and installation (1992 and later models)

1 On later models the oil filter is mounted on the crankcase front portion. Steel lines, running under the gearcase, connect the filter mount to the oil pump and tank/pan. The filter can be serviced as described in Chapter 1.

Removal

2 Drain the engine oil as described in Chapter 1.
3 Disconnect the oil lines at their unions with the mount and pump connectors. The mount is secured to the crankcase by two bolts and washers.

Installation

4 Install in the reverse order of removal, replacing all O-rings and seals as a matter of course. Tighten the filter mount bolts to 13 to 17 ft-lbs (18 to 23 Nm).

13 Engine oil pan – removal and installation (1993 and later FLT and 1991 and later Dyna models)

Removal

1 Drain the engine oil (see Chapter 1).
2 Drain the transmission oil (see Chapter 1).
3 Disconnect the hose which runs from the oil pump to the oil pan union, having made note of its fitted position and routing (on certain models there may be identification marks on the hose ends). Discard hose clips; new ones must be fitted on installation.
4 On Dyna models remove the oil filler top section (two bolts) and pull the dipstick housing out of its O-ring in the pan. On FLT models, release the dipstick housing and gasket from the transmission case (four bolts).
5 Loosening them evenly, remove all nine (Dyna models) or twelve (FLT models) oil pan bolts. On FLT models access holes are provided in the frame crossmember for the awkwardly-positioned bolts. If stuck firmly in place, tap around the joint with a soft-faced mallet to release the gasket – do not lever with a screwdriver.
6 Maneuvering the oil pan past the crossmember on FLT models will be difficult due to limited clearance. Do not force it free; either remove the rear wheel and swingarm so that it can be removed rearwards, or disconnect the engine front mounting and raise the engine using a jack to obtain the necessary clearance.
7 Once removed, peel off the old gasket and clean the pan interior thoroughly.

Installation

8 Apply a smear of sealant to the oil pan gasket face and position the gasket on the pan. Offer the pan up to the transmission case and fit the retaining bolts loosely while taking care that the gasket remains in place.
9 Tighten all bolts evenly until the specified torque of 7 to 9 ft-lbs (9 to 12 Nm) is reached; on FLT models following the sequence **(see illustration)**.
10 Refit the dipstick assembly using new O-rings (Dyna models) or a new gasket (FLT models). If components were removed for access on FLT models, refit them.
11 Refit the pump-to-oil pan hose and secure with a new clamp.
12 Refit both drain plugs and replenish the engine and transmission oils.

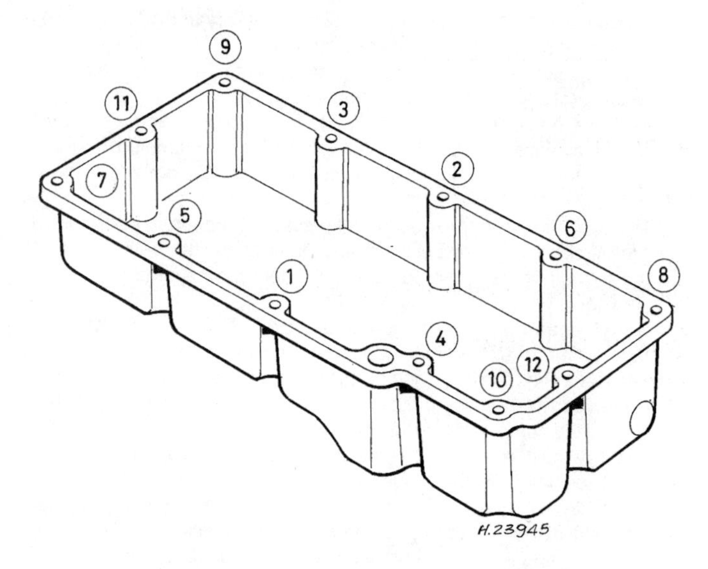

13.9 Oil pan bolt tightening sequence – 1993 and later FLT models

Chapter 4 Ignition system

Contents

Specifications

Note: *Additional ignition system specifications are included in Chapter 1, Tune-up and routine maintenance.*

Ignition system type

1970 through 1978	Mechanical contact breaker point
1979-on	Breakerless inductive discharge (electronic)

Air gap *(late 1978 and 1979 only)*

Air gap (late 1978 and 1979 only)	0.004 to 0.006 in (0.102 to 0.152 mm)

Ignition coil resistance

1970 through early 1978	
Primary	4.7 to 5.7 ohms
Secondary	16,000 to 20,000 ohms
Late 1978 and 1979	
Primary	4.5 to 5.7 ohms
Secondary	16,500 to 20,000 ohms
1980 through 1983	
Primary	3.3 to 3.7 ohms
Secondary	16,500 to 19,500 ohms
1984-on	
Primary	2.5 to 3.1 ohms
Secondary (except 1993-on FLT, FXR, Dyna)	11,250 to 13,700 ohms
Secondary (1993-on FLT, FXR, Dyna)	10,000 to 12,500 ohms

1 General information

In order for the engine to run correctly, an electrical spark must ignite the fuel/air mixture in the combustion chambers at exactly the right moment in relation to engine speed and load. The ignition system operation is based on feeding low tension (primary) voltage from the battery to the coil where it's converted to high tension (secondary) voltage by a process known as induction. The secondary voltage is powerful enough to jump the spark plug gap in the cylinders many times a second under high compression pressures, provided the system is in good condition and all adjustments are correct.

The ignition system installed on 1970 through early 1978 models as standard equipment is a conventional mechanical contact breaker point type. The system installed on late 1978 and later models is a breakerless inductive discharge system (electronic ignition).

Both systems are divided into two circuits: the primary (low tension) circuit and the secondary (high tension) circuit. The primary circuit consists of the battery, the contact breaker points (early models), the ignition timer (late 1978 and 1979) or the computerized ignition timer (1980-on), the primary coil, the ignition switch and the wires connecting the components.

The secondary circuit consists of the secondary coil, the spark plugs and the wires.

Mechanical flyweights are incorporated in the ignition systems on pre-1980 models to advance the ignition timing mechanically. On 1980 and later models, a computer module advances the timing electronically.

2 Spark plugs – check and replacement

Refer to Chapter 1, *Tune-up and routine maintenance,* for spark plug check and replacement procedures.

3 Ignition coil – check, removal and installation

Refer to illustration 3.3

1 Detach the wires from the coil and connect an ohmmeter to the coil primary wire terminals. The primary resistance should be as listed in this Chapter's Specifications – if it isn't, the coil is probably defective. If an ohmmeter isn't available, take the coil to a dealer service department to have it checked or temporarily install a known good coil. **Caution:** *Be sure to connect the wires to the correct terminals to avoid damaging other ignition components. If the ignition system trouble is eliminated by the temporary installation of the new coil, install it permanently.*

2 Pre-1979 models with contact breaker point ignitions should be checked to make sure the condenser is good before replacing the coil. A bad condenser can act just like a defective coil, but the condenser is much more likely to fail. **Note:** *The easiest way to check for a defective condenser is to substitute a known good component. If the ignition system problem goes away when the new condenser is installed, the original condenser is defective.*

3 On early models, the coil is mounted on the frame downtube, to the left of the rear cylinder, and is protected by a chrome cover **(see illustration)**. On later models, it's mounted under the front of the fuel tank.

4 Label the primary (small) wires and coil terminals, then disconnect the wires. Detach each spark plug wire from the coil by pulling the boot back, grasping the wire as close to the coil as possible and pulling the wire out of the coil.

5 Remove the bolts and detach the coil.

6 If a new coil is being installed, make sure it's the right one. **Caution:** *1980 and later models must have a coil marked "Electronic Advance". Installing the wrong type of coil could result in failure of the electronic components.*

7 Installation of the coil is the reverse of removal. Be sure the spark plug wire boots are seated on the coil towers to keep dirt and moisture away from the terminals. Install new boots if the originals are hardened, cracked or torn.

3.3 On early models, the ignition coil is located under a chrome cover on the left side of the engine

4 Contact breaker points – check and replacement

Refer to Chapter 1, *Tune-up and routine maintenance,* for contact breaker point replacement and adjustment procedures.

5 Condenser – removal and installation

1 A condenser is included in the primary circuit to prevent arcing across the contact breaker points as they open. The condenser is connected in parallel with the points – if it fails, the ignition system will malfunction.

2 If the engine is difficult to start, or if misfiring occurs, it's possible the condenser is defective. To check it, separate the contact breaker points by hand when the ignition switch is on. If a spark occurs across the points and they appear to be discolored and burned, the condenser is defective.

3 It isn't possible to check the condenser without special test equipment. Since the cost is minimal, install a new condenser to see the effect on engine performance.

4 Because the condenser and contact breaker points supply a spark to both cylinders, it's virtually impossible for a faulty condenser to cause a misfire on one cylinder only.

5 The condenser is attached to the contact breaker baseplate by a clamp and screw. If the wire is disconnected and the screw removed, the condenser can be detached from the baseplate.

6 When installing the new condenser, be sure the clip around the body (which forms the ground connection) makes good contact with the baseplate. Tighten the mounting screw securely.

6 Air gap – check and adjustment

1 On late 1978 and 1979 models, the air gap between the trigger rotor and sensor is adjustable.

2 Remove the gearcase cover from the right side of the engine (see Section 8 in Chapter 2).

3 Remove the spark plugs and rotate the crankshaft until the wide lobe on the trigger rotor is centered in the sensor.

4 Insert a feeler gauge of the specified thickness between the lobe of the rotor and the sensor. If the air gap isn't correct, loosen the screws securing the sensor and move it until the specified gap is obtained. Rotate the crankshaft 180-degrees until the other lobe of the rotor is centered in the sensor and measure the gap to be sure it's the same.

7.1 A simple spark test tool can be made from a block of wood, a large alligator clip, some nails, screws, and wire and the end of an old spark plug – leave about 1/4-inch between the electrode tips (attach the plug wire to the protruding electrode [arrow] and grip the cylinder head fins with the alligator clip)

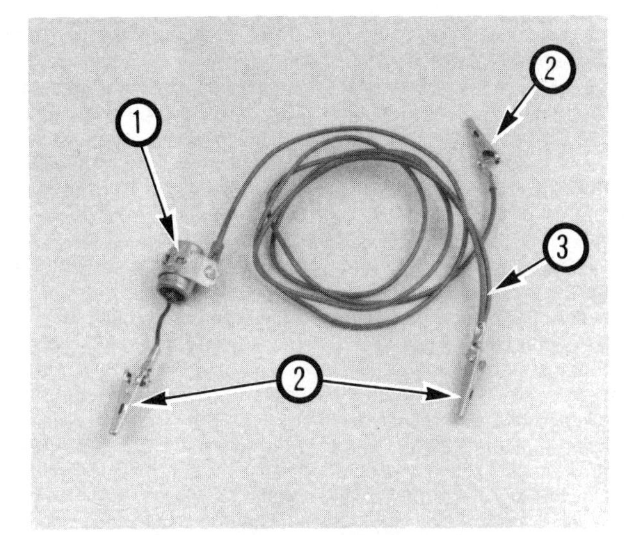

7.12 Fabricate a jumper wire like this one to check the ignition system on 1984 and later models

1 0.33 MF capacitor
2 Alligator clip
3 16-gauge wire (approx. 18-inches long)

7 Electronic ignition system – check

All models

Refer to illustration 7.1

1 Check the condition of the spark plugs and the spark plug wires as described in Chapter 1. One way to check the ability of the ignition system to produce a strong enough spark is to construct a simple home-made test tool **(see illustration)** and hook it up to the plug wires, one at a time (with the alligator clip grounded on the engine), to see if the spark will jump the gap. Another way, though less conclusive, is to attach a spark plug that's known to be good to the spark plug wires and ground the plug against the engine. Crank the engine over and check for spark. If there is a spark, the coil and ignition system are working properly. Check the choke and the carburetor and replace the spark plugs with new ones if the engine doesn't run right.

2 If no spark is obtained, attach a voltmeter to the battery (black lead to the negative terminal, red lead to positive) and turn the ignition switch and the engine stop switch to the On position. The reading on the voltmeter should be at least 11.5 volts.

3 Measure the air gap (late 1978 and 1979 models only) between the sensor and both trigger rotor lobes (refer to Section 6). If the gap between both rotor lobes can't be adjusted within the range given, replace the trigger rotor and/or the timer mechanism.

4 Make sure the ground cable from the battery is secure and making good contact.

5 Check the ground for the module at the timer plate (late 1978 and 1979 models) or at the frame (1980 and later models). It should be clean and tight to ensure a good connection.

6 Attach a voltmeter between the positive terminal of the coil and ground (red lead to the coil, black lead to ground). Rotate the crankshaft until the lobes of the trigger rotor (late 1978 and 1979 models) or the slots in the rotor (1980 and later models) are equal distances from the center of the sensor. Turn the ignition switch and the engine kill switch to the On position and read the voltage on the meter. It should be within 1/2-volt of the battery voltage. If not, the trouble is somewhere between the battery and the coil. Check the connections at the ignition switch, the engine kill switch and the circuit breaker.

7 Disconnect the blue wire from the negative terminal of the ignition coil and attach a voltmeter between the terminal and ground. With the ignition and kill switches on, the voltage should be the same as the battery. If the voltage is different, the coil should be replaced with a new one (after checking the resistance as described in Section 3).

Late 1978 and 1979 models only

8 Reconnect the blue wire to the negative terminal of the coil, then connect the voltmeter between ground and the negative terminal of the coil. The voltmeter should read 1.0 to 2.0 volts. Place the blade of a screwdriver against the face of the sensor and read the voltmeter. The reading should be between 11.5 and 13.0 volts. By removing the screwdriver, the voltage should drop to 1.0 to 2.0 volts. If not, the ignition module is defective and must be replaced with a new one.

9 Attach a known good spark plug to the plug wire and ground the plug against the engine. Check for spark each time the screwdriver is placed against the face of the sensor. If there is no spark, the ignition coil should be replaced with a new one.

1980 through 1983 models only

10 Unplug the connector between the module and the sensor plate. Connect a voltmeter between the red and black wires in the connector on the module side (positive voltmeter lead to the red wire, negative lead to black). Turn the ignition and the engine kill switches on and read the voltmeter. It should indicate 4.5 to 5.5 volts. If not, the ignition module is defective and should be replaced with a new unit.

11 A special jumper cable test adapter is needed to do the following test. It's available from a harley-davidson dealer (part number HD 94465-81). Attach the test adapter to the connector halves between the module and sensor. Be careful not to let the exposed wires touch a grounded component or each other or damage to the module will result. Recheck the voltage between the red and black wires. Connect the voltmeter between the green and black wires to test the sensor output (voltmeter positive lead to the green wire, negative lead to the black wire). The voltmeter should read 4.5 to 5.5 volts with the rotor slots away from the sensor, and 0 to 1.0 volt with the slot aligned with the sensor. If either of these voltage readings are not attained, the sensor plate must be replaced with a new unit.

1984 and later models only

Refer to illustration 7.12

12 Assemble a jumper wire from 16-gauge wire, a 0.33 microfarad capacitor and three alligator clips **(see illustration)**. A known good condenser from a breaker point ignition system can be used if a capacitor isn't available.

13 Attach a known good spark plug (or the home-made test tool) to one of the plug wires and ground the plug or tool on the engine.
14 Connect the jumper wire with the capacitor in it to the negative coil primary terminal (the blue wire was disconnected from the terminal in Step 7). Attach the jumper wire common alligator clip to a good ground.
15 Momentarily touch the remaining jumper wire alligator clip to the negative terminal on the ignition coil – when you do, a spark should occur at the plug or test tool.
16 If not, replace the ignition coil – it's defective. If a spark occurred, proceed to Step 17.
17 Follow the instructions in Step 10 above. If the voltage readings are not as specified, check the ignition module power and ground wires for loose and dirty connections.
18 If the wires and connections are okay, turn the ignition and kill switches on and momentarily connect the black and green wire connector pins in the module wire connector with a jumper wire or a screwdriver. If a spark occurs at the plug or test tool when the screwdriver or jumper wire is disconnected, the sensor is probably defective (check the sensor resistance as described below before buying a new one).
19 If no spark occurred during the test in Step 18, check the module resistance (as described below) and install a new one if the resistance isn't as specified.

Intermittent ignition problem check

20 Check the battery terminals and the module ground connection to make sure they're clean and tight.
21 Disconnect the white wire from the ignition coil primary terminal (not the white wire that goes to the module from the same terminal).
22 Connect a 16-gauge jumper wire with alligator clips to the positive battery post and the terminal on the coil the white wire was disconnected from.
23 Start the engine and see if the problem is eliminated. If it is, the problem is possibly in the starter safety switch connections (they're probably loose). **Caution:** *The engine won't stop running until the jumper wire is removed.*
24 If the intermittent problem still exists, refer to Chapter 2 (Section 8) and remove the gearcase timing covers to gain access to the ignition sensor.
25 Start the engine and spray freon (air conditioning system refrigerant available at auto parts stores) on the sensor – be sure to wear safety glasses or a face shield as this is done!
26 If the engine dies, the sensor is temperature sensitive (to cold) and should be replaced with a new one.
27 If the engine keeps running, allow it to reach normal operating temperature, then use a blow dryer to apply heat to the sensor. If the engine stops, the sensor is temperature sensitive (to excess heat) and should be replaced with a new one.
28 If the engine doesn't stop, apply heat to the ignition module. If the engine stops, the module is defective and should be replaced.

Sensor resistance check

29 Position the ohmmeter selector switch on the Rx1 scale. Unplug the sensor wire harness connector and attach the positive (red) lead from the ohmmeter to each of the sensor harness terminals in the connector, one at a time, with the negative (black) ohmmeter lead connected to a good ground (use the sensor plate). If each terminal reading indicates infinite resistance, the sensor is good. If any of the terminal connections produce a resistance reading, the sensor is bad.
30 Connect the positive ohmmeter lead to the green sensor wire and the negative lead to the black sensor wire. If the ohmmeter indicates infinite resistance, the sensor is good. If a resistance reading is produced, the sensor is bad.
31 Reverse the ohmmeter leads. If the ohmmeter reads 300-to-750 K-ohms, the sensor is good. If it indicates infinite resistance, the sensor is defective.

Ignition module resistance check

32 Position the ohmmeter selector switch on the Rx1 scale. Unplug the ignition sensor-to-module wire harness connector and attach the positive (red) lead from the ohmmeter to the black wire terminal in the module side of the connector – the negative (black) ohmmeter lead should be connected to the black module ground wire. If the ohmmeter indicates 0-to-1 ohm, the module is good. If it indicates more than 1 ohm, replace the module.
33 Connect the positive ohmmeter lead to the white module wire at the ignition coil (disconnect it from the coil first) and check the meter reading again. If it's 800-to-1300 K-ohms, the module is good. If it indicates infinite resistance, the module is defective.
34 Reverse the ohmmeter leads (red lead to ground, black lead to the white module wire at the coil). If the ohmmeter indicates infinite resistance, the module is good. If it indicates any resistance, the module is bad.
35 Attach the positive ohmmeter lead to the blue module wire at the coil (disconnect it from the coil first) and attach the negative lead to the black module ground wire. If the meter indicates infinite resistance, the module is good. If it indicates any resistance, the module is bad.
36 Reverse the ohmmeter leads. If the ohmmeter indicates 400-to-800 K-ohms, the module is good. If infinite resistance is indicated, the module is defective.
37 If the module fails any of the tests, replace it with a new one.

8 Ignition components – removal and installation

Refer to Chapter 2, Section 8, for procedures to remove and install the engine-mounted ignition components. The electronic ignition system module is bolted to the frame near the front of the engine. The VOES is attached to a bracket behind the intake manifold.

9 Ignition timing – check and adjustment

Refer to Chapter 1 for the ignition timing check and adjustment procedures.

Chapter 5 Frame and suspension

Contents

Specifications

Front forks

Oil capacity	See Chapter 1
Oil type	See Chapter 1

Torque specifications

All models

Axle nuts (front and rear)	See Chapter 1	
1970 through early 1978	**Ft-lbs**	**Nm**
Pinch bolts	22 to 26	30 to 35
Late 1978 through 1983 (four-speed)		
FL models		
Lower pinch bolts	30 to 35	41 to 47
Upper pinch bolts	22 to 26	30 to 35
FX models – pinch bolts		
FXWG only	30 to 35	41 to 47
All other FX models	20 to 25	27 to 34
Front fender mounting bolts	16 to 20	22 to 27
1980 through 1983 (five-speed)	**Ft-lbs**	**Nm**
FLT models		
Fork stem nuts	45	61
Slider cap nuts	11	15
Rear chain housing rubber boot clamps	3 to 4	4 to 5.4
Front fender mounting bolts	18 to 29	24 to 40
Swingarm pivot shaft nuts	45	61
Footrest mounting nuts	20 to 25	27 to 34

Torque specifications (continued)

1980 through 1983 (five-speed) continued	Ft-lbs	Nm
FXR models		
Slider cap nuts	11	15
Front fender mounting bolts	10 to 14	14 to 19
Swingarm pivot shaft nuts	45	61
Swingarm pivot shaft mounting bracket bolts	34 to 42	46 to 57
Footrest mounting nut	20 to 25	27 to 34
1984-on		
FLT models		
Fork stem nut	35 to 45	47 to 61
Slider cap nuts	9 to 13	12 to 18
Rear chain rubber boot fasteners (1984 only)	3 to 4	4 to 5.4
Front fender mounting bracket bolts	16 to 20	22 to 27
Rear shock absorber mounting bolts		
Upper		
1984 through early 1988	35 to 40	47 to 54
Late 1988-on	33 to 35	45 to 47
lower	35 to 40	47 to 54
Swingarm pivot shaft nut/bolt		
Late 1986 through 1988 (12-point bolt head)	85	115
All others	45	61
FXR models		
Slider cap nuts		
FXLR (all) and 1987 FXRSE	7 to 9	9.4 to 12
All others	9 to 13	12 to 18
Front fender mounting bracket bolts	16 to 20	22 to 27
Swingarm pivot shaft nut/bolt	Same as FLT models	
Swingarm pivot shaft mounting bracket bolts	34 to 42	46 to 57
FX/Softail models		
Slider cap nuts	9 to 13	12 to 18
Rear fender supports		
1985 FXST/C only	40 to 45	54 to 61
All others	30 to 33	41 to 45
Springer forks		
Steering stem bearing retainer	6	8
Rigid fork leg studs	60 to 65	81 to 88
Upper triple clamp pinch bolt	45 to 50	61 to 68
Fork stem acorn nut	20 to 25	27 to 34
Shock absorber acorn nuts	45 to 50	61 to 68
Handlebar riser locknuts	25 to 35	34 to 47
FXD Dyna models:		
Fork pinch bolts		
FXDB, FXDC, FXDL	25 to 30	34 to 41
FXDWG	30 to 35	41 to 47
Upper triple clamp-to-steering stem pinch bolt	21 to 27	28 to 37
Rear shock absorber mountings	25 to 40	34 to 54

1 General information

The Harley-Davidson models covered by this manual have a duplex tube, full cradle frame with a single top tube running from the steering head to the seat tube. Unlike many other designs, the twin tubes are very close together so the engine tends to straddle the frame rather than sit in it.

The front suspension on most models is conventional, consisting of oil-damped, telescopic forks. On 1989 and later FXSTS models, "springer forks" consisting of rigid fork and spring legs, external springs and a separate hydraulic shock absorber, serving both fork legs, have been revived.

The rear suspension on most models is also conventional and is composed of a swingarm and two hydraulically-damped shock absorbers with adjustable spring rates. Some 1985 and later models have air-assisted rear shocks. Softail models have a large, triangulated, pivoting rear frame section that resembles a rigid frame, but is actually suspended by two horizontal shock absorbers. On 1988 and earlier models, separate fluid reservoirs, with hoses connected to the forward shock mounts, are used to store excess fluid and prevent fluid cavitation due to the horizontal shock position.

Some 1984 through 1987 models (FLT, FXRD and FXRT) are equipped with air forks that include an anti-dive feature. The anti-dive operation is controlled by a solenoid activated by either brake switch (front or rear). The FXRT and FXRD have an air accumulator attached to the lower triple clamp. On FLT models, the engine guard serves as an air reservoir as well. Some 1988 and later models (FLT, FXRT, FXRS-SP, and FXRS-CONV) also have air forks and anti-dive. The system is very similar to the earlier one, but the handlebar serves as the air reservoir on all models and the FXRT doesn't have an accumulator.

Several frame and suspension-related routine maintenance procedures are covered in Chapter 1. They include *Suspension – inspection, Steering head bearings – check, Fork oil – change, Sidestand – check and maintenance* and *Springer fork rocker bearings – adjustment.*

2 Frame – inspection and repair

1 The frame is unlikely to require attention unless accident damage has occurred. In most cases, frame replacement is the only satisfactory reme-

dy for such damage. A few frame specialists have the jigs and other equipment necessary for straightening the frame to the required standard of accuracy, but even then there's no sure way of determining exactly how much the frame was overstressed.

2 After the machine has accumulated many miles, it's a good idea to examine the frame closely for cracks at the welded joints. Rust can also cause weakness at the joints. Loose engine mount bolts can cause enlargement of the holes and cracks at the mounting tabs. Minor damage can often be repaired by welding, depending on the extent and nature of the damage.

3 Remember an out-of-alignment frame will cause handling problems. If misalignment is suspected as the result of an accident, you will have to strip the machine completely so the frame can be thoroughly checked.

3 Instrument cluster – removal and installation

1970 through 1983 models

Handlebar-mounted instruments

1 Before either of the instruments can be removed, you must disconnect the drive cable. Unscrew the coupling nut from the underside of each instrument and pull the cables off. Some later models have an electronic tachometer so there will be no cable to remove – instead, the wire will have to be disconnected.

2 The light bulbs will pull away from the underside of each instrument head, along with the bulb holders. The detachable baseplate must be removed first, as it acts as a retainer.

3 Apart from defects in either the drive or the drive cable, a speedometer or tachometer that malfunctions is difficult to repair. Install a new or used one.

Fuel tank-mounted instruments

4 Remove the screws securing the instrument panel.
5 Pull the odometer trip knob off (if so equipped).
6 On models where the choke control is in the instrument panel, unscrew the choke knob and the retaining locknut.
7 Lift the panel off the fuel tank and unscrew the mounting hardware.
8 Label the wires at the instrument cluster and disconnect them.
9 Unscrew the speedometer and, if equipped, the tachometer cable.
10 If the instruments are defective, they must be replaced with new ones.
11 Installation is the reverse of the removal procedure. Be sure to connect the wires and cables.

1984 and later models

Note: *The instruments are not repairable, but before replacing a defective component, make sure the electrical connections are clean and tight. Lubricate the speedometer cables with graphite grease every 5,000 miles. Do not turn the instruments upside-down after they are removed – the damping oil used in the fuel gauge will leak out and stain the gauge face.*

FLT/C (1984-on) and FLHT/C (1984 and 1985)

12 To replace the bulbs, remove the two screws from the instrument panel. Remove the Phillips screw and pull the odometer trip knob out, then lift up on the panel to expose the bulbs.
13 To remove the instrument panel, disconnect the wire harness and detach the speedometer cable.
14 Remove the fasteners and detach the bracket.
15 Mark all wires, then unplug them and replace the malfunctioning instrument.
16 Installation is the reverse of removal.

FLHT/C (1986-on) and FLHS (1987-on)

17 To replace the bulbs, remove the light bar and outer fairing. The instrument bulbs and all gauges are now accessible from the front of the inner fairing.
18 Unplug all wire connectors from the instrument housing. Remove the four Allen-head screws holding the panel to the fairing.
19 Remove the round knob attaching the trip odometer knob to the fairing.
20 Disconnect the speedometer cable and detach the instrument panel.
21 Remove the fasteners and detach the mounting bracket.
22 Mark all wires, then unplug them and replace the malfunctioning instrument.
23 Installation is the reverse of removal.

Models with fuel tank-mounted instruments

24 On most models, access to the instruments and bulbs is gained by removing the odometer knob (loosen the set screw first), unscrewing the choke knob and removing the choke cable housing locknut or the acorn nut holding the instrument panel to the gas tank **(see illustration 2.22** in Chapter 3, if necessary).
25 On some models the instrument panel is attached to the tank with three Allen-head screws.
26 Disconnect any wires and remove the speedometer cable before replacing the instruments.

4 Shock absorbers – removal and installation

Note: *Check if replacement parts are available before working on the shock absorber – in many cases only the mounting bushings can be purchased. Air-assisted units should not be dismantled.*

Refer to illustrations 4.6 and 4.7

1 Raise the rear of the motorcycle and support it securely on blocks. If blocks aren't available, remove one shock absorber at a time.
2 Remove the saddlebags, if necessary, to gain access to the shock absorbers.
3 On some models the mufflers must be turned out of the way of the bottom shock absorber mounting bolts (loosen the clamps first). On others (FLT, FXR, FXRS Sport and FXRT), the mufflers must be completely removed.
4 On models with air shocks, the air line must be detached from the shock absorber fitting.
5 On Softail models (with horizontal shock absorbers), remove the reservoir clamp bolt (it holds both reservoirs in place) and detach the reservoirs. **Warning:** *Do not detach the hoses from the reservoirs or the shock absorbers for any reason! The left shock absorber hose on 1985 and earlier models is attached to the bottom of the transmission mounting plate with a clamp that must be removed as well.*
6 Remove the mounting nuts securing the shock absorbers to the swingarm and frame (later models use bolts and nuts and some models have bolts with internal threads that mate with a frame stud at the upper mount). Carefully separate the shock absorber from the frame and swingarm. When the shock absorber is removed, the rear wheel and swingarm will drop to the ground if not supported **(see illustration)**. **Note:** *Softail models require adaptor no. SRES24 or SRES28 (as applicable), available from Snap-on Tools Corporation, to remove and refit the shock absorber mounting bolts.*

4.6 The shock absorbers are attached to the frame mounts by large bolts/nuts, washers and rubber bushings

5

7 Where dismantling is possible, compress the spring with a spring compressor and remove the spring keeper(s) at the top of the shock absorber assembly (early models have a split-key type of keeper, while later models have a one-piece keeper) **(see illustration)**. Lift off the shroud and the spring seat (if equipped) from the top of the spring.

8 Carefully release the spring pressure until the compressor can be removed. Lift the spring off the shock absorber assembly and remove all of the mounting hardware. Note the positions of the mounting components for reassembly.

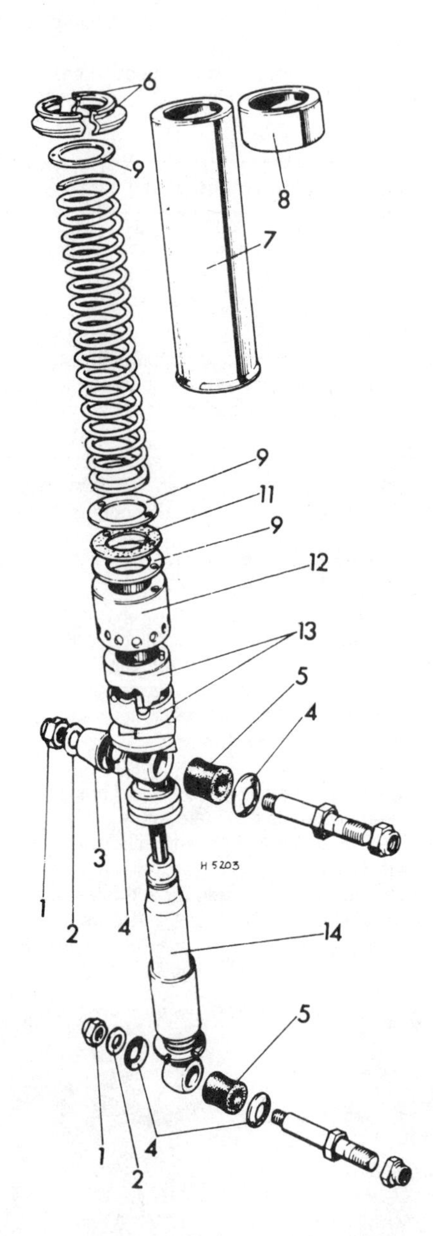

4.7 Rear shock absorber components – exploded view

1	*Mounting nut*	*8*	*Shroud (short)*
2	*Washer*	*9*	*Washer*
3	*Stud cover*	*10*	*Spring*
4	*Cup washer*	*11*	*Seal washer*
5	*Rubber bushing*	*12*	*Adjusting cup*
6	*Spring keeper*	*13*	*Cam*
7	*Shroud (long)*	*14*	*Damper unit*

9 Clean and inspect all of the components for wear and damage (especially the rubber parts). Replace any defective components with new parts.

10 Reassemble the shock absorber and spring in the reverse order of disassembly. Be sure the adjustment cams on both shock absorbers are positioned at the same level. Compress the spring until the spring keeper(s) can be installed, then slowly release the spring compressor.

11 Have an assistant lift the rear wheel and swingarm to the correct level to install the shock absorber. With one shock absorber in place, the swingarm should be in the correct position to install the other shock.

12 Tighten the nuts (and bolts) to the torque listed in this Chapter's Specifications and lower the motorcycle off the support blocks.

5 Swingarm bearings – check

1 The swingarm pivots on bearings, which rarely wear out.

2 Raise the rear of the motorcycle off the ground and support it securely on blocks. Grab the swingarm in front of the axle and move it from side-to-side. If there's any discernible play in the swingarm pivot, the bearings may be worn out.

3 The swingarm should be removed and inspected as described in the following Section.

6 Swingarm – removal, inspection and installation

Removal

Refer to illustrations 6.5, 6.8, 6.9 and 6.12

1 With the motorcycle on a level surface, raise the rear wheel. Support it securely on blocks under the frame where the engine rests.

2 Disconnect the final drive from the rear wheel sprocket. On chain drive models, remove the master link from the chain and separate it from the rear sprocket (see Chapter 1, if necessary). On models with fully enclosed chains, you'll have to remove the rubber boots from the rear sprocket housing before the master link can be disconnected. On belt drive models, the rear wheel must be removed in order to disconnect the belt from the rear wheel sprocket.

3 Remove the rear wheel as described in Chapter 6.

4 On early models with rear drum brakes, remove the brake components from the swingarm, as described in Chapter 6, and tie them up securely, out of the way. **Caution:** *Do not let the brake components hang by the hydraulic brake line or the line may be damaged.*

5 Separate the disc brake caliper and hose from the swingarm, on models so equipped, and securely tie the caliper to the frame **(see illustration)**. **Caution:** *Do not allow the caliper to hang by the hydraulic brake line or the line may be damaged.*

6.5 On some models, the rear brake hose is secured to the swingarm with a clamp – remove it before attempting to detach the swingarm

6.8 Tab washers are used on some models to secure the swingarm pivot bolt – flatten the ears before removing the bolt

6 On some models the exhaust system must be disconnected from the swingarm. Usually the brackets can be unbolted, but on some models the exhaust pipe and muffler or even the entire exhaust system must come off.
7 Support the swingarm and remove the lower shock absorber mounting bolts (rear bolts on Softail models). Remove the shock reservoir clamp bolt on 1988 and earlier Softail models.
8 Flatten the tab washer ears securing the swingarm pivot shaft **(see illustration)**.
9 On models other than those described in Steps 10 and 11, unscrew and withdraw the pivot shaft from the swingarm **(see illustration)**.
10 On Softail models, remove the bolt threaded into each side of the swingarm axis tube and detach the lock washer (right side), spacer (left side) and axis tube (inside the swingarm). The swingarm can now be pulled out of the frame mounts.
11 On five-speed transmission-equipped models, proceed as follows:

a) On models through early 1986 – Remove the nut and spacer from the right end of the pivot shaft.
b) On late 1986 through 1988 models – Remove the right pivot bolt while holding the left pivot bolt.
c) On 1989 and later models – Hold the right side nut (11/16-inch six

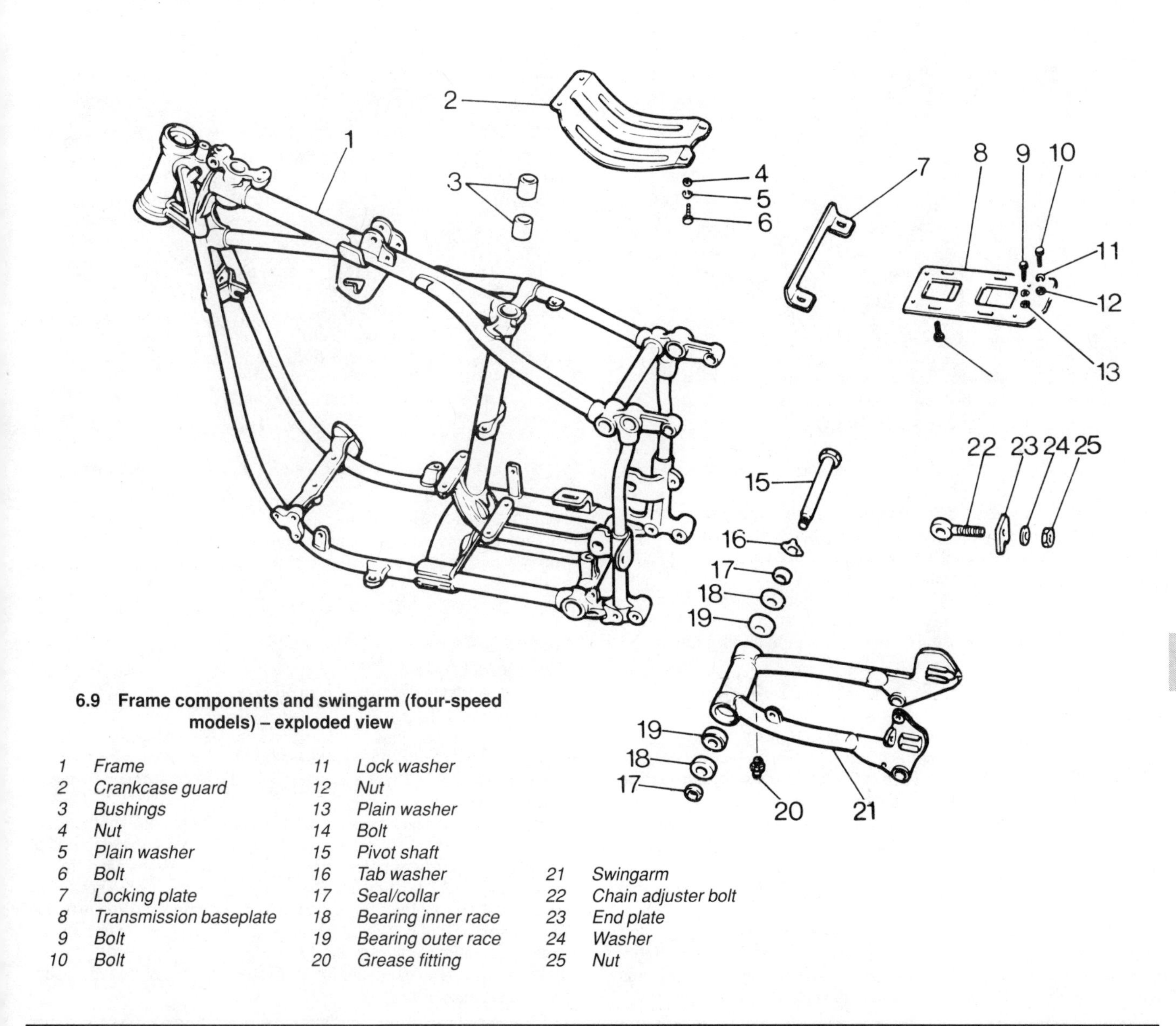

6.9 Frame components and swingarm (four-speed models) – exploded view

1	*Frame*	11	*Lock washer*	21	*Swingarm*
2	*Crankcase guard*	12	*Nut*	22	*Chain adjuster bolt*
3	*Bushings*	13	*Plain washer*	23	*End plate*
4	*Nut*	14	*Bolt*	24	*Washer*
5	*Plain washer*	15	*Pivot shaft*	25	*Nut*
6	*Bolt*	16	*Tab washer*		
7	*Locking plate*	17	*Seal/collar*		
8	*Transmission baseplate*	18	*Bearing inner race*		
9	*Bolt*	19	*Bearing outer race*		
10	*Bolt*	20	*Grease fitting*		

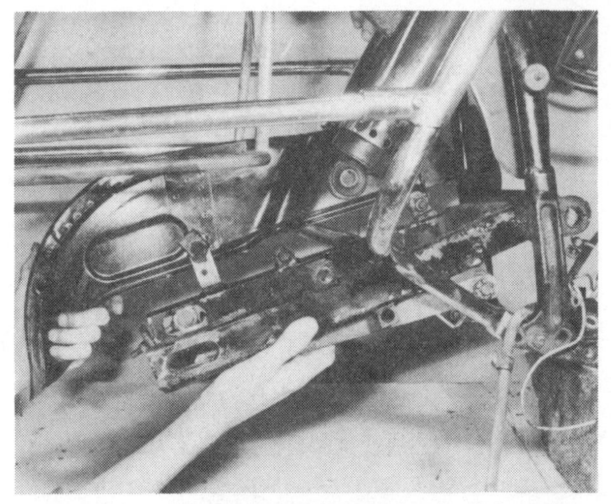

6.12 Remove the swingarm by withdrawing it from the frame mounts to the rear

6.17a Pack the swingarm bearing and race with grease (early models only)

6.17b Install a new seal after the bearing is in place (early models only)

6.17c Install the seal collar (early models only)

point head) and remove the left side locknut (3/4-inch six-point head) and cup washer.

d) On all five-speed models – Remove the passenger footrest brackets (FLT) and the pivot shaft mounting brackets (FXR).
e) On all models through 1988 – Tap the pivot shaft out of position from the right side (on models through early 1986, it isn't necessary to remove the nut and washer from the left side of the shaft).
f) On 1989 and later models – Tap the pivot shaft out from the left side.

12 Pull the swingarm out of the frame mounts **(see illustration)**.

Inspection

Four-speed models

Note: *The inspection procedure applies only to 1970 through early 1978 models and later models equipped with tapered roller swingarm bearings. Late 1978 through 1984 models have spherical bearings that are lifetime lubricated and require no service other than cleaning.*

13 Mount the swingarm securely in a vise and pull the seal collars off each end. The inner bearing races are obscured under the seals. The inner bearing races can be driven out with a drift punch and hammer inserted through the swingarm crossmember from the opposite side. This will also drive the seals out of position.

14 Clean and examine the tapered roller bearings for signs of wear and damage, which should be self-evident. If there is any doubt about the condition of the bearings, replace them both as a precaution. Most bearings fail as a result of pitting, which occurs on the inner and outer races.

15 It's also a good idea to replace the seals with new ones every time they're tampered with.

Five-speed models

16 Clean and inspect the swingarm around the pivot shaft. Check the pivot shaft bushings for signs of wear and deterioration. If the Clevebloc or bushings are worn, the swingarm must be taken to a Harley-Davidson dealer to have the defective components replaced.

Installation

Four-speed models

Refer to illustrations 6.17a, 6.17b and 6.17c

17 Reassemble and install the swingarm by reversing the above procedures. Be sure the bearings are liberally packed with grease and absolutely clean **(see illustration)**. Even a small amount of grit on a bearing will accelerate wear. Install the seals and collars in each side of the swingarm **(see illustrations)**.

18 After the pivot bolt has been installed, set the preload on the roller bearings. This is done by attaching a spring scale to the extreme end of one of the swingarm legs before the shock absorbers are installed. Take a reading (in pounds) required to make the swingarm turn on the pivot. The pivot bolt must be tightened only slightly so the swingarm will move freely. Now tighten the pivot bolt until the drag is increased by about two pounds, as measured on the spring scale.

19 When the proper preload is attained, bend the ears of the tab washer up to secure the pivot bolt.

Five-speed models

20 Installation of the swingarm is the reverse of the removal procedure. Special tool (no. HD-33805) may be required to spread the swingarm Cleveblocs.

21 On early 1986 and earlier models, coat the pivot shaft with anti-seize compound. Insert the shaft through the left side rubber mount (small diameter side of mount facing OUT), then install the left side washer and nut on the shaft. Tighten the nut until it bottoms on the shaft threads. Slide the left side nylon washer onto the shaft (small diameter side of washer facing the swingarm), then insert the pivot shaft through the swingarm and transmission from the left side of the machine. Install the nylon washer on the right side of the shaft (small diameter side facing the swingarm), followed by the rubber mount (small diameter side facing OUT), washer and nut. Position the dowel pin holes in the nylon washers at the top. Tighten the right side nut finger-tight.

22 On late 1986 through 1988 models, thread the left side pivot bolt onto the center stud, then coat the shank of the pivot bolt with anti-seize compound and insert the bolt through the left side rubber mount (small diameter side of mount facing the bolt head). Slide the left side nylon washer onto the bolt (small diameter side of washer facing the swingarm), then insert the left side pivot bolt and center stud assembly through the swingarm and transmission from the left side of the machine. Coat the right side pivot bolt shank with anti-seize compound, then install the right side rubber mount (small diameter side of mount facing the bolt head) and the right side nylon washer (small diameter side facing the swingarm). Insert the right side bolt into the swingarm and thread it onto the center stud finger-tight.

23 On 1989 and later models, coat the pivot shaft with anti-seize compound. Insert the shaft through the right side rubber mount (small diameter side of mount facing the shaft bolt head), then slide the right side nylon washer onto the shaft (small diameter side of washer facing the swingarm). Insert the pivot shaft through the swingarm and transmission from the right side of the machine. Install the nylon washer on the left side of the shaft (small diameter side facing the swingarm), followed by the rubber mount (small diameter side facing OUT), cup washer and nut. Tighten the nut finger-tight.

24 When installing the footrest or the pivot shaft mounting brackets, be sure the roll pins engage the locating hole in the rubber mount. Also be sure the flat on the pivot shaft registers with the flat on the rubber mount on the right.

25 Position the footrests so they will fold up at a 45-degree angle to the rear before tightening them.

26 Tighten the pivot shaft nut/bolt(s) to the specified torque. **Note:** *If the pivot shaft is a three-piece bolt and stud type, hold one bolt while tightening the other to the specified torque.* Make sure the clips are installed in the shaft mounting brackets. The flat side of the clip must be at the bottom. On 1989 and later models, hold the right side nut and tighten the left side nut to the specified torque.

27 On FLT models, the rear chain boots and their mating surfaces on the inner primary chaincase and transmission must be coated with RTV-type sealant.

7 Forks – removal and installation

All models

Refer to illustration 7.4

Note: *Detach the negative battery cable from the battery before beginning this procedure. On models with air-assisted forks, bleed the air out of the system and remove the banjo bolts from the fork caps. Loosen or remove the air tubes.*

1 The forks can be removed very simply by separating the fork tubes from the triple clamps. Do not disassemble the steering head components unless the bearings require attention (Section 9).

2 Raise the front of the machine until the front wheel is off the ground. Support it securely on blocks. This will require a little thought and planning because the lower frame tubes are close together and the machine won't be very stable unless it has some side support.

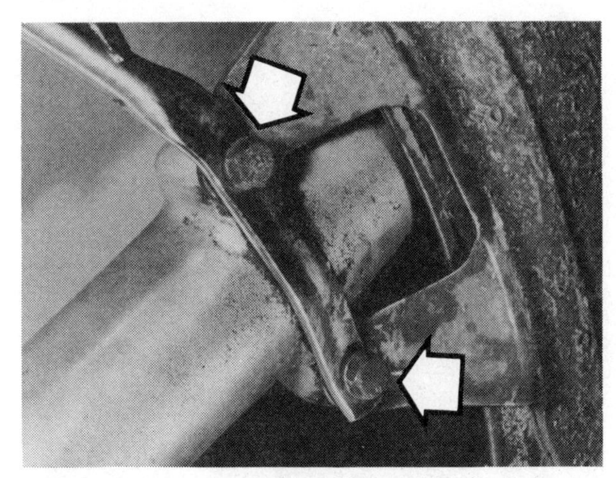

7.4 Remove the bolts securing the fender to the fork sliders

3 Remove the front wheel as described in Chapter 6.

4 Remove the front fender, which is attached to the inside of the fork legs by two bolts on each side **(see illustration)**. On most 1984 and later models, the front fender and headlight bracket must be removed. Let the headlight hang by the wire harness.

5 On models with front disc brakes, remove the clamp securing the hydraulic lines to the fork legs or fender. Disconnect the caliper(s) from the fork leg(s) and tie the caliper(s) to the frame, out of the way.

FL models

Refer to illustrations 7.8, 7.9a, 7.9b, 7.10 and 7.12

6 Remove the fairing, instrument panel and light bar (1986 and later FLHT/C models) from the forks. The mounting method will depend on the type of fairing installed.

7 Remove the screw securing the sealed beam in position. The screw passes through the bottom of the chrome rim. Pull the headlight away and disconnect the wires from the rear of the sealed beam.

8 The headlight shell is retained by screws on each side. When the screws are removed, the headlight shell can be removed from the forks. Some models have switches in the headlight shell for spotlights. These switches can be removed by unscrewing the flat nuts **(see illustration)**.

9 Remove the cap that secures the handlebars. It's held in place with four bolts. On some models, the handlebars are equipped with a bracket

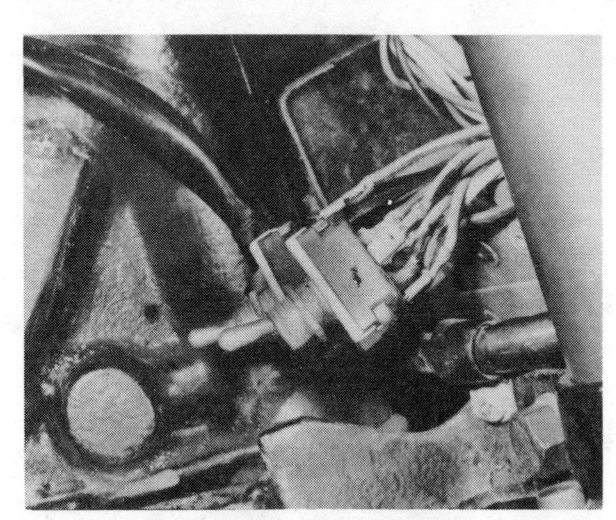

7.8 The spotlight switches (not all models) are retained in the headlight shell by flat nuts

7.9a Bolts (arrow) screw into the bottom of the handlebar brackets on some models

7.9b Be sure to label all wires and terminals if they must be disconnected

that's secured with two bolts **(see illustration)**. The handlebars can usually be moved out of the way without disconnecting any of the wires or control cables. If necessary, disconnect the cables and wires from the handlebar controls or remove the cables and wires with their respective controls still attached. The shape of the handlebars and the length of the cables will dictate the best approach. Be sure to keep the front brake master cylinder level (on disc brake models) to keep air from entering the system. **Note:** *If the wires are disconnected, be sure to label them or draw a sketch to be sure they are reconnected properly during reassembly* **(see illustration)**.

10 Loosen the pinch bolts securing the fork legs to the upper and lower triple clamps **(see illustration)**.

11 Remove the caps from the top of the fork legs.

12 Carefully remove each fork leg from the triple clamps **(see illustration on page 155)**. You may have to twist the fork leg to break it loose from the clamps. Be careful (if the forks have not been drained) or the fluid will be ejected out of them. If the fork legs are stuck in the triple clamps, screw the caps back onto the fork legs and strike the caps sharply with a soft-faced hammer. Do not use excessive force or the internal threads may be damaged. On 1984 and later FLT models, remove the rubber fork stop from each fork leg.

1 *Front forks (complete)*
2 *LH fork leg (complete)*
3 *Fork tube*
4 *Damper rod*
5 *Damper component set*
6 *Shim*
7 *Circlip*
8 *Sealing washer*
9 *Damper rod seat*
10 *Tapered bushing*
11 *Sealing washer*
12 *Slider*
13 *Axle cap*
14 *Stud*
15 *Nut*
16 *Oil seal set (complete)*
17 *Shroud*
18 *Screw*
19 *Fork spring*
20 *Drain plug*
21 *Circlip*
22 *Backing washer*
23 *Felt seal*
24 *Oil seal*
25 *Upper bushing*
26 *Lower bushing*
27 *Fork stem nut*
28 *Tab washer*
29 *Upper triple clamp*
30 *Adjusting nut*
31 *Tapered roller bearing*
32 *Outer race*
33 *Bearing housing*
34 *Washer*
35 *Lower triple clamp/steering stem*
36 *Bolt*
37 *Upper triple clamp (adjustable fork)*
38 *Bolt*
39 *Washer*
40 *Lower triple clamp/steering stem*
41 *Special tab washer*
42 *Cotter pin*
43 *Bolt*
44 *Tab washer*
45 *Castellated nut*
46 *Pinch bolt*
47 *Lower triple clamp*
48 *Damper adjusting screw*
49 *Trim*
50 *Damper knob*
51 *Adjusting screw*
52 *Coil spring*
53 *Diaphragm spring*
54 *Friction plate*
55 *Friction washer*
56 *Anchor pin*
57 *Anchor plate*
58 *Friction washer*
59 *Lower friction plate*
60 *Top bolt/breather (adjustable fork)*
61 *Top bolt*

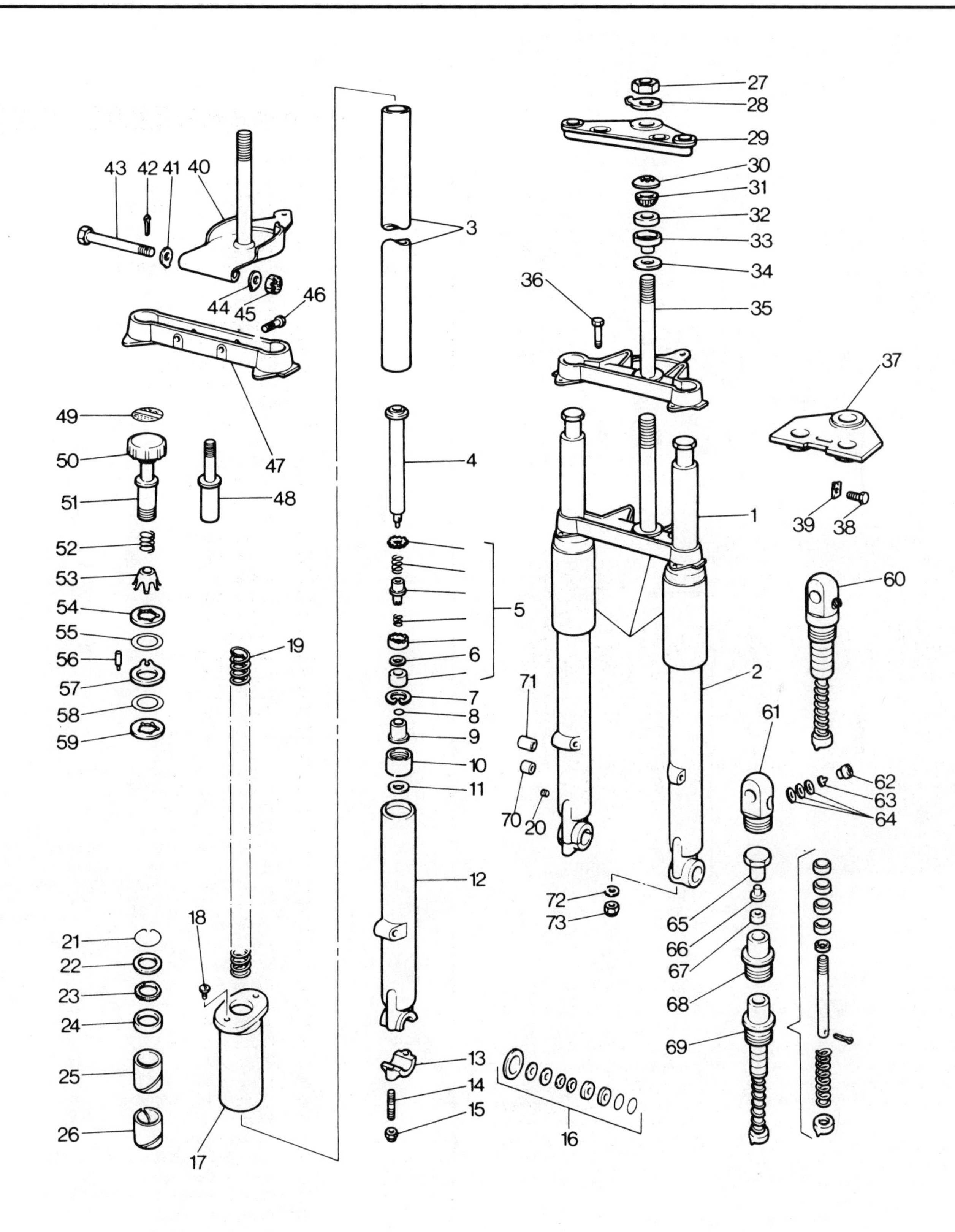

7.10 Typical FL model fork components – exploded view (early models)

62	*Oil fill plug (adjustable fork)*	*66*	*Breather valve*	*70*	*Caliper upper bushing*
63	*Insert*	*67*	*Oil seal*	*71*	*Caliper lower bushing*
64	*Sealing washer*	*68*	*Breather body*	*72*	*Plain washer*
65	*Top bolt (non-adjustable fork)*	*69*	*Breather (complete)*	*73*	*Nut*

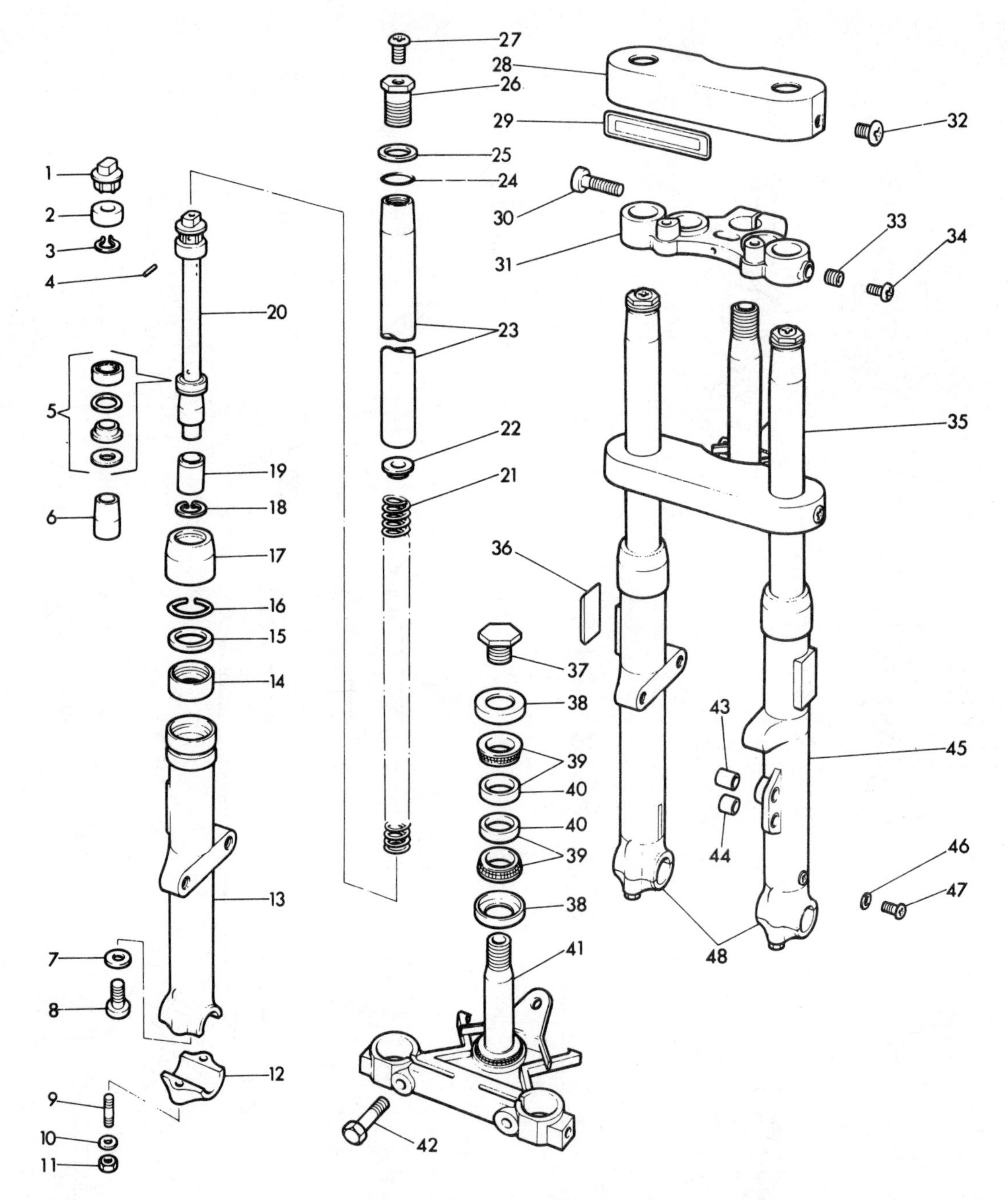

7.16 Typical FX model fork components – exploded view

1	*Upper piston seat*	17	*Dust boot*	33	*Threaded insert*
2	*Upper damper piston*	18	*Circlip*	34	*Cover screw*
3	*Snap-ring*	19	*Lower piston*	35	*LH fork leg (complete)*
4	*Roll pin*	20	*Damper rod (complete)*	36	*Reflector*
5	*Damper valve assembly*	21	*Fork spring*	37	*Fork stem retaining bolt*
6	*Damper rod seat*	22	*Upper spring seat*	38	*Dust shield*
7	*Washer*	23	*Fork tube*	39	*Tapered roller bearing*
8	*Allen-head screw*	24	*O-ring*	40	*Outer bearing race*
9	*Stud*	25	*Washer*	41	*Lower triple clamp/steering stem*
10	*Plain washer*	26	*Cap*	42	*Pinch bolt*
11	*Nut*	27	*Screw*	43	*Caliper upper bushing*
12	*Axle cap*	28	*Cover*	44	*Caliper lower bushing*
13	*Fork slider*	29	*Trim*	45	*Slider*
14	*Oil seal*	30	*Upper pinch bolt*	46	*Sealing washer*
15	*Backing washer*	31	*Upper triple clamp*	47	*Drain plug*
16	*Circlip*	32	*Screw*	48	*Axle clamp*

7.12 Slide the fork legs down through the triple clamps to remove them

FX models

Refer to illustration 7.16

13 Remove the cap that retains the handlebars. It's held in place with four Allen-head bolts.

14 The handlebars can usually be moved out of the way without disconnecting any of the control cables or wires. If necessary, disconnect the control cables and wires from the handlebar controls or remove the control cables and wires with their respective controls still attached. The shape of the handlebars and the length of the cables will dictate the best approach. Be sure to keep the front brake master cylinder level (on front disc brake models) to keep air from entering the system.

15 Unscrew the drive cables from the speedometer and tachometer. Remove the instruments and mounts by unscrewing the bolt that retains each mount to the top of each fork leg. Disconnect the wires and store the instruments in a safe place.

16 Remove the screws and lift the cover up to expose the upper triple clamp. Do the same for the lower triple clamp cover (if equipped). Unscrew the cap at the top of each fork leg, but don't remove them at this point unless they retain the fork leg in the upper triple clamp **(see illustration)**.

17 Loosen the pinch bolts in the triple clamps, under the plated covers (some models only have pinch bolts in the lower triple clamp). Remove the two bolts that secure the mounting bracket between the upper and lower triple clamps and detach the bracket (not used on all models).

18 On some models, the upper ends of the fork tubes are tapered and mate with a matching taper in the upper triple clamp. To break the taper, give the loosened fork caps a sharp tap with a hammer and block of wood – do not use excessive force or the internal threads of the fork tube may be damaged. If the fork legs are tight, try tapping on the upper triple clamp at the same time. On models which have parallel-sided fork tube ends take note of the tube's installed height in the triple clamp before removing; some are fitted flush with the triple clamp surface, whereas others project 0.42 to 0.50 inches (11 to 13 mm) – measured from top of cap bolt to clamp surface.

19 After the forks are free, remove the caps from the top of each fork leg.

20 Remove the fork legs from the lower triple clamp by pulling only on the fork tubes. **Caution:** *Do not compress the fork tube and slider or oil may spurt out the open end of the tube. Keep the fork leg upright until the caps are installed (also, store them in an upright position).*

All models

21 Install the fork legs in the triple clamps by reversing the removal procedure. Be sure to push only on the fork tubes as the fork legs are slipped into place and do not compress the fork tube and slider until the caps are in place.

22 Tighten the lower pinch bolts temporarily, then add the specified amount of fork oil to each fork leg (see Chapter 1).

23 Install the fork caps, tighten them securely, then tighten the lower pinch bolts to the specified torque.

24 Install the anti-dive air control system (where fitted).

8.3 Unscrew the caps from the top of the fork tubes . . .

25 Install the instruments and handlebars, then attach the fender to the forks. Refer to Chapter 6 and install the front wheel and related components.

Springer forks

Note: *Do not attempt to disassemble the springer forks. If problems are encountered with the springs, fork legs or rocker assemblies, take the machine or the forks to a Harley-Davidson dealer.*

26 Detach the front brake caliper and remove the wheel (see Chapter 6).

27 On models through 1992 the front fender can be released from the forks after removing its four retaining nuts and bolts. On 1993 and later models the fender is not fixed rigidly to the forks, but floats with wheel movement via a mounting on the brake reaction link and short link rods to the forks. **Note:** *The fender-to-reaction link pivot can be dismantled, while taking note of the exact fitted position of the washers and spacers, but the fender-to-fork links require the use of a Harley-Davidson service tool to hold the links in position while the fender is removed.*

28 Remove the headlight and mounting bracket. Let the headlight hang by the wire harness.

29 Remove the acorn nuts and washers, then pull out the bolts and detach the shock absorber.

30 Remove the fork (steering) stem acorn nut and washer, then loosen the upper triple clamp pinch bolt.

31 Remove the handlebars and detach the risers.

32 Remove the studs from the upper ends of the rigid fork legs, then lift off the upper triple clamp.

33 Remove the hex-shaped bearing retainer and dust shield from the top of the fork stem.

34 Withdraw the fork stem from the bottom and detach the forks from the steering head on the frame.

35 Installation is the reverse of the removal procedure. When installing the hex-shaped bearing retainer, tighten it to 6 ft-lbs (8 Nm). When installing the rigid fork leg studs, start both of them, then tighten them evenly to the specified torque. Use anti-seize compound on the upper triple clamp pinch bolt threads. Use thread locking compound on the shock absorber mounting bolt/nut threads. Use new locknuts when installing the handlebar risers. Center the bosses in the washer cutouts as the riser locknuts are tightened.

8 Forks – disassembly, inspection and reassembly

1 Remove the fork legs as described in Section 7.

2 Always disassemble one fork leg at a time to avoid mixing up parts.

Disassembly

All models

Refer to illustrations 8.3, 8.4, 8.5, 8.6a and 8.6b

3 Remove the cap, turn the fork leg upside-down and drain the oil into a container **(see illustration)**. Keep in mind when the fork is inverted the

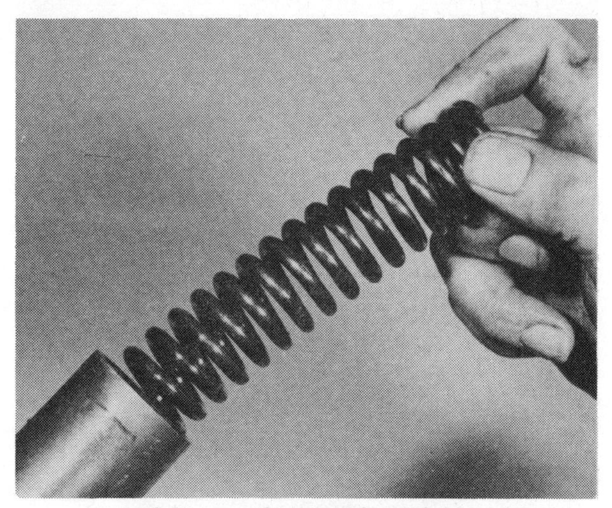

8.4 . . . and remove the springs

8.5 Remove the nut (arrow) or Allen-head bolt from the bottom of each fork slider

8.6a Push the circlip (if equipped) out of the slider groove, then . . .

8.6b . . . remove the serrated washer and felt dust seal

8.8 Remove the circlip . . .

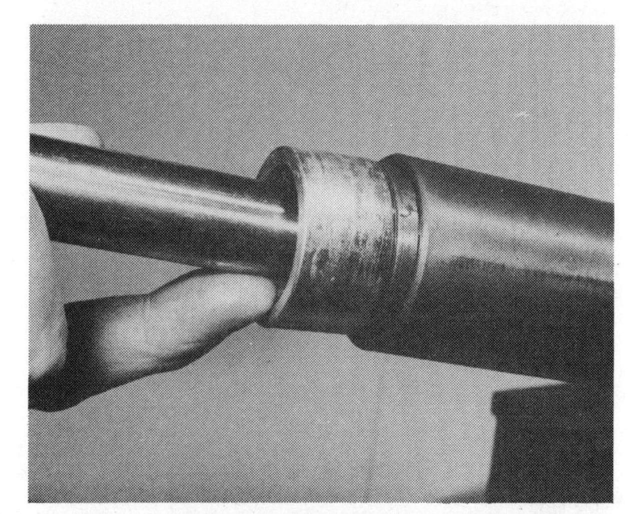

8.9 . . . to allow the damper rod to be pulled out (1970 through early 1977 models)

spring will slide out – don't let it fall. Compress the fork tube and slider a few times to ensure it drains completely.

4 When the majority of the oil has drained, slide out the spring and wipe off as much oil as possible **(see illustration)**.

5 Secure the slider in a vise equipped with soft jaws. Clamp it very lightly – just enough to prevent it from rotating while the Allen-head bolt is removed from the end of the slider. On some models a locknut is used in place of the bolt **(see illustration)**. If the nut or bolt won't loosen because

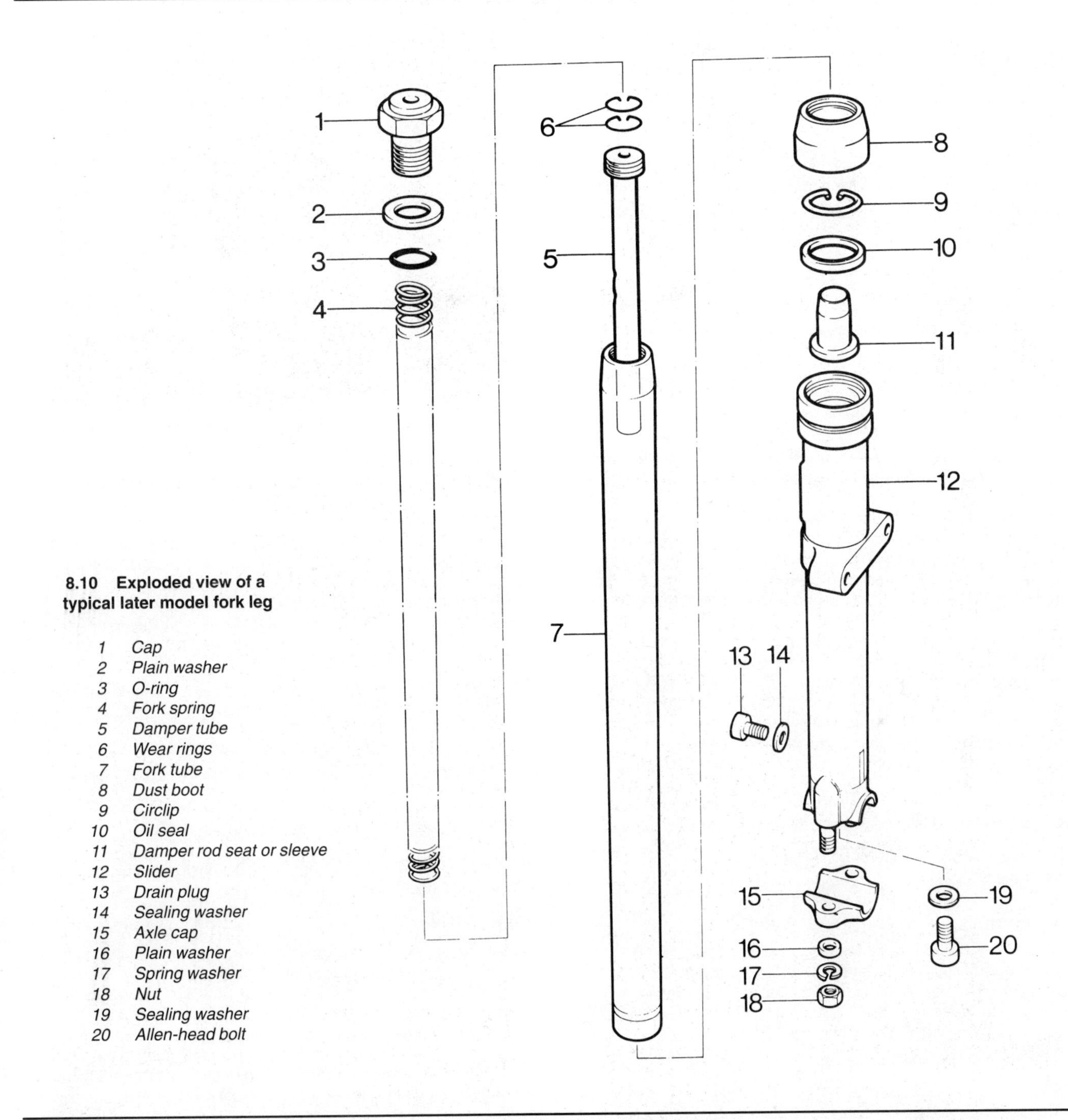

8.10 Exploded view of a typical later model fork leg

1 *Cap*
2 *Plain washer*
3 *O-ring*
4 *Fork spring*
5 *Damper tube*
6 *Wear rings*
7 *Fork tube*
8 *Dust boot*
9 *Circlip*
10 *Oil seal*
11 *Damper rod seat or sleeve*
12 *Slider*
13 *Drain plug*
14 *Sealing washer*
15 *Axle cap*
16 *Plain washer*
17 *Spring washer*
18 *Nut*
19 *Sealing washer*
20 *Allen-head bolt*

the damper rod turns, try to lock the rod by installing and applying pressure to the fork spring with the cap. On most models, the top of the damper rod has a slotted or milled head. If rotation can't be prevented by compressing the fork spring, an improvised tool or a long screwdriver can be used to hold the rod stationary.

6 Pull the dust boot off the slider (if used) and remove the circlip/retaining ring **(see illustration)**, washer(s) and felt dust seal (if used). Note the installed order of the washers so they can be reinstalled in the same locations during reassembly **(see illustration)**.

7 Pull the fork tube out of the slider. Many models have a bushing on the fork tube and in the top of the slider. Due to the fork tube bushing being slightly wider in OD than the slider bushing ID, when the fork tube and slider are pulled apart in a 'slide-hammer' action the fork tube bushing will displace the slider bushing.

1970 through early 1977

Refer to illustrations 8.8 and 8.9

8 Remove the internal circlip/retaining ring from the fork tube **(see illustration)**.

9 The damper unit can now be removed **(see illustration)**. It can be disassembled if parts require replacement. There are a variety of dampers used in the forks on these models **(see illustrations 7.10 and 7.16).** The pistons and bushings are the parts most likely to wear, which will be greatly accelerated if the forks are run without enough oil or with contaminated oil.

Late 1977-on

Refer to illustration 8.10

10 Remove the spring (if still in place) and damper tube from the fork tube **(see illustration)**.

11 Detach the wear rings from the slots in the damper tube **(see illustration 8.10).**
12 Remove the sleeve or the damper rod seat from the slider.

All models

Refer to illustration 8.13

13 The oil seal in the top of the fork slider can be pried out after the circlip (not used on early models) has been removed **(see illustration).** This seal should not be disturbed unless it's worn or if the bushings (early models only) in the slider are worn and must be replaced with new ones.

Inspection

14 The parts most likely to wear over an extended period of time are the fork bushing(s) and oil seals. Worn bushings cause a shudder when the front brake is applied, and with the forks compressed and the front brake held on it will be possible to detect an amount of play between the slider and fork tube (but don't confuse this with head bearing play).
15 Replacing bushings on early models may prove difficult due to the need to ream them to size after installation, a task that is best left to a Harley-Davidson dealer. On later models (with bushings on the fork tube and slider) replacement is fairly straightforward. Pry the fork tube bushing apart just enough to ease it off the tube end and install the new bushing in the same manner. The slider bushing will have been dislodged on dismantling and the new bushing must be installed using a tubular drift of exact ID and OD to ensures that the bushing is tapped evenly into its location.
16 The oil seal in each slider can be pried out after the circlip has been removed. New oil seals should be installed with great care and located correctly before the circlip is replaced. Be very careful when the fork tube is being reinstalled during reassembly of the forks – the seal lip is easily damaged. Coat the seal lip with grease first.
17 If the fork damping is inadequate, the damper pistons (pre-1978 models) of the damper assembly should be replaced (these are the parts most likely to wear).
18 Before reassembling the forks, check the sliding surfaces of the fork tubes. If they're pitted, scuffed, scored or badly worn, they should be replaced, along with the fork bushings (if equipped).
19 It isn't possible to straighten forks bent in an accident, even if a repair service is available. There is no way of knowing whether the parts have been overstressed and will be subject to sudden fatigue failure. The tubes should be checked for straightness by rolling them on a flat surface.
20 The triple clamps are also prone to twist or distort in an accident. They also should be replaced, not repaired, especially since they're more difficult to align correctly.

Reassembly

21 Reassembly is the reverse of the dismantling procedure. Always fit a new oil seal and install with the lettered side facing upwards; coat the seal lips with fork oil to prevent damage as the fork tube is installed.
22 Be sure to tighten the slider Allen-head bolt or the locknut securely. Use thread locking compound on the threads.
23 Refer to Section 7 and reinstall the forks (don't forget to add oil).

9 Steering head bearings – maintenance

1 If the steering head bearing check (Chapter 1) reveals excessive play in the steering head bearings, the entire front end must be disassembled and the bearings and races (cups/cones) replaced with new ones.
2 Refer to Chapter 6 and remove the front wheel.
3 Remove the forks as described in Section 7.
4 Remove the fork stem retaining nut or bolt **(see illustrations 7.10 and 7.16).** Some models use a tab washer to lock the nut in position. On these models, bend the ears of the tab washer down before loosening the nut. On early FL models with adjustable forks, remove the steering damper adjusting screw. Detach the adjuster components from the upper bracket. **Note:** *Remove the parts and set them out, in order, to be sure they're installed in the proper order during reassembly. Some of the parts may have to be carefully pried apart with screwdrivers.*

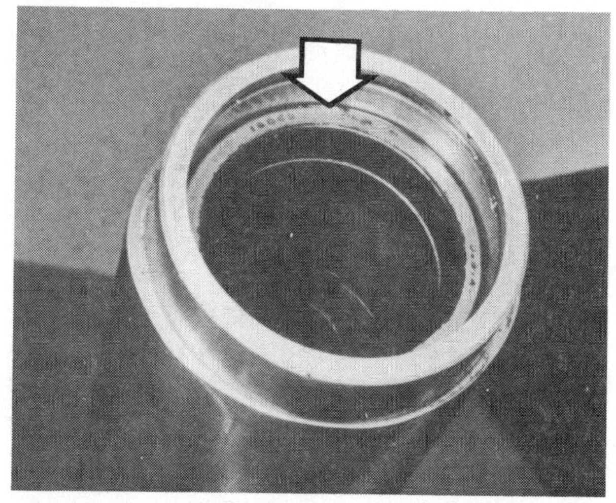

8.13 The oil seal (arrow) can be removed after the circlip

FX models

5 Remove the pinch bolt from the upper triple clamp. Be sure to support the lower triple clamp.
6 Remove the upper bearing dust shield, followed by the upper bearing. The bearing may be easier to remove if the lower triple clamp is partially removed from the bottom of the steering head **(see illustration 7.16).**
7 Remove the lower triple clamp from the steering head and lift the bearing and the dust shield off the stem. Do not mix the upper and lower bearings up. They should be reinstalled in their original locations.

FL models

8 Remove the upper triple clamp from the steering head. You may have to tap it up with a soft-face hammer.
9 Use special tool no. HD 96219-50 or a punch and hammer to remove the head bearing nut at the top of the stem. Support the lower triple clamp while the nut is removed.
10 Remove the upper bearing dust shield and the upper bearing from the stem. You may have to partially lower the lower triple clamp to remove the upper bearing.
11 Remove the lower triple clamp from the bottom of the steering head. Lift the lower bearing and dust shield off the stem.

All models

12 Clean and examine the bearing races. They should have a polished appearance with no pitting or indentations.
13 If replacement is necessary, the bearings and races should be replaced as matched sets. The races are a tight fit in the steering head and will have to be driven out of position.
14 Clean and examine the bearings, which should also be highly polished and show no signs of wear, pitting or surface cracks. Again, if the bearings are replaced, the bearing races should also be replaced as a matched set.
15 When reassembling the steering head bearings, pack the races and the bearing cones with grease.
16 Install the forks as described in Section 7 and the front wheel as described in Chapter 6.
17 Adjustment of the steering head bearings is critical. They should be tightened sufficiently to eliminate all play but they must not be overtightened. Note that it's possible to place a load of several tons on the bearings by overtightening and yet still be able to turn the handlebars. Adjustment is done by turning the adjuster nut or bolt at the top of the steering head. **Note:** *Loosen the lower triple clamp pinch bolts before making any adjustments. This will allow the fork legs to move in the lower triple clamp as the bearings are tightened or loosened. Be sure to tighten the pinch bolts when the adjustment is complete. When the adjustment is correct, there should be no play in the bearing and, when the front wheel is raised off the ground, a light tap on the end of the handlebar should cause the forks to swing to the full lock position.*

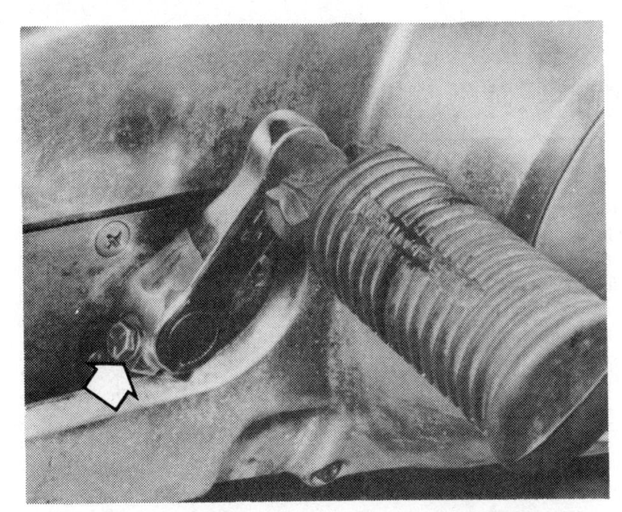

11.1 The footrests are retained by pinch bolts (arrow)

18 Overtightened bearings in the steering head will cause the machine to roll at slow speeds, while loose bearings will cause front fork shudder when the front brake is applied.

10 Adjustable front forks – altering trail angle

1 On machines originally equipped with a sidecar or supplied for sidecar use, adjustable front forks are installed, which have provision for alteration of the trail angle to suit either solo or sidecar use.
2 To alter the trail angle from solo to sidecar use, remove the nut from the adjustment bracket bolt. A cotter pin secures the nut. Tap the bolt out enough so the special tanged washer can be pried away from the bracket. Grasp the fork legs and pull them forward sharply. Rotate both tanged washers so the tangs or pins face forward and locate in the holes in the bracket. Install the nut and tighten it completely. DO NOT forget the cotter pin. Altering the forks for solo use is carried out by reversing the procedure so the tangs on the washer face to the rear.

11 Footboards and footrests – inspection and repair

Refer to illustration 11.1

1 The rider's footrests fold up and are attached by a pinch bolt. If accident damage occurs, the folding portion can be removed from the main body by removing the pivot bolt that serves as the hinge **(see illustration)**.
2 Footrests should never be straightened while they're attached to the frame. They can be straightened by heating (after the footrest rubber has been removed) and bending in the desired direction, but it's a good idea to install new components, especially if only the folding portion has been damaged.
3 On FL models, the rider's footboards pivot on shouldered bolts, allowing each board to fold up if the machine falls over or is involved in an accident. When excessive weight is applied to a footboard, the components most likely to be damaged are the two hinge brackets, which are retained by two rivets. The brackets can be straightened in position, or if damaged beyond repair, removed after drilling out the rivets.

12 Anti-dive system – check

General check

1 Check the system lines and fittings for leaks and make sure everything is tight.
2 Attach a no-loss air gauge to the system air valve on the engine guard, accumulator or handlebar.
3 Make sure the ignition switch is off, then apply the front brake and bounce the front end of the machine.
4 Watch the air gauge – it should fluctuate as the front end moves up-and-down.
5 Turn the ignition switch on and repeat the check. This time the air pressure must remain constant as the front end moves.

Solenoid check

6 The solenoid is mounted just below the upper fork assembly triple clamp.
7 With the ignition switch on, apply either brake and listen for a faint click from the solenoid. The sound is very faint, so it may help to touch the solenoid as the brake is applied and released. You should feel the solenoid operate.
8 If the solenoid doesn't seem to be working, check the winding resistance by connecting an ohmmeter to the solenoid wires. The meter should indicate 10-to-20 ohms on models through 1992 and 10-to-30 ohms on models from 1993. If it doesn't, replace the solenoid.
9 Attach one ohmmeter lead to one of the wires and the other lead to the solenoid metal case. The ohmmeter must indicate infinite resistance. If it doesn't, replace the solenoid.
10 To remove the solenoid, proceed as follows:

a) FLT models – On 1984 through 1987 FLT models, unscrew the hose fitting and remove the O-ring from the bottom of the solenoid. On 1988 and later FLT models, remove the banjo bolt, O-rings and hose fitting from the bottom of the solenoid. On all FLT models, remove the nut (if used) and withdraw the case, solenoid, spacer and rubber washer. The plunger body can be unscrewed with a spanner wrench, by locking two thin nuts together on the plunger body threads or with a screwdriver (1985-on). Remove the spring, plunger and O-ring.
b) 1984 through 1987 FXRT and FXRD models – On 1985 and earlier models, detach the accumulator hose at the solenoid right-angle fitting (the hose clamp will have to be cut or pried off), then unscrew the fitting and remove the O-ring, washer and nut. Unscrew the large nut at the bottom of the solenoid case. Remove the case bolts/nuts and detach the case, solenoid and spacer. On 1986 and later models, remove the banjo bolt, fitting and O-rings from the bottom of the solenoid, then detach the case, solenoid and spacer. The plunger can be removed as described in Paragraph a).
c) 1988 and later FXRT, FXRS-SP and FXRS-CONV models – Remove the banjo bolt or hex hose fitting from the bottom of the solenoid and detach the fitting and O-rings. Pull down on the case and detach the solenoid, the rubber washer, the metal washer and the O-ring from the plunger body. Remove the plunger as described in Paragraph a).

Chapter 6 Wheels, brakes and tires

Contents

Specifications

Wheel bearing end play

FLT models	
1980 and 1981	0.004 to 0.014 inch (0.1 to 0.35 mm)
1982 on	0.002 to 0.006 inch (0.05 to 0.15 mm)
All other models	
Through mid-1991	0.004 to 0.008 inch (0.1 to 0.2 mm)
Mid 1991 on	0.002 to 0.006 inch (0.05 to 0.15 mm)

Wheel runout

All wheels (cast and spoke)	
Lateral	3/64 in (0.040 in, 1.0 mm)
Radial	1/32 in (0.030 in, 0.8 mm)

Brakes

Minimum brake shoe thickness	See Chapter 1
Minimum brake pad thickness	See Chapter 1
Disc runout (warpage) limit	3/32 in (2.38 mm)
Rear pedal height setting	See Section 16 of this Chapter
Rear master cylinder free play	See Section 16 of this Chapter
Front master cylinder free play (at end of lever)	1/4 in (6.35 mm)
Caliper piston retraction (minimum)	
Front	0.020 to 0.025 in (0.5 to 0.6 mm)
Rear	0.033 to 0.038 in (0.8 to 1.0 mm)

Torque specifications

	Ft-lbs (unless otherwise indicated)	Nm
Axle nut (front and rear)	See Chapter 1	
Brake disc-to-hub	See Chapter 1	
Disc brake caliper mounting bolts	See Chapter 1	
Brake drum-to-hub	35	47
Rear master cylinder mounting bolts	155 to 190 in-lbs	17.5 to 21
Rear master cylinder mounting nut (late 1987 and later FXR models)	30 to 40	41 to 54
Sprocket-to-rear wheel hub bolts (through 1983)		
FLT only	65 to 75	88 to 102
All others	35 to 50	47 to 68
Sprocket-to-rear hub bolts (1984 through 1990)		
FLT with enclosed chain	65 to 75	88 to 102
FLT with belt drive and all FX/Softails		
Grade 5 bolts	45 to 50	61 to 68
Grade 8 bolts	65 to 70	88 to 95
FLT/FXR with open chain drive	50 to 55	68 to 75
Sprocket to rear hub bolts (1991 on)		
FLT (1991)		
Spoked wheel	45 to 50	61 to 68
Cast wheel		
Grade 5 bolts	45 to 50	61 to 68
Grade 8 bolts	65 to 70	88 to 95

Torque specifications (continued)

	Ft-lbs (unless otherwise indicated)	Nm
FLT (1992)		
Spoked wheel	45 to 50	61 to 68
Cast wheel	45 to 55	61 to 75
FLT/FXR (1993-on)		
Spoked wheel	45 to 55	61 to 75
Cast wheel	55 to 65	75 to 88
FXR (1991 through 1992)	45 to 55	61 to 75
FX/Softails		
1991 with disc wheel		
Grade 5 bolts	45 to 50	61 to 68
Grade 8 bolts	65 to 70	88 to 95
1991-92 spoked and 1992 disc wheel	45 to 55	61 to 75
1993-on	55 to 65	75 to 88
FXD Dynas		
1991	50 to 55	68 to 75
1992	65 to 70	88 to 95
1993-on		
Spoked wheel	45 to 55	61 to 75
Cast wheel	55 to 65	75 to 88
Rear brake anchor nut	50	68
Rear brake reaction pin nut (FXST/C only)	20	27
Brake hose-to-caliper banjo fitting bolts (1985 on)		
With copper washers	30 to 35	41 to 47
With steel and rubber washers	17 to 22	23 to 30

1 General information

Depending on the model, either wire spoke wheels or cast alloy wheels are standard equipment. Wire wheels require frequent inspection and maintenance, while cast wheels are virtually maintenance-free.

Brakes are various combinations, depending on model and year; 1970 through 1972 FX models and 1970 and 1971 FL models are equipped with drum brakes on both front and rear wheels. Beginning with 1972 FL models and 1973 FX models, the front and rear wheels are both equipped with disc brakes (dual discs are installed on the front wheels of 1978 models). Most 1979 and later models are equipped with dual front discs and a single rear disc. **Warning:** *Dust created by the brake system contains asbestos, which is harmful to your health. Never blow it out with compressed air and don't inhale any of it. An approved filtering mask should be worn when working on the brakes. Do not, under any circumstances, use petroleum-based solvents to clean brake parts. Use brake system cleaner or denatured alcohol only.*

2 Wheels – inspection and repair

Wire wheels

1 Wire wheels should be inspected frequently to ensure the wheel runs true and to prevent potential damage from loose or broken spokes.

2 Clean the wheels thoroughly to remove mud and dirt, then make a general check of the wheels and spokes as described in Chapter 1.

3 Raise the motorcycle so the front wheel is off the ground and support it securely with blocks. Because the frame is very narrow under the engine, be sure to support the motorcycle so it can't fall over sideways. Attach a dial indicator to the fork slider and position the stem against the side of the rim. Spin the wheel slowly and check the side-to-side (axial) runout of the rim. In order to accurately check the radial runout with the dial indicator, the wheel would have to be removed from the machine and the tire removed from the wheel. With the axle clamped in a vise, the wheel can be rotated to check the runout.

4 An easier, though slightly less accurate, method is to attach a stiff wire pointer to the fork slider and position the end a fraction of an inch from the wheel (where the wheel and tire join). If the wheel is true, the distance from the pointer to the rim will be constant as the wheel is rotated. Repeat the procedure to check the rear wheel.

5 A wheel that wobbles from side-to-side can be trued by loosening the spokes that lead to the hub from the high side of the rim and tightening the spokes that lead to the hub from the low side. This in effect will pull the bulge out of the rim. Always tighten/loosen spokes in small increments to avoid distorting the rim and make sure all the spokes are uniformly tight (see Chapter 1).

6 An out-of-round wheel can be trued by loosening the spokes (both sides of the hub) that lead to the low area, and tightening the spokes that lead to the high or bulged-out area.

7 Generally, the wheels will probably have a combination of side-to-side wobble and out-of-roundness. Keep in mind that tightening and loosening spokes will affect wheel runout in both directions. Wheel truing requires patience and practice to develop any degree of skill, so it's best left to a dealer service department or motorcycle repair shop.

8 If the inspection reveals a bent, cracked, or otherwise damaged rim, the entire wheel will have to be rebuilt using a new rim and spokes. This is a complicated task requiring experience and skills beyond those of the average home mechanic and should be done by a dealer service department or motorcycle repair shop.

Cast wheels

9 The cast alloy wheels should be visually inspected for cracks, flat spots on the rim and other damage. Check the axial and radial runout as described previously for wire wheels.

10 If damage is evident, or if runout in either direction is excessive, the wheel will have to be replaced with a new one. Never attempt to repair a damaged cast wheel.

3 Front wheel – removal and installation

Caution: *On disc brake equipped models, do not operate the brake lever while the front wheel is removed – the piston in the caliper might be forced out. If the piston is forced out of the bore, the caliper will have to be completely disassembled and rebuilt. Also, after the wheel is reinstalled, the wheel bearing end play must be checked with a dial indicator. If it falls outside the range listed in this Chapter's Specifications, the long spacer in the wheel hub or thin spacer washer (as applicable) will have to be replaced by another of different length – see Chapter 1.*

1 Raise the front wheel off the ground and support the machine securely on blocks. Be sure the motorcycle is stable from side-to-side. The frame under the engine is very narrow, so it may be necessary to support the side of the motorcycle.

6

3.5 Loosen the slider cap nuts before attempting to remove the axle (1970 and 1971 FL models)

1970 and 1971 FL models

Refer to illustration 3.5

2 Straighten the cotter pin at the end of the axle and pull it out.
3 Unscrew the axle nut and remove the washer from the axle.
4 Remove the screws securing the brake drum to the wheel hub.
5 Loosen the nuts securing the caps at the bottom of the fork sliders and pull the axle out **(see illustration)**.
6 Apply the front brake to hold the brake drum in place while the wheel is removed from the forks.
7 Installation of the front wheel is done by reversing the removal procedure. Be sure the mating surfaces of the wheel hub and brake drum are clean to ensure a tight fit.
8 Tighten the screws securing the brake drum to the torque listed in this Chapter's Specifications. Use a criss-cross pattern to avoid warping the drum.
9 Tighten the axle nut to the specified torque, followed by the fork slider cap nuts. Install a new cotter pin in the axle.

1970 through 1972 FX models

10 Disconnect the front brake cable from the front wheel by removing the clevis pin from the brake operating arm. Remove the brake backing plate anchor bolt and the lock washer.
11 Unscrew the nut from the end of the axle.
12 Loosen the pinch bolts (at the bottom of each fork leg) that secure the axle to the forks **(see illustration 3.5)**.
13 Tap the axle out of position and remove the front wheel.
14 Installation is the reverse of the removal procedure. Inject an ounce of multi-purpose grease into the wheel hub before installing the axle. Tighten the axle nut to the torque listed in this Chapter's Specifications, followed by the axle pinch bolts.

1973 FX models

Refer to illustrations 3.16 and 3.17

15 Remove the axle nut and washer from the front axle.
16 Loosen the nuts securing the axle cap to the bottom of the right fork slider **(see illustration)**.
17 Tap the axle out of position and lower the front wheel. Pull the speedometer drive out of the front hub **(see illustration)**. Leave the speedometer drive unit connected to the cable and tie them up out of the way.
18 Reverse the removal procedure for installation of the front wheel. Align the brake pads on each side of the brake disc while installing the wheel.
19 Engage the speedometer drive gear with the hole in the wheel hub.
20 Tighten the axle nut to the specified torque, followed by the slider cap nuts.

1974 through 1977 FX models

21 Remove the brake caliper mounting bolt, the washers and the locknut.
22 Remove the axle nut from the axle, along with the washer and lock washer.
23 Loosen the bolts securing the axle caps to the bottom of each fork slider.
24 Tap the end of the axle to loosen it, then pull the axle out of the forks and front hub.
25 Lower the front wheel until the speedometer drive unit can be disengaged from the front hub. Tie the speedometer drive unit and cable out of the way.
26 Remove the front wheel.
27 Installation is the reverse of the removal procedure. Align the brake pads on each side of the brake disc while installing the front wheel.
28 Engage the speedometer drive gear with the hole in the wheel hub. Tighten the axle nut to the specified torque, followed by the slider cap nuts.

1972 through early 1978 FL models

29 Straighten the ends of the cotter pin in the end of the axle and remove it.
30 Remove the axle nut and flat washer from the axle.
31 Loosen the slider cap nuts and remove the axle. You may have to tap the end of the axle with a soft-faced hammer to remove it from the wheel.
32 Remove the front wheel.
33 Installation is the reverse of the removal procedure. Align the brake pads on each side of the brake disc while placing the wheel in position.

3.16 Loosen the axle cap nuts before attempting to remove the axle (1973 FX models)

3.17 On 1973 FX models, lower the front wheel and disengage the speedometer drive unit from the wheel hub

34 Tighten the axle nut to the specified torque and install a new cotter pin to secure the nut.
35 Tighten the two slider cap nuts.

1978 and later FX models (except springer forks)

Refer to illustration 3.38

36 Detach the front brake calipers from the fork legs and tie them up, out of the way.
37 Remove the nut, lock washer and flat washer from the axle.
38 Loosen the bolts/nuts securing the axle caps to the bottom of the fork slider(s) **(see illustration)**. **Note:** *On 1987 and later FXR models and FXDB/C/L Dyna models, loosen the pinch bolt locknut on the right side fork slider.*
39 Tap the end of the axle to loosen it, then pull it out of the wheel.
40 Lower the wheel from the forks and disconnect the speedometer drive unit from the hub (some models don't have a speedometer drive unit on the front wheel). Remove the front wheel. On FLST models, the hub cap will come off with the wheel.
41 Installation of the front wheel is the reverse of the removal procedure. Align the brake pads on each side of the brake discs while placing the wheel in position.
42 A spacer is installed between the left fork leg and the wheel on some FXWG models; a spacer with a groove also fits between the right fork leg and the wheel on some FXWG models.
43 Engage the speedometer drive gear with the hole in the wheel hub (except FXWG models).
44 Tighten the axle nut to the specified torque before tightening the slider cap nuts.
45 Securely tighten the caliper mounting hardware.

Late 1978 through 1980 FL models

46 Remove the caps from the ends of the axle (they're held in place with a small set screw).
47 Remove the axle nut, lock washer and flat washer.
48 Loosen the nuts securing the slider cap and pull out the axle.
49 Remove the front wheel, along with the hub cap and spacer.
50 Reverse the removal procedure to install the front wheel. Align the brake pads on each side of the brake disc.
51 Apply multi-purpose grease to the axle and insert it through the right fork leg, through the wheel, spacer, hub cap and the left fork leg.
52 Tighten the axle nut to the specified torque before tightening the slider cap nuts.
53 Install the caps on the ends of the axle.

1981 and later FL models (except FLT)

54 Remove the caps from the ends of the axle (they're held in place with a small set screw).
55 Remove the axle nut, lock washer and flat washer from the left side of the wheel.
56 Loosen the slider cap nuts and remove the axle.
57 Remove the wheel, along with the spacer, speedometer drive and hub cap.
58 Installation is the reverse of the removal procedure. Be sure to align the brake pads on each side of the brake disc.
59 Coat the axle with multi-purpose grease before installing it.
60 Insert the axle through the right fork leg, through the speedometer drive, hub cap, wheel, spacer and the left fork leg.
61 Tighten the axle nut to the specified torque, followed by the slider cap nuts.
62 Install the caps on the ends of the axle.

1980 and later FLT models

63 Remove the brake caliper mounting nuts and bolts and tie the calipers out of the way.
64 Remove the axle nut and washers from the left side of the axle.
65 Loosen the slider cap nuts on the bottom of the right fork leg. Tap the axle out of the left fork leg with a soft-faced hammer.
66 Lift the wheel and speedometer drive out of position. **Note:** *The speedometer drive on 1987 and later models is mounted on the left side instead of the right side as on previous years. The speedometer drive is also equipped with a rubber washer-type seal between the drive unit and the front wheel.*

3.38 On later models, remove the caliper mounting bolts (1), the axle nut (2) and washer and the slider cap nuts (3) to detach the front wheel

67 Reverse the removal procedure to install the front wheel. The valve stem must be on the left side of the machine when the wheel is installed.
68 Install the speedometer drive between the right side of the forks and the wheel (left side on 1987 and later models). The tab on the drive must engage with the slot in the brake disc.
69 Apply multi-purpose grease to the axle and insert the axle through the wheel and forks from the right side. Install the wheel spacer on the correct side, between the fork leg and the wheel.
70 Tighten the axle nut to the specified torque before tightening the slider cap nuts. **Note:** *The calipers on later models are secured with locknuts. When the locknuts are removed, they're destroyed and must be replaced with new ones.*

Springer forks

71 Remove the brake caliper and suspend it out of the way.
72 If necessary, remove the speedometer cable.
73 Remove the axle locknut and flat washer. Discard the nut.
74 Carefully slide the axle out of the hub, then separate the wheel from the fork rockers. The washers and spacers will fall out as the axle and wheel are removed – try to note how they're installed.
75 Position the wheel between the rockers with the brake disc on the right side.
76 Insert the axle from the right side until it's just barely through the rocker. **Note:** *The thrust washers are not the same. The one with the large ID goes inside the brake bracket and the one with the small ID goes outside the bracket. The teflon coated side of the washers must be against the brake bracket. If the teflon coating is worn off, install new washers.*
77 Slide the wave washer and the large ID thrust washer onto the spacer, then slide the spacer and washer assembly into the hole in the brake bracket.
78 Position the small ID thrust washer over the axle, then slide the axle through the brake bracket/spacer and into the hub.
79 Continue sliding the axle through the hub while positioning the seal and speedometer drive assembly in the left side. Make sure the speedometer drive unit engages the notch in the hub correctly.
80 Install the washer and a NEW locknut. Tighten the locknut to the torque listed in the Chapter 1 Specifications.

4 Front drum brake – inspection and brake shoe replacement

Refer to illustration 4.8

1 Drum brakes don't usually require frequent maintenance, but they

should be checked periodically to ensure proper operation. If the linkage is properly adjusted, the brake shoes aren't contaminated or worn out and the return springs and cables are in good condition, the brakes should work fine.

2 Check the cable ends to make sure they aren't frayed and check the lever pivot for binding and excessive play. As a general rule, the cable should be adjusted so the brake shoes don't drag when the lever is released and the lever doesn't touch the handlebar when the brake is applied.

3 If the lever doesn't operate smoothly, lubricate the cable, the cable ends and the pivot (refer to Chapter 1). If the brakes still don't operate smoothly, the problem is in the shoe actuating mechanism.

4 If the brake shoe wear check (refer to Chapter 1) indicates the shoes are near the wear limit, refer to Section 3 and remove the front wheel. Measure the thickness of the brake shoe lining and compare it to the Specifications in Chapter 1. If the shoes have worn beyond the allowable limits, or if they're worn unevenly, they must be replaced with new ones.

5 If the linings are acceptable as far as thickness is concerned, check them for glazing, high spots and hard areas. A light touch-up with a file or emery paper will restore them to usable condition. If the linings are extremely glazed, they have probably been dragging. Be sure to properly adjust the lever free play to prevent further glazing.

6 Occasionally the linings may become contaminated with grease from the wheel bearings or brake cam. If this happens, and it's not too severe, cleaning the shoes with a brake system solvent (available at auto parts stores) may restore them. Better yet, replace the shoes with new ones – the cost is minimal.

7 To remove the shoes from the brake plate on FX models, remove the pivot stud bolt, the operating shaft nut and the operating lever. Tap on the operating shaft and lift out the shaft, the pivot stud, the brake shoes and the springs as an assembly.

8 On FL models, grab both brake shoes simultaneously and release them from the brake backing plate. See the accompanying illustration for the proper technique.

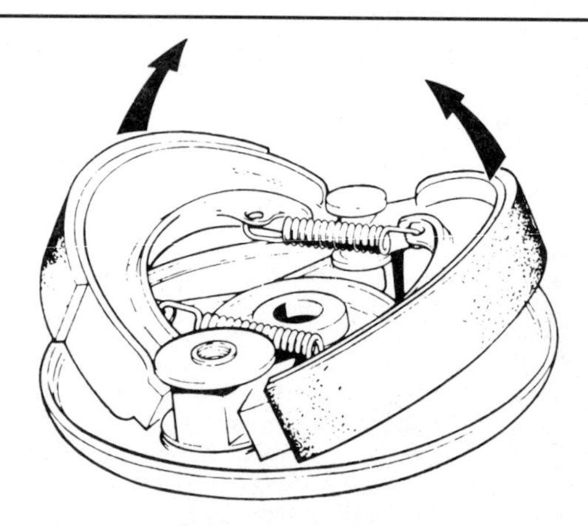

4.8 On FL models, tilt both brake shoes up, without removing the springs, to separate them from the backing plate – reverse the procedure to install them

9 Remove the cotter pin from the end of the cam lever stud and lift off the cam lever washer.

10 Loosen the clamp nut on the cable clevis and depress the brake lever. The cable ferrule can now be disconnected from the anchor pin in the hand lever. When the cable is disconnected, the cam lever can be removed from the cam lever stud.

11 On all models, remove the springs from the shoes and check the springs for cracks and distortion. Replace them with new ones if defects are noted.

12 Clean the brake plate with solvent to remove brake dust and dirt. Also, clean the operating shaft and the pivot stud. If compressed air is available, use it to dry the parts thoroughly.

13 Check the operating shaft and the hole in the brake plate for excessive wear. Slide the operating shaft back into the hole and make sure it turns smoothly without binding. If excessive side play is evident, the brake plate will have to be replaced with a new one. Check the shoe contact areas of the cam for wear also.

14 Before installing the new shoes, file a taper on their leading edges. Install the springs, then apply a thin coat of high-temperature grease to the shoe contact areas of the cam and pivot stud.

15 Clean the brake drum out with a wet rag. Don't use solvent in the brake drum because the rubber seals in the hub will be damaged by it. **Caution:** *Don't blow the brake drum out with compressed air – the brake dust may contain asbestos, which can damage your lungs if inhaled.*

16 Check the drum for rough spots, rust and evidence of excessive wear. If the outer edge of the drum has a pronounced ridge, excessive wear has occurred. Measure the diameter of the drum at several places to determine if it's worn out-of-round. Excessive wear and out-of-roundness indicate the need for a new hub/drum. Slight rough spots and roughness can be removed with fine emery paper. Use one of the brake shoes as a sanding block so low spots aren't created in the drum.

17 On FX models, slip the brake shoe assembly into position in the brake plate and install the pivot stud bolt and washer and the operating lever and shaft nut.

18 Insert the brake shoe assembly into the brake drum and install the front wheel as described in Section 3.

19 On FL models, assemble the brakes in the reverse order of the disassembly procedure. Connect the brake shoes with the top spring only and set the shoes in position on the pivot stud and cam lever. Be sure the spring hooks are in the shoe spacer notch nearest the side cover.

20 Attach the cable ferrule to the anchor pin of the hand lever with the slot in the anchor pin facing in. On earlier models with slotted end anchor pins, the open end of the pin should face down when the cable is installed.

21 Install the front wheel.

22 Check the adjustment of the brake as described in Step 2 of this Section. If the brake needs adjusting, loosen the locknut on the adjusting sleeve, then turn the adjusting sleeve nut until the lever moves freely for about one-quarter of its full movement before the brakes begin to drag. Tighten the locknut against the adjusting sleeve.

23 If the adjustment is correct, but the brakes drag, the brake shoes must be centered in the brake drum. Loosen the pivot stud bolt and the axle nut, then spin the front wheel. While the wheel is spinning, apply the brake and tighten the pivot stud bolt and the axle nut. Recheck the adjustment.

5 Front disc brake – inspection and brake pad replacement

Inspection

Refer to illustration 5.4

1 Carefully examine the master cylinder, the hoses and the caliper unit for evidence of brake fluid leakage. Pay particular attention to the hoses. If they're cracked, abraded, or otherwise damaged, replace them with new ones. If leaks are evident at the master cylinder or caliper, they should be rebuilt by referring to the appropriate Sections in this Chapter.

2 Check the lever for proper operation. It should feel firm and return to its original position when released. If it feels spongy, or if lever travel is excessive, the system may have air trapped in it. Refer to Chapter 1 and bleed the brakes.

3 Check the brake pads for excessive wear by referring to Chapter 1.

4 Examine the brake disc for cracks and score marks. Measure the thickness of the disc. If its worn beyond the allowable limit **(see illustration)**, it must be replaced with a new one.

5 If the brake lever pulsates when the brake is applied during operation of the machine, the disc may be warped. Attach a dial indicator set up to the fork slider and check the disc runout. If the runout is greater than specified, replace the disc with a new one. If a dial indicator isn't available, a dealer service department or motorcycle repair shop can make this check for you.

6 If the brake pads are worn out or contaminated with brake fluid or dirt, they must be replaced with new ones. Failure to replace the pads when

5.4 The minimum brake disc thickness is stamped near one of the mounting bolts

5.7 Remove the screw securing the brake hose to the fender or fork leg

necessary will result in damage to the disc and severe loss of stopping power.

Pad replacement

Caution: *Do not operate the brake lever while the caliper is apart – the piston will be forced out of the caliper. Always replace all pads in the front brake caliper(s); never replace only one pad or the pads in only one caliper on dual disc models.*

1972 through 1983 FL models (except FLT) and 1973 FX models

Refer to illustrations 5.7 and 5.8

7 Remove the clamp securing the brake hose to the fork leg **(see illustration)**.

8 Unscrew the bolts holding the caliper together, then separate the outer half and the damper spring from the rest of the caliper **(see illustration)**.

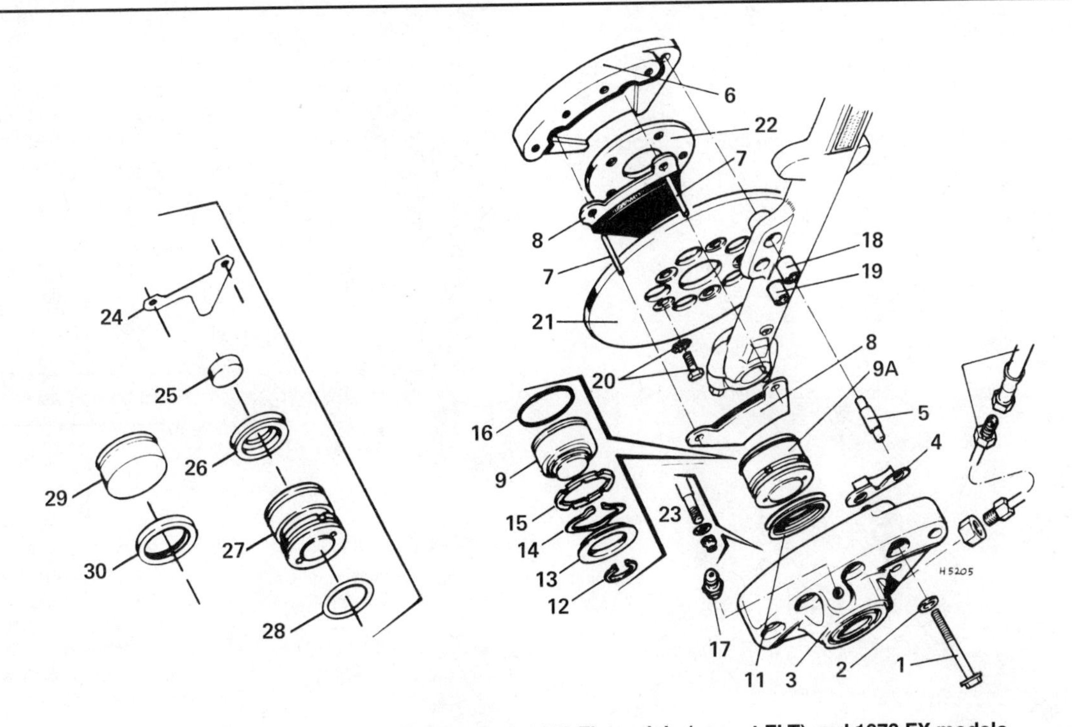

5.8 Front wheel disc brake – 1972 through 1983 FL models (except FLT) and 1973 FX models

1 *Bolt*
2 *Washer*
3 *Outer caliper half*
4 *Damper spring*
5 *Mounting pin*
6 *Inner caliper half*
7 *Brake pad mounting pins*
8 *Brake Pads*
9 *Brake piston*
9A *Brake piston assembly*
10 *Brake line*
11 *Piston boot*
12 *Snap-ring*
13 *Backing plate*
14 *Wave spring*
15 *Friction ring*
16 *O-ring*
17 *Bleeder valve*
18 *Bushing*
19 *Bushing*
20 *Bolt and lock washer*
21 *Brake disc*
22 *Brake disc spacer*
23 *Replacement mounting pin*
24 *Plate (late 1978 through 1983)*
25 *Insulator (late 1978 through 1983)*
26 *Dust boot*
27 *Piston (late 1978 through early 1980)*
28 *O-ring (late 1978 through early 1980)*
29 *Piston (late 1980 through 1983)*
30 *Seal (late 1980 through 1983)*

9 Remove the mounting pin and the inner half of the caliper.

10 Disengage the brake pad mounting pins and detach the brake pads. Late 1978 through 1983 models are equipped with a plate and an insulator between the piston and the brake pad.

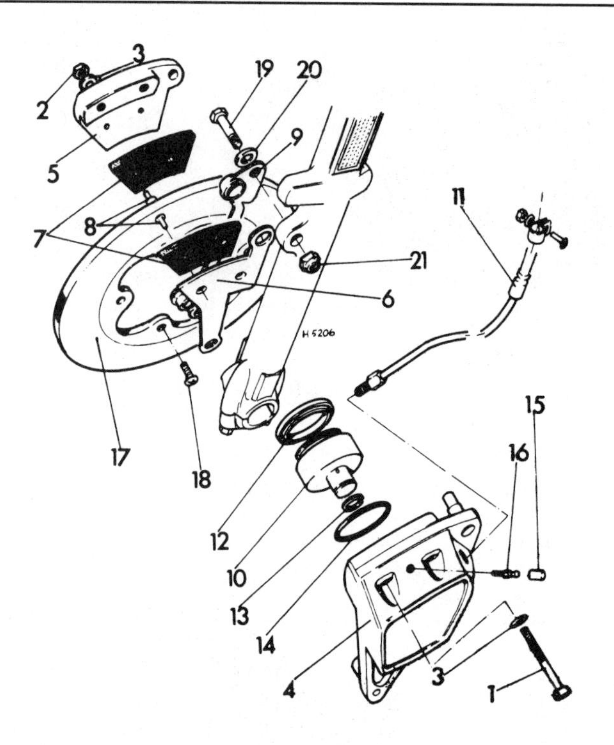

5.11a Front wheel disc brake – 1973 through 1977 FX models

1 Allen-head bolt
2 Locknut
3 Washer
4 Outer caliper half
5 Inner caliper half
6 Backing plate
7 Brake pad
8 Rivet
9 Torque arm
10 Piston
11 Brake hose
12 Rubber boot
13 Friction ring
14 O-ring
15 Bleeder valve cap
16 Bleeder valve
17 Brake disc
18 Mounting bolt
19 Torque arm mounting bolt
20 Washer
21 Lockout

5.11c ... and separate the caliper sections to get at the brake pads on 1972 through 1983 FL models and 1973 FX models

1974 through 1977 FX models

Refer to illustrations 5.11a, 5.11b, 5.11c and 5.12

11 Remove the Allen-head bolts and locknuts holding the caliper together **(see illustration)**. Separate the two caliper halves **(see illustrations)**.

12 Remove the brake pads and check them for wear **(see illustration)**.

1978 through 1983 FX and 1980 through 1983 FLT models

Refer to illustrations 5.13, 5.15a and 5.15b

13 Using a socket, universal joint and extension, loosen the bolt securing the two halves of one of the calipers together **(see illustration)**. You have to work from the back side of the caliper to loosen the bolt.

14 Remove the Allen-head bolts and nuts attaching the caliper to the lower fork leg.

15 Detach the caliper from the forks and separate the two halves. Remove the brake pads from the guide pins **(see illustrations)**.

16 Repeat the procedure for the remaining caliper.

1984 and later models (except FXSTS)

Refer to illustrations 5.17, 5.18 and 5.20

17 Loosen the pad retainer screw at the back (inner) side of the caliper **(see illustration)**.

18 Remove the upper mounting bolt and the lower mounting pin **(see illustration)**. Move the caliper to the rear and down slightly, away from the fork slider, then remove the outer pad, pad holder and spring clip from the caliper as an assembly. Pull out the bushing the upper mounting bolt threads into, then slide the caliper off the brake disc. You may have to cut the plastic tie-wrap holding the wire to the brake hose to produce enough slack in the hose to remove the caliper.

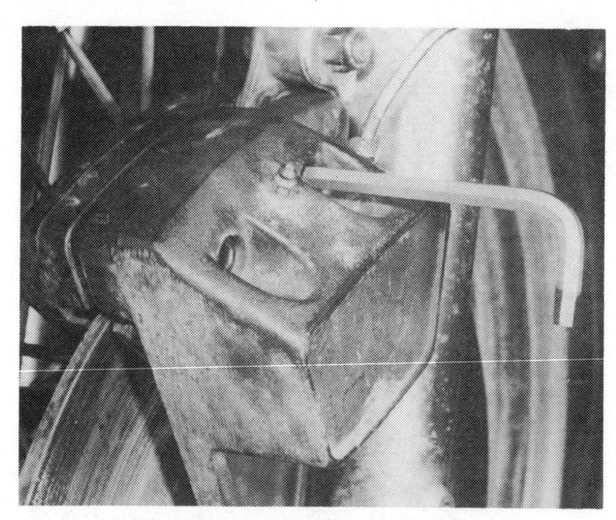

5.11b Remove the Allen-head bolts, ...

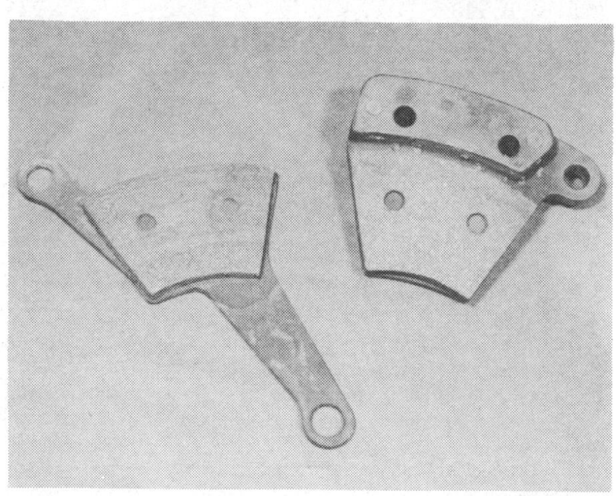

5.12 The amount of lining material remaining on the brake pads can be determined once they're out of the caliper

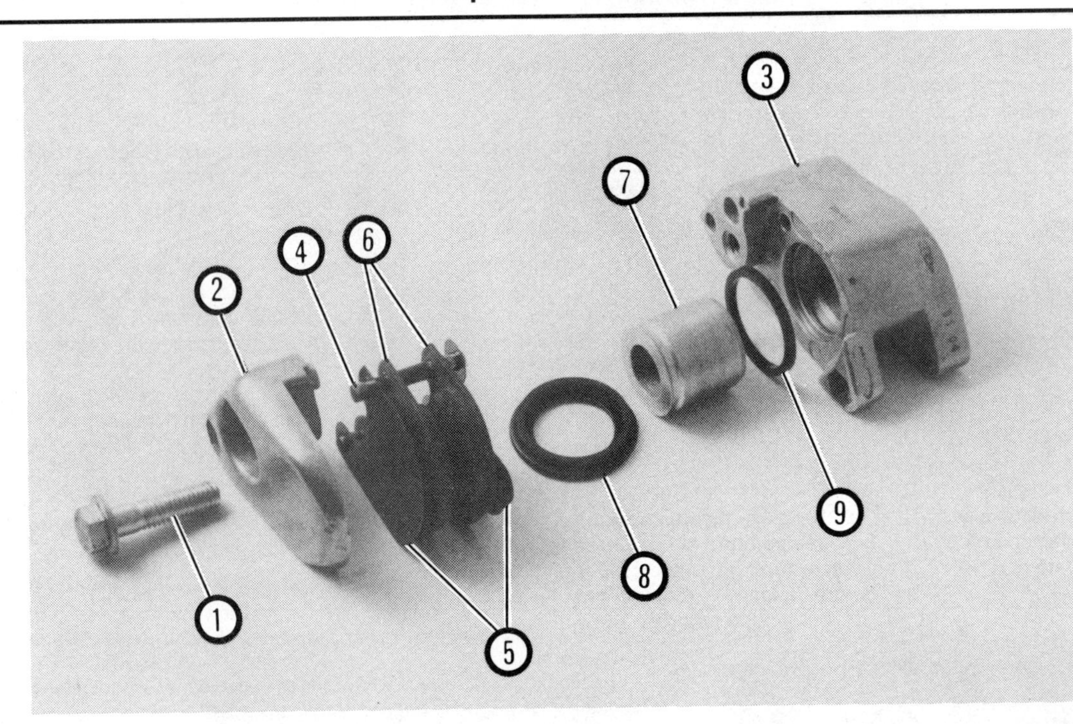

5.13 Front disc brake caliper – 1978 through 1983 FX and 1980 through 1983 FLT models

1 *Bolt (holds caliper sections together)*
2 *Inner caliper half*
3 *Outer caliper half*
4 *Pad guide pin (2)*
5 *Pad shims*
6 *Brake pads*
7 *Piston*
8 *Boot*
9 *Seal*

5.15a Remove the bolt holding the caliper sections together, . . .

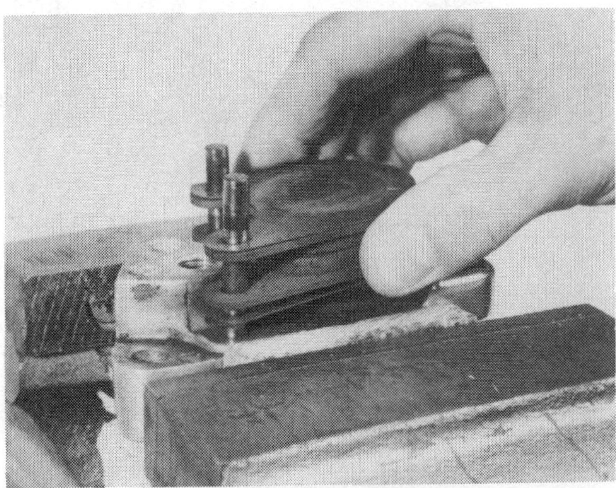

5.15b . . . then lift off the inner caliper half and remove the pads and shims

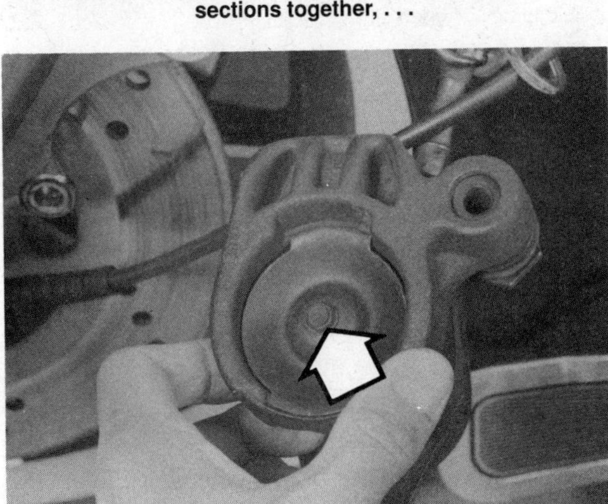

5.17 On all 1984 and later models, loosen the pad retainer screw on the back of the caliper (the caliper has been removed to clearly show the screw location), . . .

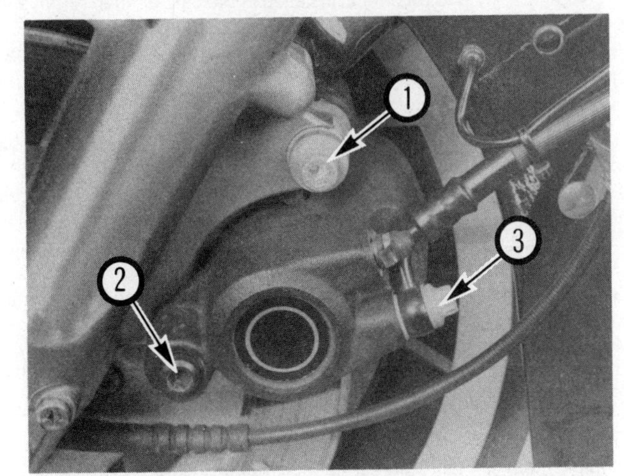

5.18 . . . then remove the caliper mounting bolt and pin

1 *Caliper mounting bolt*
2 *Caliper mounting pin*
3 *Brake hose-to-caliper banjo fitting bolt (12-point head)*

5.20 This is what the outer brake pad and spring clip look like when correctly installed in the pad holder

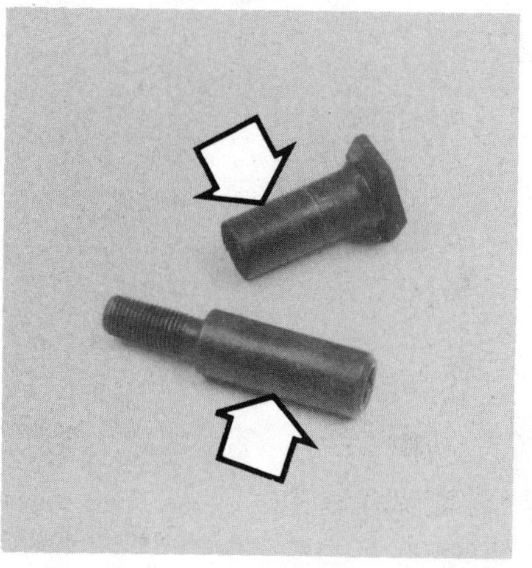

5.28 Clean them thoroughly, then apply high-temperature brake grease to the sliding surfaces (arrows) of the caliper pins/bushings before installation (components for 1984 and later models shown)

5.29 Make sure the master cylinder reservoir has brake fluid in it

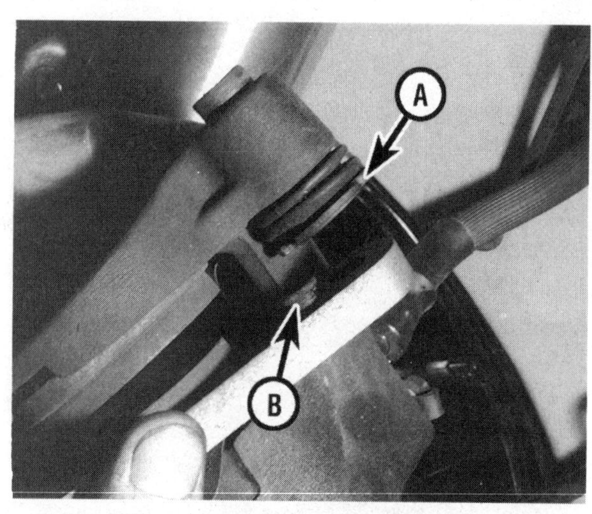

5.32 Caliper upper mounting bolt bushing flanged head (A) and rivet (B)

19 Note how the pad retainer is installed in the caliper, then remove the screw and detach the pad retainer and inner brake pad.

20 Note how the pad and spring clip are positioned in the holder **(see illustration)**. Push the pad out of the clip to remove it from the holder.

FXSTS model

21 Loosen the pad retainer screw at the back (inner) side of the caliper **(see illustration 5.17)**.

22 Remove the caliper upper mounting bolt (with washer) and lower mounting pin and lift the caliper off the disc and away from the brake reaction link bracket. Rocking the caliper gently back and forth will push the piston back into its bore and ease removal.

23 Remove the outer pad, pad holder and spring clip as an assembly. Pull out the upper mounting bolt threaded bushing.

24 Refer to Steps 19 and 20 for the rest of the procedure.

All models

Refer to illustrations 5.28 and 5.29

25 While the caliper assembly is apart, check the movement of the piston in the outer caliper half. Mount a dial indicator on the back of the outer caliper, so the plunger rests on the piston face. Apply the handlebar lever gently until the piston is extended and set the indicator to zero. Release the brake lever. If the piston movement isn't restricted it should be as specified.

26 Clean the disc surface with brake system cleaner, lacquer thinner or acetone. **Warning:** *Do not use petroleum-based solvents.*

27 The piston(s) in the caliper must be pressed into the bore as far as possible when new pads are installed.

28 Don't touch the faces of the brake pads when installing them. Make sure the pads are in the correct position and facing the right direction during reassembly.

29 Reassemble the calipers, with the new pads, in the reverse order of disassembly.

30 On 1974 through 1977 FX models, install new locknuts on the caliper.

31 On 1984 and later models, position the spring clip and the pad with the insulator backing material in the pad holder. The pad lining (friction face) and the spring clip loop must face away from the caliper piston **(see illustration 5.20).**

32 When installing the upper mounting bolt threaded bushing on 1993 and later models, note that its flanged head must locate under the mounting plate rivet so that one of its cutouts slots over the rivet body **(see illustration)**.

33 Lubricate the caliper pins with high-temperature grease **(see illustration)**.

34 Check the brake fluid level in the master cylinder after the pads are installed **(see illustration)**. The level may be too high, requiring the excess fluid to be siphoned off.

6 Front brake disc – removal and installation

1 Remove the front wheel as described in Section 3.

2 Remove the bolts/nuts and separate the disc from the hub. Later models may have Torx-head bolts which require a special tool for removal and installation.

3 Before installing the disc, be sure the threads on the bolts and nuts (or in the hub) are clean and undamaged. Use thread locking compound and be sure to tighten the bolts/nuts in a criss-cross pattern, in several steps, until the specified torque is reached.

7 Front disc brake caliper – removal and installation

1972 through 1983 FL models (except FLT) and 1973 FX models

1 The caliper is held in place by the four bolts that secure the two caliper halves together **(see illustration 5.8).**

2 Remove the clamp securing the brake hose to the fork leg, then remove the four caliper bolts.

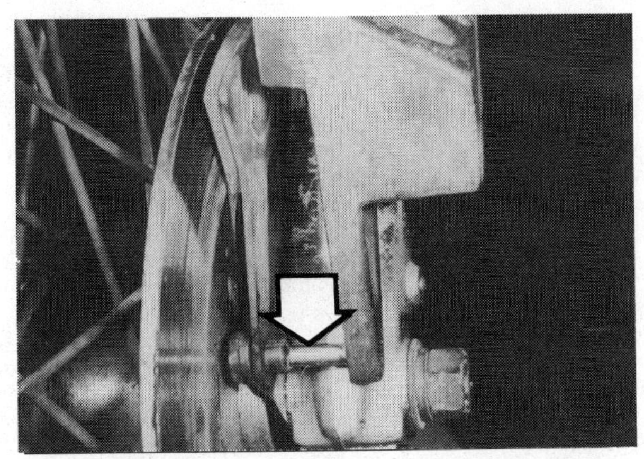

7.4 On 1974 through 1977 FX models, pull the caliper straight out until the pin (arrow) clears the torque arm

3 The caliper can be removed by sliding the mounting pins out of the fork slider bushings after separating the outer caliper half from the inner half (see Section 5).

1974 through 1977 FX models

Refer to illustration 7.4

4 Remove the two Allen-head bolts and locknuts holding the caliper halves together **(see illustration 5.11a)**, then detach the caliper by pulling out on it until the pin is disengaged from the torque arm **(see illustration)**.

1978 through 1983 FX and 1980 through 1983 FLT models

5 Loosen the bolt securing the two caliper halves together **(see illustration 5.13)**. If both calipers are being removed, loosen the bolt in each caliper.

6 Remove the Allen-head bolts and nuts attaching the caliper(s) to the fork leg(s), then detach the caliper(s).

1984-on (all models)

7 Loosen the brake hose-to-caliper banjo fitting bolt, but don't unscrew it **(see illustration 5.18)**. A 12-point socket will be required to fit the bolt head.

8 Remove the upper mounting bolt and the lower caliper pin **(see illustration 5.18)**.

9 Detach the caliper. You may have to cut the plastic tie-wraps securing the wire to the brake hose near the caliper to allow enough movement to detach the caliper.

All models

Caution: *While the caliper is apart or removed from the brake disc, do not operate the front brake lever – it will force the piston out of the caliper bore.*

10 Unscrew the brake line union or banjo fitting bolt at the caliper. Plug the end of the brake line and the opening in the caliper to prevent dirt from entering the hydraulic system.

11 Installation of the caliper is the reverse of the removal procedure. Apply Teflon tape to the threads of the brake line fitting before attaching the line to the caliper. On later models, discard the original washers used at the brake line banjo fitting and install new ones.

12 On 1974 through 1977 FX models, be sure the locating pin at the bottom of the outer caliper engages the backing plate for the brake pad as well as the torque arm **(see illustration 7.4)**.

13 Coat the caliper pins (that allow the caliper to move back-and-forth) with high-temperature grease and tighten the mounting bolts to the specified torque.

14 After the calipers are installed, the brake system must be bled as described in Chapter 1.

8.16 On 1978 through 1983 FX and FLT models, the piston can be pried out of the caliper bore (very carefully) with two screwdrivers

8 Front disc brake caliper – overhaul

1 The caliper should not be disassembled unless it leaks fluid around the piston or doesn't operate properly. If the piston travel (Section 5) is not as specified, the piston will have to be removed and inspected. Before disassembling the caliper, read through the entire procedure and make sure you have the correct caliper rebuild kit. Also, you'll need some new, clean brake fluid of the recommended type and some clean rags. **Note:** *Disassembly, overhaul and reassembly of the brake caliper must be done in a spotlessly clean work area to avoid contamination and possible failure of the brake hydraulic system components. If such a work area isn't available, have the caliper rebuilt by a dealer service department or a motorcycle repair shop.*

1972 through 1983 FL models (except FLT) and 1973 FX models

2 Remove the caliper as described in Section 7 but DO NOT disconnect the hydraulic brake line.

3 Remove the brake pads as described in Section 5.

4 Slowly pump the brake lever until the piston doesn't move any further.

5 Disconnect the brake line from the caliper and plug the line and the opening in the caliper.

6 Pull the piston boot away from the groove in the piston, then remove the piston from the bore in the caliper.

7 Remove the snap-ring from the piston (if used), then lift off the backing plate, the wave spring (1974 and earlier models), the friction ring and the O-ring. On late 1978 through 1983 models, remove the O-ring or seal from the piston.

8 Unscrew the bleeder valve from the caliper body.

1974 through 1977 FX models

9 Remove the caliper and the brake pads as described in Sections 5 and 7.

10 Disconnect the brake line from the caliper and plug it.

11 Remove the rubber boot, then, using two screwdrivers, pry the piston out of the caliper bore.

12 Check the friction ring at the end of the piston. If it's damaged, remove it and install a new one.

13 Carefully pry the O-ring out of the caliper bore with a wood or plastic tool.

14 Unscrew the bleeder valve from the caliper.

1978 through 1983 FX and 1980 through 1983 FLT models

Refer to illustration 8.16

15 Remove the caliper and brake pads as described in Sections 5 and 7. Disconnect the brake line from the caliper and plug it.

16 Carefully pry the piston out of the caliper, then remove the boot **(see illustration)**. If the piston cannot be pried out, place the caliper face down

on a clean work surface. Position a clean towel under the piston and apply low pressure air to the inlet hole to force the piston out.
17 Carefully remove the seal from the caliper bore **(see illustration 5.13)**. Use a wood or plastic tool to remove the seal (to avoid scratching the caliper bore).

1984 and later models

18 Remove the caliper and brake pads as described in Sections 5 and 7. Disconnect the brake line from the caliper and discard the washers.
19 Pry out the boot retainer by inserting a small screwdriver into the notch at the bottom of the piston bore.
20 Note how it's installed, then remove the rubber piston boot.
21 Place the caliper face down on a clean work surface. Position a clean towel under the piston and apply low pressure air to the inlet hole to force the piston out.
22 Remove the seal from the caliper bore with a wood or plastic tool.
23 Note how they're installed, then remove the threaded bushing from the caliper and pull the pin boot out.
24 Remove the O-rings from the mounting bolt/pin holes.

All models

25 Clean all the brake components (except for the brake pads) with brake system solvent (available at auto parts stores), isopropyl alcohol or clean brake fluid. **Caution:** *Do not, under any circumstances, use petroleum-based solvents to clean brake parts. If compressed air is available, use it to dry the parts thoroughly.*
26 Check the caliper bore and the outside of the piston for scratches, nicks and score marks. If damage is evident, the caliper must be replaced with a new one.
27 Reassembly of the caliper is done in the reverse order of disassembly. Be sure to lubricate all of the components with clean brake fluid during reassembly and install new O-rings.
28 Install the brake pads as described in Section 5 and connect the brake line to the caliper (see Section 7).
29 Attach the caliper assembly to the forks and bleed the system as described in Chapter 1.

9 Front disc brake master cylinder – removal, overhaul and installation

1 If the master cylinder is leaking fluid, or if the lever doesn't feel firm when the brake is applied, and bleeding the brakes doesn't help, master cylinder overhaul is recommended. Before disassembling the master cylinder, read through the entire procedure and make sure you have the correct rebuild kit. Also, you'll need some new, clean brake fluid of the recommended type, some clean rags and internal snap-ring pliers. **Note:** *Disassembly, overhaul, and reassembly of the brake master cylinder must be done in a spotlessly clean work area to avoid contamination and possible failure of the brake hydraulic system components. If such a work area isn't available, have the master cylinder rebuilt by a dealer service department or motorcycle repair shop.*

1972 through 1981 models

Refer to illustrations 9.3 and 9.4

2 Turn the handlebars so the brake master cylinder is as level as possible and remove the cover and gasket from the master cylinder. Disconnect the hydraulic brake line from the master cylinder and catch the fluid in a container.
3 Remove the handlebar switch assembly and disconnect the brake light wires **(see illustration)**.
4 Remove the snap-ring and pivot pin so the handlebar brake lever can be removed. It'll pull out with the brake lever pin, plunger, spring, two washers and the dust wiper **(see illustration)**.
5 Remove the snap-ring from the master cylinder housing bore.
6 Pull out the piston, O-ring, piston cup, spring cup and piston return spring. Be careful when removing the piston so the bore of the master cylinder and the piston don't get scratched. If they do, the entire assembly must be replaced with a new one.

1982 and later models

7 Open the bleeder valve on one of the front calipers and attach a piece of hose to the valve. Place the other end of the hose in a clean container and slowly pump the handlebar brake lever to drain the brake fluid.
8 Remove the bolt attaching the brake line to the master cylinder. Throw away the washers on either side of the brake line.
9 Remove the cover and gasket from the master cylinder.
10 Remove the snap-ring securing the pivot pin, then lift out the pivot pin. Separate the brake lever and the reaction pin from the master cylinder.
11 Remove the master cylinder clamp and detach the assembly from the handlebars.
12 Pull the pushrod and switch out, followed by the dust boot, piston and O-ring, back-up disc (not used on 1984 and later models), cup, stop and spring.
13 Remove the sight glass and grommet from the side of the master cylinder.

All models

14 Clean all of the parts with brake cleaning solvent (available at auto parts stores), isopropyl alcohol or clean brake fluid. **Warning:** *Do not, under any circumstances, use petroleum-based solvent to clean brake parts.* If compressed air is available, use it to dry the parts thoroughly. Check the master cylinder bore for scratches, nicks and score marks. If damage is evident, the master cylinder must be replaced with a new one. Be sure the vent holes are open.
15 Before reassembling the master cylinder, soak the new rubber seals in clean brake fluid for 10 or 15 minutes. Lubricate the master cylinder bore with clean brake fluid, then carefully insert the piston and related parts in the reverse order of disassembly.
16 On 1982 and later models, install the sight glass and grommet in the side of the master cylinder.
17 Apply a light coat of anti-seize compound to the reaction pin and insert it into the large hole in the brake lever. Connect the lever to the master cylinder, aligning the plunger (pushrod on 1982 and later models) with the hole in the pin.
18 Install the pivot pin and secure it with the snap-ring.
19 On 1973 through 1981 models, attach the brake light wires and assemble the handlebar switch. On later models, clamp the master cylinder to the handlebars.
20 Make sure the relief port in the master cylinder is uncovered when the brake lever is released.
21 Connect the brake line to the master cylinder (using new washers on 1982 and later models).
22 Fill the master cylinder with the recommended brake fluid and bleed the system as described in Chapter 1.

10 Disc brakes – bleeding

Refer to Chapter 1, *Tune-up and routine maintenance,* for the disc brake bleeding procedure.

9.3 On 1972 through 1981 models, detach the switch assembly (arrow) from the master cylinder and disconnect the brake light wires

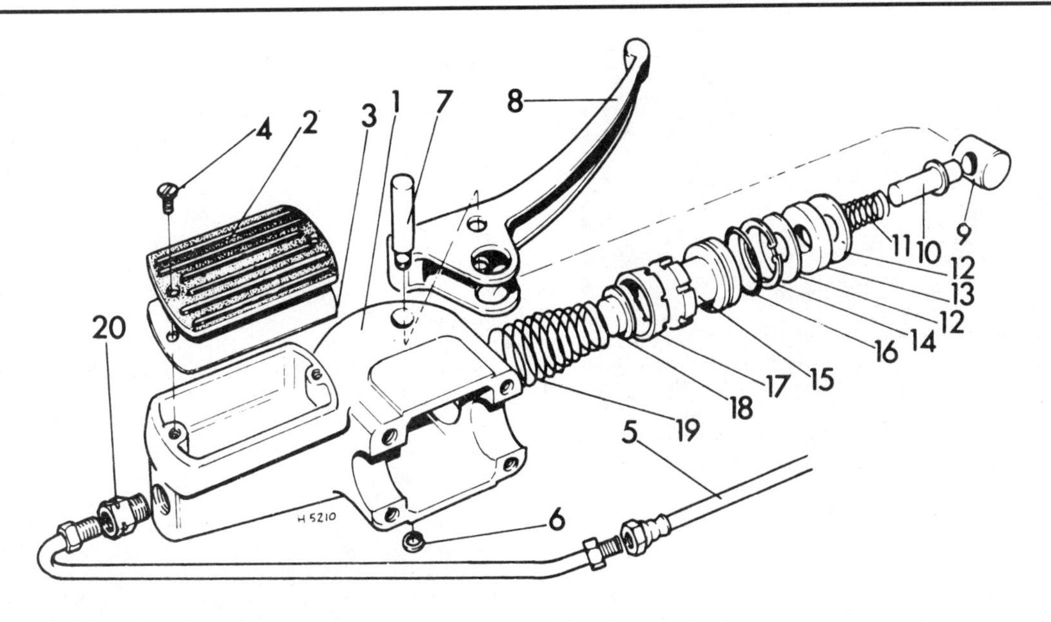

9.4 Front disc brake master cylinder components – 1972 through 1981 models

1	*Master cylinder housing*	5	*Brake line*	9	*Brake lever pin*	13	*Dust wiper*	17	*Piston cup*
2	*Master cylinder cover*	6	*Pivot pin snap-ring*	10	*Plunger*	14	*Snap-ring*	18	*Spring cup*
3	*Gasket*	7	*Pivot pin*	11	*Spring*	15	*Piston*	19	*Spring*
4	*Cover screw*	8	*Brake lever*	12	*Washer*	16	*O-ring*	20	*Union fitting*

11.9 On 1973 through early 1978 models, withdraw the axle from the right side of the rear wheel

11.10a A spacer (arrow) is used between the sprocket and swingarm on 1973 through early 1978 models

11 Rear wheel – removal and installation

Caution: *On disc brake equipped models, do not operate the brake lever while the front wheel is removed – the piston in the caliper might be forced out. If the piston is forced out of the bore, the caliper will have to be completely disassembled and rebuilt. Also, after the wheel is reinstalled, the wheel bearing end play must be checked with a dial indicator. If it falls outside the range listed in this Chapter's Specifications, the long spacer in the wheel hub or thin spacer washer (as applicable) will have to be replaced by another of different length – see Chapter 1.*

Removal

1 Raise the rear of the motorcycle and support is securely on blocks so the tire is at least four inches off the ground (it must be stable from side-to-side so it won't tip over).

1970 through 1972 models

2 Remove the two screws from the rear fender support and lift the end of the fender so the wheel will clear it.

3 Five socket head (Allen) screws secure the brake drum to the rear wheel hub. Remove the screws one at a time (they can only be removed at the rear of the axle). Remove the nut and lock washer from the end of the axle, then tap the axle out from the drum side.

4 Remove the spacer between the wheel hub and the axle clip on the right side.

5 Apply pressure to the rear brake pedal and remove the rear wheel.

1973 through early 1978 models

Refer to illustrations 11.9, 11.10a and 11.10b

6 Rotate the wheel until the master link on the drive chain is positioned on the rear wheel sprocket. Detach the master link and disengage the chain from the sprocket. Don't separate the chain from the front sprocket.

7 Remove the cotter pin securing the brake anchor nut and remove the nut.

8 Unscrew the axle nut and remove the washers from the axle.

9 Note the position of the spacer between the sprocket and the swingarm before removing the axle. Tap the axle out with a soft-face hammer from the right side **(see illustration)**.

10 With the axle removed, the wheel is free to come out the rear **(see illustrations)**.

11.10b Let the wheel drop down, then roll it out to separate it from the swingarm

Late 1978 and later models (except FLT with enclosed chain)

11 Remove the saddlebags, if applicable. On some later models the belt/chain guard must be removed as well.
12 Remove the cotter pin (if used) and discard it, then unscrew the axle nut and remove the washer(s). **Note:** *On 1990 FLSTF models and all other models from 1991 and later, the axle nut is installed on the left side for proper clearance during operation of the motorcycle. When reinstalling the axle, make sure the nut is on the left side.*
13 Pull the axle out from the left side and allow the wheel to drop down. You may have to tap the axle out with a soft-face hammer.
14 Note the position of the spacer (debris deflector on some models) between the swingarm and the sprocket, then remove the spacer.
15 Detach the belt or chain from the sprocket on the rear wheel.
16 Separate the wheel from the swingarm.

Late 1978 and later FLT models with enclosed chain

17 Remove the mufflers to gain access to the rear wheel mounting components.
18 Remove both saddlebags.
19 Remove the rubber plug from the housing to get at the rear sprocket mounting screws. Insert a 3/8-inch Allen wrench through the plug opening and remove the sprocket mounting screws. Turn the wheel after each screw is removed until the next screw is accessible.
20 Support the swingarm and remove the lower shock absorber mounting bolts. Lower the rear wheel to the ground.
21 Remove the bolts from the caliper mounting bracket and anchor.
22 Unscrew the axle nut and remove the washers from the axle.
23 Tap the axle out with a soft-face hammer far enough to clear the wheel but still support the sprocket and chain housing. Slide the brake caliper up and swing it away from the wheel.
24 Support the swingarm, then separate the wheel from the sprocket and chain housing and roll it out of the swingarm.

Installation

1970 through 1972 models

25 Reverse the removal procedure to install the rear wheel. Be sure the mating surfaces of the brake drum and wheel hub are clean, then install the mounting screws. Tighten the screws to the torque listed in this Chapter's specifications, following a criss-cross pattern.
26 Tighten the axle nut to the specified torque.

1973 through early 1978 models

27 Install the rear wheel by reversing the removal procedure. Tighten the axle nut and the brake anchor nut to the specified torque.
28 Adjust the chain as described in Chapter 1.

Late 1978 and later models (except FLT)

29 Set the rear wheel in position in the swingarm. Connect the belt or chain to the sprocket on the rear wheel.
30 Lift the wheel until the axle hole is aligned with the holes in the swingarm. Place the spacer in position on the left side of the swingarm.
31 Apply a coat of multi-purpose grease to the axle and insert it through the swingarm, spacer and rear wheel hub.
32 Install the washer and a new lock washer on the axle and thread the nut on.
33 Adjust the tension of the belt or drive chain, as described in Chapter 1, and tighten the axle nut to the specified torque. Install a new cotter pin.
34 Check the end play of the rear wheel bearings and compare it to the Specifications. If the end play is incorrect, different size spacers can be installed to change it.

Late 1978 and later FLT models

35 Position the rear wheel in the swingarm and engage it in the chain housing and sprocket. Reposition the brake caliper.
36 Apply a thin coat of multi-purpose grease to the axle, then insert it through the swingarm and wheel from the left side (right side from 1991 and later). You may have to tap it into position with a soft-face hammer.
37 Install the washer, a new lock washer and the nut on the end of the axle. Tighten the axle nut to the specified torque.
38 Install the two brake caliper mounting bolts and the anchor bolt. If it was disconnected, reattach the brake line to the front mounting bolt.
39 Raise the swingarm and reinstall the lower shock absorber mounting bolts.
40 Apply clean engine oil to the sprocket mounting screws and carefully install them. When all of the sprocket mounting screws are in place, tighten them to the specified torque in a criss-cross pattern.
41 Check the rear wheel bearing end play and compare it to the Specifications. If the end play doesn't fall within the specified limits, the bearing assembly will have to be replaced (refer to Chapter 1).
42 Adjust the chain/belt as described in Chapter 1.
43 Install the mufflers and saddlebags.

12 Rear drum brake – inspection and brake shoe replacement

Warning: *The dust created as the brake shoes wear may contain asbestos, which is harmful to your health. Never blow it out with compressed air and don't inhale any of it. An approved filtering mask should be worn when working on the brakes. Do not, under any circumstances, use petroleum-based solvents to clean brake parts. Use brake cleaner or denatured alcohol only!*

1 Drum brakes don't normally require frequent maintenance, but they should be checked periodically to ensure proper operation. If the hydraulic system is free of air, if the brake shoes aren't contaminated or worn out and if the return springs are in good condition, the brakes should work fine.
2 If the brake shoe wear check (Chapter 1) indicates the shoes are near the wear limit, refer to Section 11 and remove the rear wheel. Measure the thickness of the brake shoe lining and compare it to the Specifications. If the shoes have worn beyond the allowable limits, or if they're worn unevenly, they must be replaced with new ones.
3 If the linings are acceptable as far as thickness is concerned, check them for glazing, high spots and hard areas. A light touch-up with a file or emery paper will restore them to usable condition. If the linings are extremely glazed, they have probably been dragging. Be sure to properly adjust the pedal free play to prevent further glazing.
4 Occasionally the linings may become contaminated with grease from the wheel bearing or brake cam. If this happens, and it's not too severe, cleaning the shoes with brake system solvent (available at auto parts stores) may restore them. However, the best approach is to replace the shoes with new ones.
5 Disconnect the upper brake shoe return spring and lift the brake shoes and lower return spring off the backing plate.

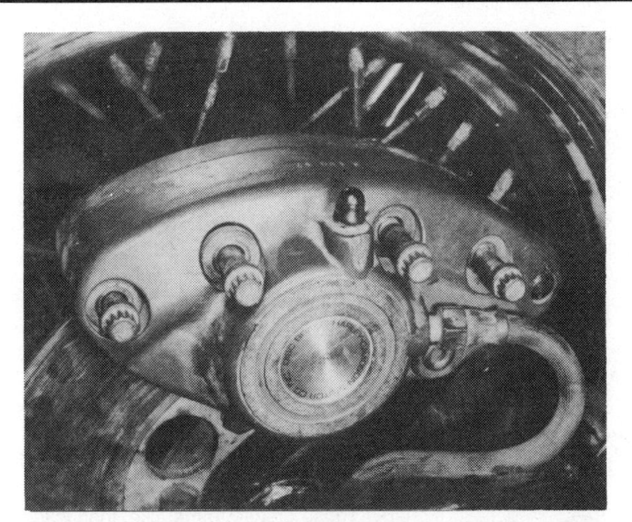

13.8a On 1972 through 1980 models (except FLT and FXR) the caliper halves are held together with four bolts – use a 12-point socket to remove them, . . .

13.8b . . . then detach the outer section of the caliper from the bracket

6 If you have to detach the wheel cylinder, remove the two mounting screws located on the outside of the backing plate.
7 Remove the lower spring from the brake shoes and check them for cracks and distortion. Replace the springs with new ones if defects are noted.
8 Clean the backing plate to remove brake dust and dirt. Also, clean the operating shaft and the pivot stud. If compressed air is available, use it to dry the parts thoroughly.
9 Examine the backing plate and the wheel cylinder for signs of leaking fluid. If leaks are evident, determine if the wheel cylinder or the brake line-to-wheel cylinder connection is leaking.
10 Clean the brake drum out with a wet rag.
11 Check the drums for rough spots, rust and excessive wear (if the outer edge of the drum has a pronounced ridge, excessive wear has occurred). Measure the diameter of the drum at several places to determine if it's worn out-of-round. Excessive wear and out-of-roundness indicate the need for a new hub/drum. Slight rough spots can be removed with fine emery paper. Use one of the brake shoes as a sanding block so low spots aren't created in the drum.
12 If the wheel cylinder is leaking or sticking, a repair kit or new wheel cylinder should be installed. Remove the wheel cylinder and mount it in a vise with soft jaws (use blocks of wood if necessary).
13 Remove the outer boots from the wheel cylinder.
14 Remove the piston, cups and piston return spring from the wheel cylinder.
15 Inspect the bore for score marks, corrosion and other damage. Small marks and corrosion can be removed with a brake cylinder hone. However, if the bore condition is questionable, replace the wheel cylinder with a new one.
16 Clean the wheel cylinder with brake solvent.
17 Coat all of the parts in the repair kit with brake assembly lubricant or clean brake fluid.
18 Install the return spring between the two cups. Make sure the cups are installed wide-side in. Insert a piston into each end of the bore with the concave side facing out. Install a boot over each end of the wheel cylinder.
19 Install the wheel cylinder on the backing plate. Carefully attach the hydraulic brake line.
20 Assemble the brake shoes and springs in the backing plate in the reverse order of disassembly. Apply a light coat of grease to the backing plate, where the shoes make contact, and to the hold-down springs. The short hook on the spring must be inserted into the elongated hole in the front brake shoe.
21 Install the rear wheel. If the wheel cylinder was disconnected, bleed the brakes as described in Chapter 1.

13 Rear disc brake – inspection and brake pad replacement

Inspection

1 Carefully examine the master cylinder, the hoses and the caliper for evidence of brake fluid leakage. Pay particular attention to the hoses. If they're cracked, abraded, or otherwise damaged, replace them with new ones. If leaks are evident at the master cylinder or caliper, they should be rebuilt by referring to the appropriate Sections in this Chapter.
2 Check the pedal for proper operation. It should feel firm and return to its original position when released. If it feels spongy, or if pedal travel is excessive, the system may have air trapped in it. Refer to Chapter 1 and bleed the brakes.
3 Check the brake pads for excessive wear by referring to Chapter 1.
4 Examine the brake disc for cracks and score marks. Measure the thickness of the disc. If it has worn beyond the limit stamped on the side of the disc, it must be replaced with a new one.
5 If the brake pedal pulsates when the brake is applied during operation of the machine, the disc may be warped. Attach a dial indicator to the swingarm and check the disc runout. If the runout is greater than specified, replace the disc with a new one. If a dial indicator isn't available, a dealer service department or motorcycle repair shop can make this check for you.
6 If the brake pads are worn out or contaminated with brake fluid or dirt, they must be replaced with new ones. Failure to replace the pads when necessary will result in damage to the disc and severe loss of stopping power.

Pad removal

1972 through 1980 models (except FLT and FXR)

Refer to illustrations 13.8a, 13.8b and 13.9

7 Remove the clamp securing the brake line to the swingarm.
8 **Note:** *Don't operate the brake pedal while the caliper is apart – the piston will be forced out of the caliper.* Unscrew the bolts holding the caliper together, then separate the outer half and the damper spring from the rest of the assembly **(see illustrations)**.

6

13.9 The inner caliper section can then be removed

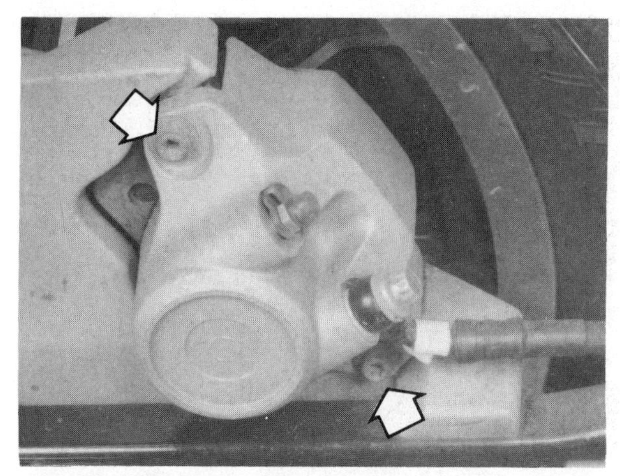

13.21 On 1986 and later FLT models and all 1987 and later models, the rear brake caliper is attached to the bracket and slides on two mounting (pin) bolts

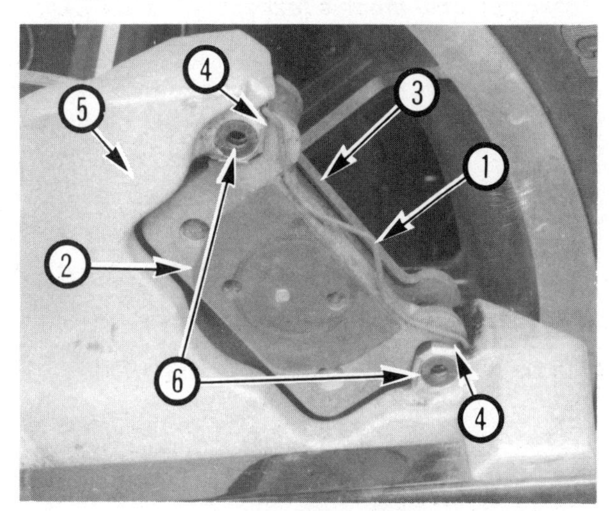

13.22 Brake pads and related components – 1986 and later FLT and all 1987 and later models

1 *Wire retainer clip*
2 *Outer brake pad*
3 *Inner brake pad*
4 *Pad shim (2)*
5 *Bracket*
6 *Rubber bushing (2)*

9 Remove the mounting pin and the inner half of the caliper **(see illustration)**.

10 Disengage the brake pad mounting pins and remove the brake pads. Late 1978 through 1980 models are equipped with a plate and insulator between the piston and the brake pad.

1980 through 1985 FLT and 1981 through 1984 FL models (except 1983 and 1984 FLHS)

11 Unscrew the bolts from the pins, then remove the pins, washers, spring (wave) washers, seals and bolts. **Note:** *Don't operate the brake pedal while the caliper is apart – the piston will be forced out of the caliper.*

12 Detach the caliper and remove the brake pads, plates and springs (springs are used only on FLT models).

1980 through early 1987 FXR, 1983 through early 1987 FX (except FXE) and 1983 through early 1987 FLHS models

13 **Note:** *Don't operate the brake pedal while the caliper is apart – the piston will be forced out of the caliper.* Remove the screws securing the caliper to the mounting bracket.

14 Detach the caliper and remove the brake pads.

15 Remove the upper and lower pins from the bracket, along with the boots, and detach the pad spring from the body of the caliper.

1981 and 1982 FX and 1983 FXE models

16 **Note:** *Don't operate the brake pedal while the caliper is apart – the piston will be forced out of the caliper.* Remove the bolts holding the caliper halves together.

17 Separate the caliper halves from the mounting bracket and remove the damper and sleeve from the outer half of the caliper.

18 Remove the brake pads, guide pins, plates, insulator and pad retaining spring clip. Note the arrangement of the components to simplify reassembly.

1986 and later FLT, late 1987 and later (all other models)

Refer to illustrations 13.21 and 13.22

19 On FLT models, detach the right saddlebag and side cover, then on all except 1993 and later models remove the battery and its carrier.

20 On FX/Softail models, remove the muffler to make room for brake pad removal.

21 Remove the caliper mounting (pin) bolts **(see illustration)**, then carefully lift the caliper up and off the brake disc and pads. **Note:** *The caliper piston may catch on the rear face of the pad and prevent removal; rock the caliper to push the piston back in its bore slightly.*

22 Carefully note how it's installed, then remove the wire retainer clip from the back side of the bracket **(see illustration)**. Slide the outer brake pad off the bracket.

23 Slide the inner pad off toward the wheel.

24 Detach the pad shims from the bracket.

All models

25 While the caliper assembly is apart, check the movement of the piston(s) in the caliper. Mount a dial indicator on the back of the outer caliper, so the plunger rests on the piston face. Gently apply the brake pedal until the piston is extended, then zero the indicator. Release the brake pedal. If the piston movement isn't restricted it should retract from 0.033 to 0.038-inch.

Pad installation

26 Clean the disc surface with brake system cleaner, lacquer thinner or acetone. Do not use petroleum-based solvents.

27 The piston(s) must be pressed into the caliper bore(s) as far as possible when new pads are installed.

28 Don't touch the faces of the brake pads when installing them. Make sure the pads are in the correct position and facing the right direction during reassembly.

29 Reassemble the calipers, with the new pads, in the reverse order of disassembly.

30 On late 1978 through 1980 models (except FLT and FXR), install the plate between the outer brake pad and the piston so the wheel will rotate into the notch.

31 Coat the screws with multi-purpose grease and secure the caliper halves to the mounting bracket. Tighten the screws to the specified torque.

13.41 Position pad shims (arrow) with looped ends outwards – mid 1991 and later

1981 and 1982 FX and 1983 FXE models

32 Set the insulator in position, then attach the brake pad pins to the outer half of the caliper. Assemble the plate and brake pads in the outer half of the caliper.
33 Attach the spring, with the arc towards the top and the prongs facing the right side of the caliper, on top of the brake pads.
34 Place the damper and sleeve on the mounting bracket, then set the left caliper half in position on the mounting bracket.
35 Apply a coat of multi-purpose grease to the screws that connect the caliper halves and tighten them to the specified torque.

1980 through early 1987 FXR, 1983 through early 1987 FX (except FXE) and 1983 through early 1987 FLHS models

36 Attach the pad spring to the top of the caliper with the long tab above the piston. The short tab must be hooked above the ridge on the caliper casting, opposite the piston, to hold the spring in place.
37 Position the pads on the bracket and place the caliper on the bracket without turning the pins. The flat sides of the pin heads should be parallel with the opening in the bracket.
38 Install the caliper mounting bolts and tighten them to the specified torque.

1980 through 1985 FLT and 1981 through 1984 FL models (except 1983 and 1984 FLHS)

39 Position the caliper, plates and brake pads on the mounting bracket.
40 Install the O-ring seals, followed by the pins, spring (wave) washers, washers and mounting bolts. Coat the pins with multi-purpose grease before installing them.

1986 and later FLT, late 1987 and later (all other models)

Caution: *Ensure that only the correct parts are fitted when replacing pads, discs or shims. Modifications to the pad and disc material from 1992 and later and redesign of the shims and pad shape from mid-1991 prevent the mixing of early and late components.*
41 On all models through mid-1991 position the pad shims on the mounting bracket with the tabs seated in the holes. From mid-1991 fit the shims so that their looped ends are positioned outwards (towards the piston) and hold them in place while the pads and wire retainer are fitted **(see illustration)**.
42 Slide the inner pad onto the shims from the wheel side. Slide the outer pad on from the outside.
43 Insert the wire retainer clip ends into the mounting bracket holes and position the clip over the outer brake pad **(see illustration 13.22)**. **Caution:** *Make sure the pads are still riding on the shims after the retainer clip is installed.*
44 Make sure the mounting (pin) bolts are clean so the caliper can move freely. Carefully lower the caliper over the pads and disc and align the holes, then install the mounting bolts.
45 Tighten the mounting bolts to the torque listed in the Chapter 1 Specifications

All models

46 Check the brake fluid in the master cylinder after the pads are installed. The level may be too high, requiring the excess fluid to be siphoned off.

14 Rear disc brake caliper – removal, overhaul and installation

1 If the caliper is leaking fluid around the piston, it should be removed and overhauled to restore braking performance. Before disassembling the caliper, read through the entire procedure and make sure you have the correct caliper rebuild kit. Also, you'll need some new, clean brake fluid of the recommended type and some clean rags. **Note:** *Disassembly, overhaul and reassembly of the brake caliper must be done in a spotlessly clean work area to avoid contamination and possible failure of the brake hydraulic system components. If such a work area isn't available, have the caliper rebuilt by a dealer service department or a motorcycle repair shop.*
2 Remove the caliper as described in Section 13.

1972 through 1980 models (except FLT and FXR)

3 Remove the brake pads.
4 Slowly pump the brake lever until the piston doesn't move any further.
5 Disconnect the brake line from the caliper and plug the line and the fitting in the caliper.
6 Pull the piston boot away from the groove in the piston, then remove the piston from the bore in the caliper.
7 On 1972 through early 1978 models, remove the snap-ring (if used) from the piston, then lift off the backing plate, the wave spring (1972 through 1974 models), the adjusting ring and the O-ring. On late 1978 through 1980 models, remove the O-ring or seal from the piston with a wood or plastic tool.
8 Unscrew the bleeder valve from the caliper body.

1980 through 1985 FLT and 1981 through 1984 FL models (except 1983 and 1984 FLHS)

9 Remove the brake pads as described in Section 13.
10 Slowly pump the brake pedal until the pistons don't move any further. Disconnect the brake line and plug the line and the fitting in the caliper. Don't lose the brake hose seat.
11 Remove the pistons, dust boots and seals from the caliper. You may have to apply low pressure air to the caliper inlet to dislodge the pistons. If so, position the caliper so the pistons are facing down and place a clean towel under them. Be careful because the pistons may come out with considerable force. Use a wood or plastic tool to remove the seals.

1980 through early 1987 FX, FXR and FXE and 1983 through early 1987 FLHS models

12 Remove the brake pads from the caliper.
13 Detach the retaining wire and remove the rubber boot from the caliper.
14 Slowly pump the brake pedal until the piston doesn't move any further.
15 Disconnect the brake line from the caliper and plug the line and the fitting in the caliper.
16 Remove the piston and seal from the caliper. You may have to force the piston out of the caliper with air pressure. Place the caliper on a clean workbench with the piston facing down. Place a clean towel under the piston and apply low pressure air to the caliper inlet. Be careful – the piston may come out with considerable force. Use a wood or plastic tool to remove the seal.

1986 and later FLT, late 1987 and later (all other models)

17 Detach the retaining wire and remove the rubber boot from the caliper.
18 Slowly pump the brake pedal until the piston doesn't move any further.
19 Disconnect the brake line from the caliper and plug the line and the opening in the caliper.

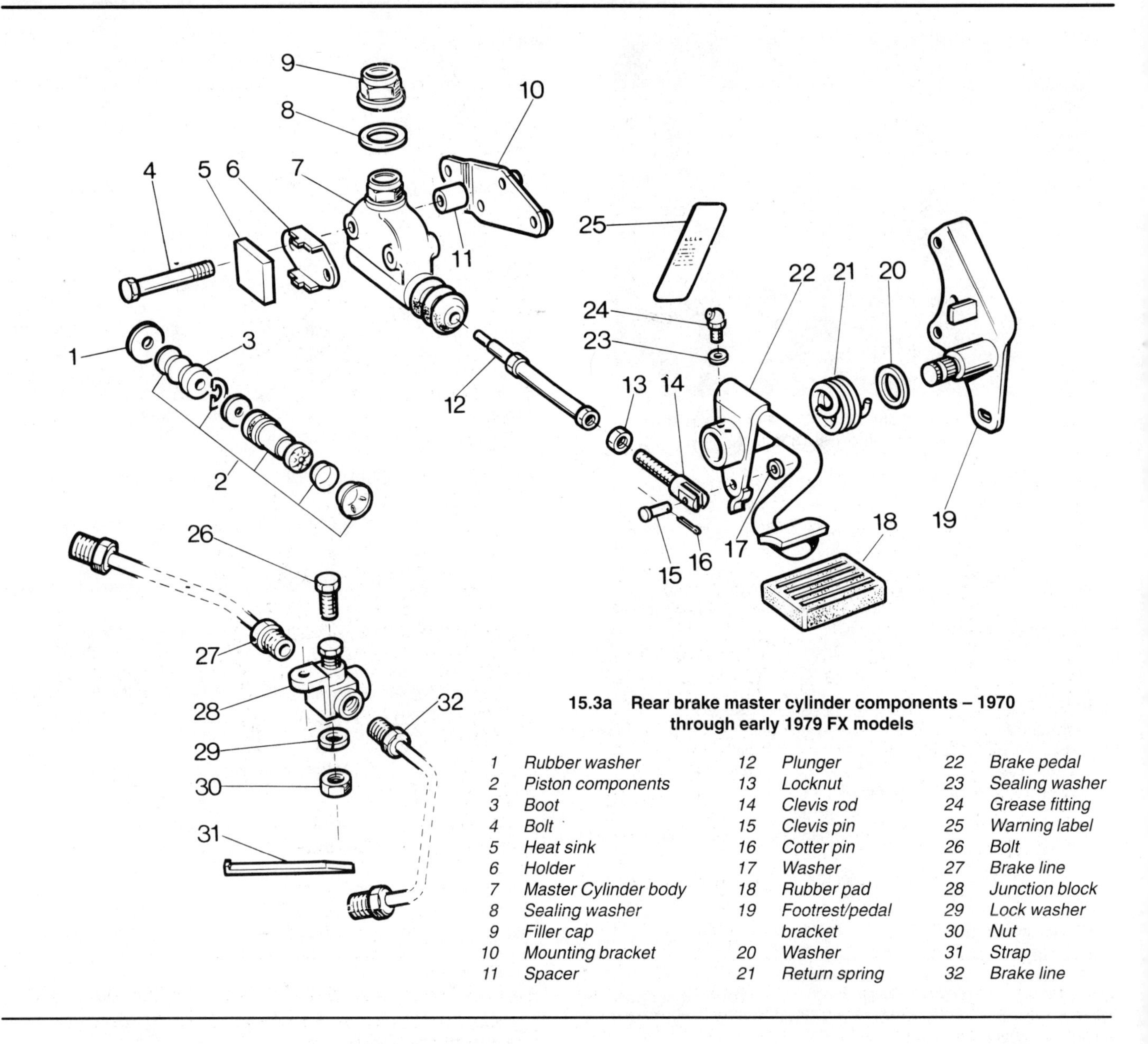

15.3a Rear brake master cylinder components – 1970 through early 1979 FX models

1	*Rubber washer*	*12*	*Plunger*	*22*	*Brake pedal*
2	*Piston components*	*13*	*Locknut*	*23*	*Sealing washer*
3	*Boot*	*14*	*Clevis rod*	*24*	*Grease fitting*
4	*Bolt*	*15*	*Clevis pin*	*25*	*Warning label*
5	*Heat sink*	*16*	*Cotter pin*	*26*	*Bolt*
6	*Holder*	*17*	*Washer*	*27*	*Brake line*
7	*Master Cylinder body*	*18*	*Rubber pad*	*28*	*Junction block*
8	*Sealing washer*	*19*	*Footrest/pedal bracket*	*29*	*Lock washer*
9	*Filler cap*			*30*	*Nut*
10	*Mounting bracket*	*20*	*Washer*	*31*	*Strap*
11	*Spacer*	*21*	*Return spring*	*32*	*Brake line*

20 Remove the piston and seal from the caliper. You may have to force the piston out of the caliper with air pressure. Place the caliper on a clean workbench with the piston facing down. Place a clean towel under the piston and apply low pressure air to the caliper inlet. Be careful – the piston may come out with considerable force. Use a wood or plastic tool to remove the seal.

21 Remove the rubber bushings from the mounting bracket bores.If they're worn or damaged, install new ones.

All models

22 Clean all of the brake components (except for the brake pads) with brake system solvent (available at auto parts stores), isopropyl alcohol or clean brake fluid. **Warning:** *Do not, under any circumstances, use petroleum-based solvents to clean brake parts.* If compressed air is available, use it to dry the parts.

23 Check the caliper bore and the outside of the piston for scratches, nicks and score marks. If damage is evident, the caliper must be replaced with a new one. **Warning:** *Don't attempt to rebore or hone the caliper.*

24 Before reassembling the caliper, soak the new rubber parts in clean brake fluid for 10 or 15 minutes. Lubricate the caliper bore with brake fluid, then install the seal. **Warning:** *Always install a new seal – do not reuse the old one if it was removed from the bore.*

25 Carefully insert the piston as far as possible into the bore. Install the boot and retaining wire (if used).

26 Reattach the brake hose and complete reassembly as described in Section 13. **Note:** *If the brake hose is attached to the caliper with a banjo fitting and bolt, use new washers when installing the bolt. The replacement washers must be the same type as the originals (some are zinc-plated copper, while others are steel with a rubber O-ring). Be sure to tighten the banjo fitting bolt to the correct torque – it's different, depending on the type of washers used.*

27 Fill the master cylinder with the recommended brake fluid and bleed the system as described in Chapter 1.

15 Rear brake master cylinder – removal, overhaul and installation

1 If the master cylinder is leaking fluid, or if the pedal doesn't feel firm when the brake is applied – and bleeding the brakes doesn't help – master

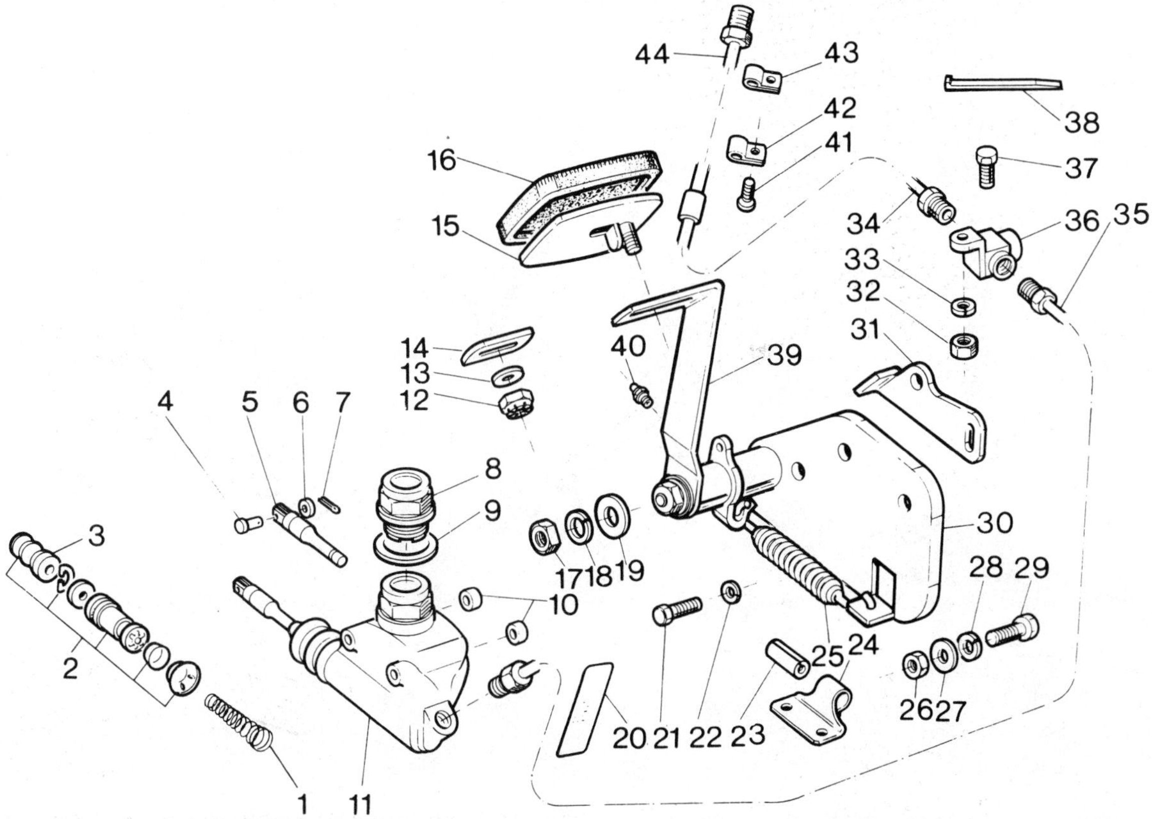

15.3b Rear brake master cylinder components – 1970 through early 1979 FL models

1. *Spring*
2. *Piston components*
3. *Boot*
4. *Clevis pin*
5. *Plunger*
6. *Washer*
7. *Cotter pin*
8. *Filler cap*
9. *Sealing washer*
10. *Spacer*
11. *Master cylinder body*
12. *Castellated nut*
13. *Washer*
14. *Spacer plate*
15. *Brake pad*
16. *Rubber pad*
17. *Nut*
18. *Lock washer*
19. *Plain washer*
20. *Warning label*
21. *Bolt*
22. *Lock washer*
23. *Rubber sleeve*
24. *Clip*
25. *Return spring*
26. *Nut*
27. *Plain washer*
28. *Lock washer*
29. *Bolt*
30. *Mounting plate*
31. *Stop plate*
32. *Nut*
33. *Lock washer*
34. *Rear brake line*
35. *Front brake line*
36. *Junction block*
37. *Bolt*
38. *Strap*
39. *Brake pedal*
40. *Grease fitting*
41. *Bolt*
42. *Clip*
43. *Clip*
44. *Rear brake hose*

cylinder overhaul is recommended. Before disassembling it, read through the entire procedure and make sure you have the correct rebuild kit. Also, you'll need some new, clean brake fluid of the recommended type, some clean rags and internal snap-ring pliers. **Note:** *Disassembly, overhaul and reassembly of the master cylinder must be done in a spotlessly clean work area to avoid contamination and possible failure of the brake hydraulic system components. If such a work area isn't available, have the master cylinder rebuilt by a dealer service department or motorcycle repair shop.*

2 Disconnect the master cylinder-to-caliper brake line from the master cylinder and plug the end. Use a flare-nut wrench, if possible, to avoid rounding off the fitting (later models have a banjo fitting and bolt, so a flare-nut wrench isn't necessary). **Caution:** *If brake fluid is spilled on a painted surface, remove it immediately or the paint will be damaged.*

1970 through early 1979 models

Refer to illustrations 15.3a and 15.3b

3 Disconnect the master cylinder plunger from the brake pedal by removing the cotter pin from the clevis pin **(see illustrations)**. Remove the washer from the clevis pin, then pull the clevis pin out.

4 Pull the plunger out of the master cylinder.

5 Remove the boot from the master cylinder.

6 Remove the snap-ring and pull out the piston assembly. Pay attention to the order in which the components are removed. After the piston assembly is out, remove the wafer, piston cup, spring seat and spring. Detach the O-ring from the piston and discard it.

Late 1979 through 1982 FX, late 1979 through 1983 FXE, late 1979 through 1984 FL and 1980 through 1982 FLT models

Refer to illustrations 15.7, 15.8 and 15.10

7 Remove the two mounting bolts and pull the master cylinder off the pushrod **(see illustration)**.

15.7 On most models, the master cylinder is attached to the frame bracket with two bolts

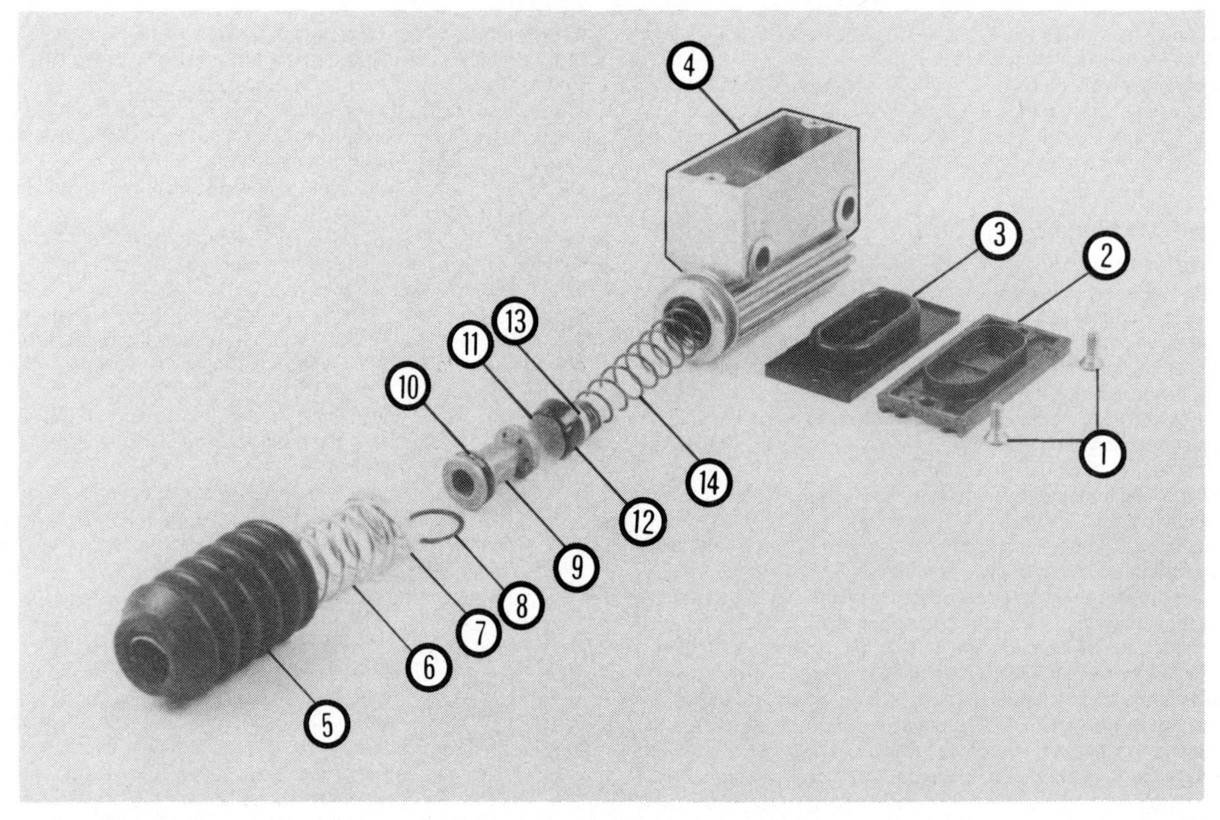

15.8 Rear brake master cylinder components – late 1979 through 1982 FX, late 1979 through 1983 FXE, late 1979 through 1984 FL and 1980 through 1982 FLT models

1	*Screws*	*5*	*Dust boot*	*10*	*O-ring*
2	*Cover*	*6*	*Spring*	*11*	*Wafer*
3	*Gasket*	*7*	*Cupped washer (2)*	*12*	*Piston cup*
4	*Master cylinder body*	*8*	*Snap-ring*	*13*	*Spring seat*
		9	*Piston*	*14*	*Spring*

8 Remove the cupped washers, spring and dust boot **(see illustration)**.
9 Remove the screws and detach the cover and gasket from the top of the master cylinder. Drain the brake fluid.
10 Remove the snap-ring **(see illustration)**, followed by the piston assembly, wafer (if equipped), piston cup, spring seat and spring from the master cylinder bore. Remove the O-ring from the piston **(see illustration 15.8)** and discard it.

1980 through early 1987 FXR and 1983 through 1991 FLT models

11 Remove the cover from the master cylinder fluid reservoir.
12 Disconnect the brake reservoir hose from the master cylinder fitting and drain the brake fluid into a container.
13 Disconnect the wire from the brake light switch.
14 Remove the mounting bolts and detach the master cylinder from the pushrod.
15 Remove the snap-ring, followed by the piston and seal (FLT models have an O-ring in place of the seal), piston cup, piston stop and spring.
16 Remove the seal or O-ring from the piston and discard it.

1985 through early 1987 FLST models

17 Remove the two bolts that attach the master cylinder to the bracket, then pull the master cylinder off the pushrod and remove the cupped washer, spring and rubber dust boot **(see illustration 15.8)**.
18 Remove the screws and detach the cover and gasket, then drain the brake fluid from the master cylinder.
19 Remove the snap-ring **(see illustration 15.10)**, then pull out the piston, piston cup, spring seat and spring.
20 Remove the seal from the piston and discard it.

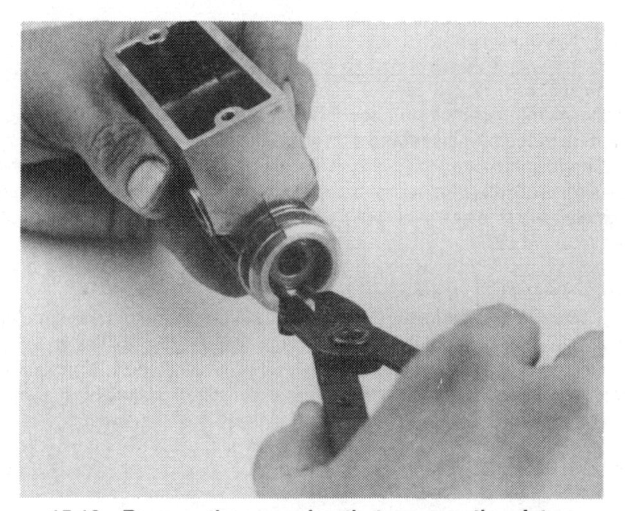

15.10 Remove the snap-ring that secures the piston assembly in the master cylinder bore

1983 through early 1987 models (except those listed above)

21 Remove the cover from the master cylinder fluid reservoir.

22 Disconnect the brake reservoir hose from the master cylinder fitting and drain the brake fluid into a container.
23 Remove the mounting bolts and detach the master cylinder.
24 Remove the snap-ring **(see illustration 15.10)**, followed by the piston, washer, piston cup, spring retainer and spring.
25 Remove the seal from the piston and discard it.

Late 1987 and later FXR, FXST/C, FXSTS, all FXD, and 1992 and later FLT models

26 Where necessary, remove the exhaust pipe/muffler for access to the master cylinder connections. Disconnect the clevis pin from the brake pedal (it's held in place by a cotter pin). Remove the banjo fitting bolt and detach the brake line from the fitting on the master cylinder. Discard the washers – new ones must be used on installation.
27 On all except FXD models, carefully remove the hose clamp and detach the reservoir hose from its fitting on the master cylinder. The fitting is easily damaged, so don't use excessive force.
28 Drain the fluid from the reservoir while taking care not to spill any on the surrounding parts.
29 Release the lockplate tabs, then remove the large nut that attaches the master cylinder to the mounting bracket.
30 Loosen the jam nut on the pedal pushrod, then raise the pedal so the master cylinder can be moved forward and out of the bracket.
31 Turn the master cylinder to unscrew the pushrod from the pedal rod. There are flats on the pushrod so a wrench can be used if necessary.
32 Clean the exterior of the master cylinder, then thread the banjo bolt into the cartridge body. The bolt will protect the sealing surface in the cartridge body during disassembly. **Note:** *To avoid confusion during reassembly, pay close attention to how the parts fit together as the master cylinder is disassembled – lay the parts out in the proper order and make a simple sketch if necessary.*
33 Remove the rubber boot from the reservoir adapter, then hold the master cylinder upright with the banjo bolt resting on the workbench and push the reservoir adapter off the cartridge body.
34 Don't allow the cartridge body to be contaminated with dirt or grease. Do not disassemble the cartridge body – it contains the piston and related components, which are not sold separately. If the piston seal is leaking, install a new cartridge body.
35 Press down on the large washer to compress the return spring, then remove the snap-ring from the pushrod groove.
36 Release the spring and remove the washer, rubber boot, spring retainer (inside the boot) and spring.
37 Remove and discard the snap-ring inside the cartridge body, then pull out the pushrod and washer.
38 Check the reservoir adapter bore for scratches. Inspect the threads on the cartridge body, pushrod and banjo bolt.
39 Carefully remove the O-rings from the cartridge body and clean the grooves with a soft cotton cloth saturated with alcohol. Check the grooves for scratches and nicks. Replace the cartridge body if damage is evident.

Late 1987 and later FLST/C/F/N models

40 Remove the exhaust pipe/muffler to gain access to the master cylinder. Also remove the chromed cover over the reservoir.
41 Remove the master cylinder mounting bolts and detach the clevis pin from the brake pedal (it's held in place by a cotter pin).
42 Remove the banjo fitting bolt and detach the brake line from the master cylinder. Discard the washers.
43 Remove the large nut that attaches the master cylinder to the mounting bracket.
44 Follow the procedure in Steps 32 through 39 above.

All models

45 Clean all of the components with brake system solvent (available at auto parts stores), isopropyl alcohol or clean brake fluid. **Warning:** *Do not, under any circumstances, use petroleum-based solvents to clean brake parts.* If compressed air is available, use it to dry the parts. Make sure the tiny fluid passages in the master cylinder aren't clogged.
46 Always use a new seal or O-ring when the master cylinder is reassembled – never reinstall the old one (not applicable to late 1987 and later models).
47 Inspect the master cylinder bore for scratches, nicks and score marks. If damage is evident, the master cylinder must be replaced with a new one. Do not attempt to hone the bore.
48 Inspect all of the components for wear and damage and replace defective parts with new ones.
49 Before reassembling the master cylinder, soak the new rubber seal/O-ring in clean brake fluid for 10 or 15 minutes. Lubricate the master cylinder bore with clean brake fluid.
50 Reassemble and install the master cylinder by reversing the removal and disassembly procedures. **Note:** *On late 1987 and later models, pay close attention to the following important points during reassembly and installation:*

a) Install new O-rings in the cartridge grooves and lubricate them with DOT 5 brake fluid. Lubricate the reservoir adapter or reservoir bore with the same brake fluid.
b) When installing the cartridge body in the reservoir adapter (FXR, FXST/C, FXSTS, FXD Dynas and 1992 and later FLT models), align the reservoir hose fitting on the adapter with the notch in the threaded end of the cartridge body.
c) The notch in the threaded end of the cartridge body must engage the lug on the inside of the adapter or reservoir.
d) Make sure the snap-ring is seated in the groove in the cartridge bore after installing the pushrod. The pushrod must rotate freely.
e) Make sure the drain hole in the rubber boot is facing down.
f) On FXR, FXST/C, FXSTS, FXD Dynas and 1992 and later FLT models, the square on the reservoir adapter must engage in the square hole in the mounting bracket with the reservoir hose fitting on top. The lockplate lip should fit over the bracket. Tighten the large mounting nut to 30 to 40 ft-lbs, then bend the tab on the lockplate over one of the nut flats.
g) On FLST/C/F/N models, don't tighten the large mounting nut until after the brake hose banjo fitting bolt is tightened. Bend the tab on the lockplate over one of the nut flats.

51 Carefully attach the hose/line to the master cylinder. If a banjo fitting is used, be sure to install new washers of the correct type (DO NOT interchange zinc-plated copper washers with steel/rubber washers) and tighten the bolt to the torque listed in this Chapter's Specifications – the torque is different, depending on the type of washers used. Fill the reservoir with the recommended brake fluid.
52 Connect the wire to the brake light switch, if so equipped.
53 Bleed the brakes as described in Chapter 1.
54 Check the brake pedal height setting and master cylinder pushrod free play as described in Section 16.

16 Rear brake pedal – check and adjustment

Warning: *On models with hydraulic disc brakes, rear brake pedal adjustment is very important. Insufficient pedal clearance may result in interference with brake operation – inadequate master cylinder pushrod free play could result in dragging brakes.*

FLT models

1980 through 1982

1 The rear brake master cylinder pushrod must have 3/32-to-1/8 inch of free play between the stop bolt and brake pedal. To adjust it, loosen the locknut and turn the stop bolt. Once the free play is correct, hold the stop bolt and tighten the locknut securely.

1983-on

2 The rear brake pedal must have a minimum of 2-1/4 inches of clearance above the footboard (measured from the upper surface of the footboard to the lower corner of the pedal pad). Check the clearance whenever the footboard, the brake pedal height or master cylinder pushrod free play is adjusted. **Warning:** *Insufficient rear brake pedal clearance will cause the pedal to contact the footboard and interfere with rear brake operation.*

3 To change the pedal height, remove the screw from the clevis assembly, remove the clevis assembly from the brake pedal shaft and turn the pedal to the desired position.
4 Install the clevis assembly on the shaft. Install the screw and tighten it to 30-to-35 ft-lbs. Adjust the footboard to obtain the correct clearance.
5 On 1992 and later models fine pedal height adjustment can be made by loosening the pushrod jam nut, disconnecting the clevis pin from the pedal and turning the pushrod to adjust (the linkage should not be adjusted to the point where more than 1/2 inch (13 mm) of thread is exposed otherwise there is danger of the pushrod being detached from the piston).
6 On 1983 through 1991 models check the pushrod free play adjustment, measured between the stop bolt and clevis assembly; it should be 3/32-to-1/8 inch (2.4 to 3.1 mm). To adjust, loosen the locknut and turn the pedal stop bolt in or out as required. Tighten the locknut while holding the stop bolt to keep it from turning. Pushrod free play is preset from 1992 onwards.

FXR models (except FXRD)

1984 through early 1987

7 There are two important adjustments relating to proper FXR rear brake operation: Brake pedal height and pushrod free play, which should be adjusted together.
8 The top of the rear brake pedal should be 4-1/8 to 4-3/8 inches above the center of the pivot shaft with the machine upright. Measure from the floor to the center of the pivot shaft, then measure from the floor to the top of the brake pedal – the difference should be 4-1/8 to 4-3/8 inches.
9 To adjust the pedal, proceed as follows:

a) Make sure the pedal doesn't contact the footpeg bracket. The center of the footpeg rubber should be 7/8 to 1-3/16 inch ABOVE the center of the pivot shaft. Adjust the footpeg if necessary.
b) Loosen the locknut and turn the brake pedal stop bolt in or out to obtain the specified pedal height (Step 7).
c) Tighten the locknut while holding the brake pedal stop bolt to keep it from turning.
d) The pushrod must have 1/16-inch of free play before actuating the master cylinder piston. Measure the free play between the brake pedal and stop bolt.
e) Loosen the locknut and turn the pushrod to obtain the specified free play.
f) Tighten the locknut while holding the pushrod.

Late 1987 and later

10 The top of the rear brake pedal should be 4-1/8 to 4-3/8 inches above the center of the pivot shaft with the machine upright. Measure from the floor to the center of the pivot shaft, then measure from the floor to the top of the brake pedal. The difference should be 4-1/8 to 4-3/8 inches. Adjust it by loosening the jam nut and turning the pushrod (use a wrench on the pushrod flats). **Warning:** *Don't lengthen the pushrod excessively or insufficient thread engagement between the pushrod and brake rod could cause brake failure and possible personal injury!*
11 Make sure the pedal doesn't contact the footpeg bracket. The center of the footpeg rubber should be 7/8 to 1-3/16 inch ABOVE the center of the pivot shaft. Adjust the footpeg if necessary.
12 Pedal free play is built into the master cylinder and no adjustment is required. When the pedal is pushed down by hand, a small amount of free play must be felt.

FXRD models

13 Brake pedal height and pushrod free play are two important adjustments for proper rear brake operation. Because one adjustment affects the other, the brake pedal adjustment is followed immediately by pushrod adjustment. **Warning:** *When adjusting the brake pedal, it should never be positioned to provide less than 1/16-inch of pushrod free play. It's also important that the pedal is positioned where full leverage can be applied with good foot-to-pedal contact. The pedal must have full travel to bottom the master cylinder without interference by the footboard. An improperly adjusted brake pushrod could cause the brakes to drag.*
14 The brake pedal can be positioned to suit the leg reach of individual riders. The rider should sit on the machine and apply the brake at different pedal positions to establish which position is most comfortable and effective.
15 Loosen the locknut and turn the brake pedal stop bolt in or out to obtain the desired pedal position.
16 Tighten the locknut while holding the stop bolt to keep it from turning.
17 The pushrod must have 1/16-inch of free play before actuating the master cylinder piston. Measure the free play between the brake pedal and stop bolt. Loosen the locknut and turn the pushrod to obtain the specified free play.
18 Tighten the locknut while holding the pushrod.

All other models

Late 1978 through 1984

FL models

19 The rear brake master cylinder pushrod must have 1/16-inch of free play. Work the brake pedal up-and-down by hand to check the free play (slight movement before the pushrod contacts the master cylinder piston).
20 On late 1978 and early 1979 models, loosen the master cylinder rear bolt and the brake pedal stop plate bolt. Move the front end of the stop plate down to decrease free play or up to increase free play.
21 On late 1979 through 1984 models, loosen the locknut and turn the stop bolt (clockwise to decrease free play, counterclockwise to increase it). Tighten the locknut to 10 ft-lbs.

FX models

22 The rear brake master cylinder pushrod should have 1/16-inch of free play.
23 On FXWG models through 1983, loosen the locknut and turn the stop bolt (clockwise to decrease free play, counterclockwise to increase it). Hold the stop bolt to keep it from turning and tighten the locknut to 10 ft-lbs.
24 On FX models (except FXDG and 1984 FXST/FXWG), loosen the locknut and turn the pushrod on the clevis threads (forward to increase free play, to the rear to decrease free play). Adjustment should be made with the linkage loose, then turn the pushrod to the rear until the correct free play is obtained.
25 On FXDG and 1984 FXST/FXWG models, loosen the jam nut and adjust the pedal stop bolt to obtain the desired pedal-to-footpeg relationship. Loosen the locknut and turn the master cylinder pushrod on the clevis threads (forward to increase free play, to the rear to decrease free play).

1985 and later

FXEF and FXSB

26 Loosen the locknut and turn the pushrod on the clevis (forward to increase free play, to the rear to decrease free play). Adjustment should be done with the linkage very loose, then turn the pushrod to the rear until 1/16-inch of free play is obtained.

FXST/C, FXSTS and FXWG

27 Measure the pedal-to-footpeg clearance; it should be within 1/4-to-1/2 inch (6.4 to 13 mm). On early models adjustment is made via the stop bolt and locknut which bears on the clevis assembly, whereas on later models it will be necessary to loosen the pushrod jam nut, disconnect the clevis pin and turn the pushrod to make adjustment.
28 On models through early 1987 pushrod free play should be 1/16 inch (1.6 mm). On later models freeplay is preset.

FXDB/C/L Dyna

29 Pedal height should be 1-to-1.2 inches (25 to 30 mm), measured from the front tip of the pedal foot to the top of the footrest rubber. To adjust, first loosen the pushrod jam nut, then disconnect the clevis pin at the pedal. Turn the pushrod to make adjustment. **Note:** *Do not lengthen the rod to the extent that it is in danger of becoming detached from the master cylinder piston.*
30 Pushrod freeplay is preset on these models.

FLST/C/F/N and FXDWG Dyna

Note: *The rear brake pedal position is non-adjustable. On models through early 1987, adjust the stop bolt until there's 1/16-inch of free play between the stop bolt and brake pedal. On late 1987 and later models, DO NOT MAKE ANY ADJUSTMENTS! The free play is built into the master cylinder assembly.*

TIRE CHANGING SEQUENCE - TUBED TIRES

A
Deflate tire. After pushing tire beads away from rim flanges push tire bead into well of rim at point opposite valve. Insert tire lever next to valve and work bead over edge of rim.

Use two levers to work bead over edge of rim. Note use of rim protectors

B

C
Remove inner tube from tire

When first bead is clear, remove tire as shown

D

E
To install, partially inflate inner tube and insert in tire

Work first bead over rim and feed valve through hole in rim. Partially screw on retaining nut to hold valve in place.

F

G
Check that inner tube is positioned correctly and work second bead over rim using tire levers. Start at a point opposite valve.

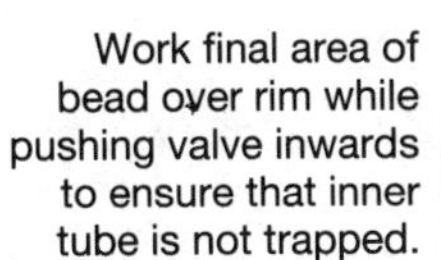
Work final area of bead over rim while pushing valve inwards to ensure that inner tube is not trapped.

H

17 Rear wheel sprocket – removal and installation

1 Remove the rear wheel as described in Section 11.

Drum brake models

Note: *Special tools are required to install a new sprocket on drum brake models. Read the procedure before beginning removal. It may be necessary to have a Harley-Davidson dealer service department do the job.*

2 The rear wheel sprocket is riveted to the brake drum. If examination indicates some of the teeth are chipped, broken or hooked, the brake drum must be removed from the wheel. It's attached to the wheel by five bolts.
3 Remove the bolts securing the brake drum to the hub.
4 the sprocket is attached to the brake drum with rivets. Using a sharp chisel, cut the heads off the rivets and dowel pins.
5 If the rivet holes aren't worn or elongated, the new rivets can be installed in the same holes. However, if they're worn or elongated, a new set of holes should be drilled midway between the existing holes and the dowel pin holes. A 0.1935-inch (no. 10) drill bit is required to bore the holes.
6 New dowel pin holes should be drilled also (use a 3/16-inch drill bit). The dowel pins must be a tight fit. Use the new sprocket as a template.
7 Drill a rivet hole, then place the sprocket in position on the drum. Insert a rivet through the hole but don't head the rivet. Drill another hole across from the first rivet and insert another rivet. Again, don't head the rivet.
8 Drill the remaining holes, then drill the four dowel pin holes.
9 Remove the sprocket from the brake drum and deburr the newly drilled holes.
10 Position the drum and sprocket on the center support of the riveting jig (special tool no. 95600-33B).
11 The dowel pins are installed first, followed by the rivets. The dowel pins and the rivets must be installed through the brake drum side.
12 Using a hollow driver, seat the dowel pin and rivet simultaneously, driving the sprocket and hub flange together.
13 Using a concave punch, flare the end of the dowel pin. Head the end of the rivet until it extends 3/32-inch above the face of the sprocket.
14 Repeat Steps 11 through 13, seating the rivets and dowel pins on the opposite side of the hub until all of them are in place. This will prevent distortion of the sprocket as the rivets are installed.
15 Install the brake drum on the hub and reinstall the rear wheel as described in Section 11.

Disc brake models

All models except FLT

16 The sprocket is attached to the rear wheel hub with five bolts.
17 If examination of the rear sprocket indicates the teeth are chipped, broken or hooked, the sprocket should be replaced with a new one.
18 Remove the bolts securing the sprocket to the hub. Lift the sprocket off and place the new one in position.
19 Install the sprocket mounting bolts and washers and tighten the bolts to the specified torque.
20 Install the rear wheel as described in Section 11.

FLT models

Refer to illustration 17.39

21 Raise the rear wheel off the ground and support the motorcycle so it'll remain steady.
22 Remove the saddlebags and the muffler from the left side of the motorcycle.
23 Disconnect the rubber boots from the enclosed chain housing.
24 Turn the wheel until the master link on the chain is located, then remove the master link.
25 Remove the bolts securing the swingarm bracket to the chain housing and detach the bracket.
26 Unscrew the axle nut and remove the washers from the right end of the axle.
27 Support the rear wheel and remove the axle. You may have to drive the axle out with a soft-face hammer.
28 Carefully lower the rear wheel to the ground and separate it, with the chain housing attached, from the swingarm. **Caution:** *Do not operate the rear brake pedal while the wheel is removed – the piston may be forced out of the brake caliper.*
29 Remove the nuts and bolts that hold the chain housing together.
30 Remove the bolts securing the axle bracket, then remove the two screws from the left side of the housing. Pull the housing apart.
31 Remove the bolts and detach the spacer and sprocket assembly from the wheel.
32 Remove the seal and bearing from the left side of the sprocket assembly. On 1982 and 1983 models, a sleeve and spacer washer must also be removed from the left side.
33 Remove the dust shield from the right side of the sprocket. On 1980 and 1981 models, a bearing and inner spacer must also be removed from the right side of the sprocket.
34 Pack the bearings with multi-purpose grease and assemble the bearings, spacers and other components in the reverse order of disassembly.
35 Coat the sprocket mounting bolt threads with engine oil and mount the sprocket assembly on the rear wheel. Tighten the bolts in a criss-cross pattern to the specified torque.
36 Place the spacer in the sprocket. Coat the mating surfaces of the two halves of the chain housing with RTV-sealant and attach the upper half to the lower half. Install the nuts and bolts in their original locations and tighten them securely.
37 Attach the axle bracket to the left side of the chain housing with the screws and bolts. Be sure the opening in the spacer is aligned with the sprocket bolt hole in the lower housing before the screws and bolts are tightened.
38 Place the rear wheel and chain housing in position in the swingarm and insert the axle from the left side. Be sure the brake pads are in the proper position in relation to the brake disc and the sprocket housing is facing forward.
39 Connect the drive chain with the master link. Be sure the closed end of the spring clip on the master link faces the direction of chain travel **(see illustration)**.

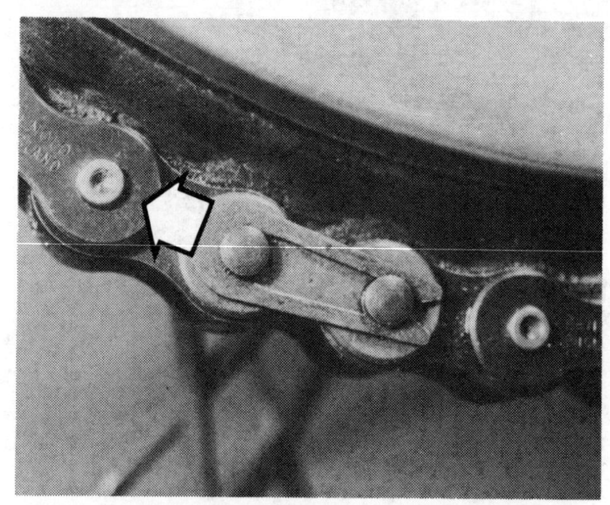

17.39 The closed end of the master link spring clip must point in the direction of normal chain travel

40 Install the washers and nut on the end of the axle but don't tighten the axle nut.
41 Check the chain tension as described in Chapter 1. Adjust it if necessary.
42 Tighten the axle nut to the specified torque.
43 Apply a coat of RTV-sealant to the mating surfaces of the chain housing and the rubber boots. Attach the boots to the housing with the mounting screws.
44 Install the housing bracket components and attach the chain housing to the swingarm.

All models

45 Note that if the rear sprocket is worn, the final drive sprocket on the transmission is probably worn also. It's a good idea to replace both sprock-

ets at the same time, as well as the chain. This will alleviate rapid wear that results from mixing old and new parts.

18 Rear brake disc – removal and installation

1 Remove the rear wheel as described in Section 11.
2 The brake disc is attached to the rear hub with five bolts. Remove the bolts and detach the brake disc from the hub.
3 Before installing the disc, be sure the threads on the bolts and in the hub are clean and undamaged. Use thread locking compound on the bolts and tighten them in small increments, in a criss-cross pattern, until the specified torque is reached.

19 Tires – removal and installation

Note: *This procedure does not apply to tubeless tires!*
1 To properly remove and install tires, you'll need at least two motorcycle tire irons, some tire mounting lubricant (available at motorcycle dealers and accessory stores) or a solution of water and liquid soap, some talcum powder and a tire pressure gauge.
2 Begin by removing the wheel from the machine. If the tire is going to be reused, mark it next to the valve stem or wheel balance weight.
3 Deflate the tire by removing the valve stem core. When it's completely deflated, push the tire bead away from the rim on both sides. In some extreme cases, and on all 16-inch wheels, this can only be accomplished with a bead breaking tool, but usually it can be done with tire irons. Riding on a deflated tire to break the tire bead is not recommended – damage to the rim and tire will result.
4 Dismounting a tire is easier when the tire is warm, so an indoor tire change is recommended in cold climates. The rubber gets very stiff and is difficult to manipulate when cold.
5 Place the wheel on a thick pad or old blanket. This will help keep the wheel and tire from slipping around. On disc brake equipped wheels, place the disc down or remove it from the hub.
6 Once the bead is completely off the rim, lubricate the inside edge of the rim and the tire bead with a solution of water and liquid soap or tire mounting lubricant. Remove the locknut and push the tire valve stem through the rim.
7 Insert one of the tire irons under the bead at the valve stem and lever the bead up over the rim. This should be fairly easy. Take care not to pinch the tube as this is done. When removing tires from cast wheels, cushion the rim with a shop towel or some other type of pad to avoid gouging the rim. If it's difficult to pry the bead up, make sure the rest of the bead opposite the valve stem is in the dropped center section of the rim.
8 Hold the tire iron down with the bead over the rim, then move about three or four inches to either side and insert the second tire iron. Be careful not to cut or slice the bead or the tire may split when inflated. Also, don't catch or pinch the inner tube as the second tire iron is levered over (this is why tire irons are recommended over screwdrivers or other implements).
9 With a small section of the bead up over the rim, one of the levers can be removed and reinserted a few inches around the rim until about one-quarter of the tire bead is above the rim edge. Make sure the rest of the bead is in the dropped center of the rim. At this point, the bead can usually be pulled up over the rim by hand.
10 Once all of the first bead is over the rim, the inner tube can be withdrawn from the tire and rim. Push in on the valve stem, lift up on the tire next to the stem, reach inside the tire and carefully pull out the tube. It usually isn't necessary to completely remove the tire from the rim to repair the inner tube. It's sometimes recommended though because checking for foreign objects in the tire is difficult while it's still partially mounted on the rim.
11 To remove the tire completely, make sure the bead is broken all the way around on the remaining edge, then stand the tire and wheel up on the tread and grab the wheel with one hand. Push the tire down over the same edge of the rim while pulling the rim away from the tire. If the bead is correctly positioned in the dropped center of the rim, the tire should roll off and separate from the rim very easily. If tire irons are used to work the last bead over the rim, the outer edge of the rim may be marred. If a tire iron is necessary, be sure to pad the rim as described earlier.
12 Refer to Section 20 for inner tube repair procedures.
13 Mounting the tire is basically the reverse of removal. Some tires have a balance mark and/or directional arrows molded into the tire sidewall. Look for these marks so the tire can be installed properly. The dot should be aligned with the valve stem.
14 If the tire wasn't removed completely to repair or replace the inner tube, the tube should be inflated just enough to make it round. Sprinkle it with talcum powder, which acts as a dry lubricant, then carefully lift up the tire edge and install the tube with the valve stem next to the hole in the rim. Once the tube is in place, push the valve stem through the rim and start the locknut on the stem.
15 Lubricate the tire bead, then push it over the rim edge and into the dropped center section opposite the inner tube valve stem. Work around each side of the rim, carefully pushing the bead over it. The last section may have to levered on with tire irons. If so, be careful not to pinch the inner tube as this is done.
16 Once the bead is over the rim edge, check to see if the inner tube valve stem is pointing to the center of the hub. If it's angled slightly in either direction, rotate the tire on the rim to straighten it out. Run the locknut the rest of the way down the stem but don't tighten it completely.
17 Inflate the tube to approximately 50 psi and check to make sure the guidelines on the tire sidewalls are the same distance from the rim around the entire circumference of the tire.
18 After the tire bead is correctly seated on the rim, allow the tire to deflate. Replace the valve core and inflate the tube to the recommended pressure, then tighten the valve stem locknut securely and install the cap.

20 Tubes – repair

1 Inner tube repair requires a patching kit that's usually available from motorcycle dealers, accessory stores and auto parts stores. Be sure to follow the directions supplied with the kit to ensure a safe, long lasting repair.
2 If the inner tube has been patched previously, or if there's a tear or large hole, it should be replaced with a new one. Sudden deflation can cause loss of control and an accident, particularly if it occurs on the front wheel.
3 To repair a tube, remove it from the tire, then inflate and immerse it in a tub or sink full of water to pinpoint the leak. Mark the position of the leak, then deflate the tube, Dry it off and thoroughly clean the area around the puncture.
4 Most tire patching kits have a buffer to rough up the area around the hole for proper adhesion of the patch. Roughen an area slightly larger than the patch, then apply a thin coat of the patching cement to the roughened area. Allow the cement to dry until tacky, then apply the patch.
5 If may be necessary to remove a protective covering from the top surface of the patch after its been attached to the tube. Keep in mind that inner tubes made from synthetic rubber may require a special type of patch and adhesive to achieve a satisfactory bond.
6 Before replacing the tube, check the inside of the tire to make sure the object that caused the puncture isn't lying inside it. Also check the outside of the tire, particularly the tread area, to make sure nothing is projecting through the tire that may cause another puncture. Check the rim for sharp edges and damage. On wire spoke wheels, make sure the rubber rim band is in good condition and properly installed before inserting the tube.

21 Wheel bearings – repack

Refer to Chapter 1 for wheel bearing inspection and maintenance procedures.

Chapter 7 Electrical system

Contents

Specifications

Battery

Voltage	12-volts
Capacity	
FL models	
FLT	
Through 1983	22 amp/hr.
1984-on	20 amp/hr. at 10 hr. rate/22 amp/hr. at 20 hr. rate
All others	32 amp/hr.
FX models	
1970 through early 1978	7 amp/hr.
Late 1978 through 1983	7.5 amp/hr.
1984-on (FX/Softail)	19 amp/hr.
FXE, FXS, FXR, FXEF, FXB, FXSB and FXWB models	19 amp/hr.
FXD Dyna models	19 amp/hr.
Specific gravity at 80-degrees F	
State of charge	Specific gravity
100%	1.25 to 1.27
75%	1.22 to 1.24
50%	1.19 to 1.21
25%	1.16 to 1.18
Ground connection	Negative

Alternator

Type	Permanent magnet rotor
Regulated output	
1985-on FX/Softail models and all FXD Dyna models	13.8 to 15-volts
All others	14-volts

Bulbs*

1970 through 1983

Headlight	
1970 through early 1978	35/45 watt or 50/60 watt
Late 1978 through 1983	
1980 and 1981 FLT	30/30 watt
All others	35/50 watt or 50/60 watt
Tail/brake light	
1970 through early 1978	4/32 CP
Late 1978 through 1983	3/32 CP
FLHT passing lights	30 watt
Instrument lights	1 CP or 2 CP
Spotlights	30 watt
Turn signals	32 CP
1984-on	Consult a Harley-Davidson dealer parts department

**All bulbs are rated at 12-volts*

Starter motor brush length (minimum)

1970 through 1983	
Prestolite	1/4 in (6.35 mm)
Hitachi	7/16 in (11.1 mm)
1984 through 1988	7/16 in (11.1 mm)
1989-on	0.413 in (10.5 mm)

Torque specifications

	Ft-lbs (unless otherwise indicated)	Nm
All models		
Battery cable-to-starter motor terminal nut	65 to 80 in-lbs	7 to 9
1970 through 1983		
Rear bracket-to-engine	12	16
Rear bracket-to-starter motor or battery carrier	6	8
Starter-to-primary chaincase (FLT and FXR models)	10 to 12	14 to 16
Starter through-bolts		
Late 1978 through 1982 FL models	60 to 80 in-lbs	6.7 to 9
All others	20 to 25 in-lbs	2.2 to 2.8
1984 through 1988		
Starter bracket-to-transmission stud nut	13 to 16	18 to 22
Rubber mount nut	6	8
Starter through-bolts	20 to 25 in-lbs	2.2 to 2.8
1989-on		
Starter motor mounting bolts	13 to 20	18 to 27
Starter through bolts		
1989 to 1990	20 to 25 in-lbs	2.2 to 2.8
1991-on	39 to 65 in-lbs	4.4 to 7
Jackshaft bolt	7 to 9	9 to 12

1 General information

All models covered in this manual are equipped with a 12-volt electrical system. The system includes an alternator, mounted on the left end of the crankshaft, a voltage regulator and a battery. Since the output of the alternator is alternating current, the regulator is combined with a rectifier to convert the alternating current (AC) to direct current (DC), which is needed to charge the battery and operate the electrical components on the motorcycle.

A large capacity battery is used for starting the engine, to provide additional current, when necessary, over the amount being generated and to operate accessories when the engine isn't running.

On models equipped with an electric starter, a solenoid relay provides power to the starter motor directly from the battery. The solenoid is controlled by a switch on the handlebars.

Keep in mind that electrical parts, once purchased, cannot normally be returned. To avoid unnecessary expense, make very sure the defective component has been positively identified before buying a replacement part.

Caution: *Always disconnect the battery when working on the electrical system to avoid causing an accidental short circuit. Always disconnect the negative cable first, followed by the positive cable.*

2 Electrical troubleshooting – general information

Refer to illustration 2.5

A typical electrical circuit consists of an electrical component, any switches, relays, motors, fuses or circuit breakers related to that component and the wiring and connectors that link the component to both the battery and the frame. To help pinpoint electrical circuit problems, wiring diagrams are included at the end of the manual.

Before tackling any troublesome electrical circuit, first study the appropriate wiring diagram to get a complete understanding of what makes up the circuit. Trouble spots, for instance, can often be narrowed down by noting if other components related to the circuit are operating properly. If several components or circuits fail at one time, chances are the problem is in a fuse or ground connection, because several circuits are often routed through the same ones.

Electrical problems usually stem from simple causes, such as loose or corroded connections, a blown fuse or a bad relay. Visually check the condition of all fuses, wires and connections in a problem circuit before troubleshooting it.

If test instruments are going to be utilized, use the diagram to plan ahead of time where to make the connections in order to accurately pinpoint the trouble spot.

The basic items needed for electrical troubleshooting include a 12-volt test light, a voltmeter, a self-powered continuity tester (which includes a bulb or buzzer, battery and set of test leads), and a jumper wire, preferably with a fuse incorporated, which can be used to bypass electrical components **(see illustration)**.

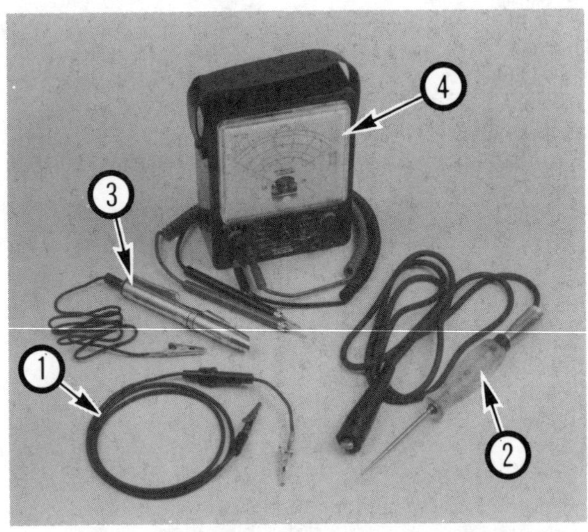

2.5 Basic electrical troubleshooting can be done with an assortment of tools from a simple jumper wire to a multimeter (volt/ohm/ammeter)

1 Jumper wire (with in-line fuse)
2 12-volt test light
3 Continuity tester
4 Multimeter

Voltage checks

A voltage check should be done if a circuit isn't functioning properly. Connect the test light alligator clip to the negative battery terminal. Use the test light probe to contact the connector wire terminal(s) in the circuit being tested, starting with the one nearest to the battery or fuse. If the test bulb lights, voltage is present, which means the part of the circuit between the connector and battery is okay. Continue checking the rest of the circuit in the same manner. When you reach a point where no voltage is indicated, the problem is between that point and the last test point with voltage. Most of the time the problem can be traced to a loose connection. **Note:** *Keep in mind that some circuits receive voltage only when the ignition key is in the On position.*

Finding a short

One method of finding a short in a circuit is to remove the fuse and connect a test light or voltmeter in its place to the fuse terminals. The circuit should be off (no voltage present in the circuit). Move the wiring harness from side-to-side while watching the test light. If the bulb lights, there's a short to ground somewhere in that area, probably where the insulation has rubbed through. The same test can be performed on each component in the circuit, even a switch.

Ground check

Perform a ground check to see if a component is properly grounded. Disconnect the battery and connect one lead of a self-powered test light, known as a continuity tester, to a known good ground. Connect the other lead to the wire or ground connection being tested. If the bulb lights, the ground is good. If the bulb doesn't light, the ground is bad.

Continuity check

A continuity check is done to determine if there are any breaks in a circuit – if it's capable of passing electricity properly. With the circuit off (no power in it), a self-powered continuity tester can be used to check it. Connect the test leads to both ends of the circuit (or to the "hot" wire and a good ground). If the test light comes on, the circuit is passing current properly. If the light doesn't come on, there's a break (called an open) somewhere. The same procedure can be used to test a switch, by connecting the continuity tester leads to the switch terminals. With the switch on, the test light should come on.

Finding an open circuit

When looking for possible open circuits, it's almost impossible to locate them by sight because oxidation and terminal misalignment are hidden by the connectors. Sometimes, wiggling a connector in the wiring harness will correct the open circuit condition temporarily. Remember this when an open circuit is indicated when troubleshooting a circuit. Intermittent problems may also be caused by oxidized or loose connections.

Electrical troubleshooting is simple if you keep in mind that all electrical circuits are basically electricity running from the battery, through the wires, switches, relays, fuses, etc. to each electrical component (light bulb, motor, etc.) and to ground, then back to the battery. All electrical problems are an interruption in the flow of electricity to and from the battery.

3 Charging system check – general information

1 If the battery isn't being recharged, the charging system should be checked first, followed by testing of the individual components (the alternator and regulator). Before beginning the checks, make sure the battery is fully charged and all system connections are clean and tight (particularly the battery cables).

2 Checking the output of the charging system and the operation of the components in the system requires special electrical test equipment. A voltmeter and ammeter or a multimeter are the absolute minimum tools required. In addition, an ohmmeter is generally required for checking the remainder of the electrical system.

3 When making the checks, follow the procedures carefully to prevent incorrect connections and short circuits – irreparable damage to electrical system components may result if short circuits occur. Because of the special tools and expertise required, checking the electrical system should be left to a dealer service department or a reputable motorcycle repair shop.

4 Battery – check and maintenance

1 Most battery damage is caused by heat, vibration, and/or low electrolyte levels, so keep the battery securely mounted, check the electrolyte level frequently, and make sure the charging system is functioning properly.

2 Refer to Chapter 1 for the electrolyte level checking procedure.

6.2 The battery retaining strap can be disconnected after loosening the nut (arrow)

3 Check around the base of the battery for sediment, which is the result of sulfation caused by low electrolyte levels. These deposits will cause internal short circuits, which can quickly discharge the battery. Look for cracks in the case and replace the battery if either of these conditions is found.

4 Check the battery terminals and cable ends for tightness and corrosion. If corrosion is evident, remove the cables from the battery and clean the terminals and cable ends with a wire brush or a knife and emery cloth. Reconnect the cables and apply a thin coat of petroleum jelly to the connections to slow further corrosion.

5 The battery case should be kept clean to prevent current leakage, which can discharge the battery over a period of time (especially when it sits unused). Wash the outside of the case with a solution of baking soda and water. **Caution:** *Do not get any baking soda solution in the battery cells.* Rinse the battery thoroughly, then dry it.

6 If acid has been spilled on the frame or battery box, neutralize it with the baking soda and water solution, dry it thoroughly, then touch up any damaged paint. Make sure the battery vent tube is directed away from the frame and chain and isn't kinked or pinched.

7 If the motorcycle sits unused for long periods of time, refer to Section 5 and charge the battery approximately once every month.

5 Battery – charging

1 If the machine sits idle for extended periods or if the charging system malfunctions, the battery can be charged from an external source.

2 To charge the battery properly, you will need a charger of the correct rating, a hydrometer, a clean rag and a syringe for adding distilled water to the battery cells.

3 The maximum charging rate for any battery is 1/10 of the rated amp/hour capacity. For example, the maximum charging rate for a 22 amp/hour battery would be 2.2 amps. If the battery is charged at a higher rate, it could be damaged.

4 Don't allow the battery to be subjected to a so-called quick charge (high rate of charge over a short period of time) unless you're prepared to buy a new battery.

5 When charging the battery, always remove it from the machine and be sure to check the electrolyte level before hooking up the charger. Add distilled water to any cells that are low.

6 Loosen the cell caps, hook up the battery charger leads (red to positive, black to negative), cover the top of the battery with a clean rag, then – and only then – plug in the battery charger. **Caution:** *Remember, the hydrogen gas escaping from a charging battery is explosive, so keep open flames and sparks well away from the area. Also, the electrolyte is extremely corrosive and will damage anything it comes in contact with.*

6.4 Always detach the negative battery cable first and hook it up last – it will prevent the positive cable from being shorted to ground by the tool being used to remove it

7 Allow the battery to charge until the specific gravity is as specified. The charger must be unplugged and disconnected from the battery when making specific gravity checks. If the battery overheats or gases excessively, the charging rate is too high. Either disconnect the charger or lower the charging rate to prevent damage to the battery.

8 If one or more of the cells do not show an increase in specific gravity after a long slow charge, or if the battery as a whole doesn't seem to want to take a charge, it's time for a new battery.

9 When the battery is fully charged, unplug the charger first, then disconnect the leads from the battery. Install the cell caps and wipe any electrolyte off the outside of the battery case.

6 Battery specific gravity – check

Caution: *Be extremely careful when handling or working around the battery. The electrolyte is very caustic and explosive hydrogen gas is given off when the battery is being charged.*

1 It may be helpful to remove the battery from the motorcycle for this check.

All models except FLT and FXR

Refer to illustrations 6.2 and 6.4

2 Loosen the nuts securing the battery retaining strap and release the strap from the bottom of the battery box **(see illustration)**.

3 On models so equipped, remove the decorative cover from the top of the battery.

4 Disconnect the cable from the negative terminal of the battery first, followed by the positive cable **(see illustration)**.

5 Lift the battery out of the box. Note how the vent hose is routed.

FLT models

6 On all models through 1992, remove the saddlebag from the right side of the motorcycle.

7 On 1993 and later models, remove the seat to gain access to the battery (retaining screw accessed from inside the tour-pak).

8 On all models disconnect the negative cable from the battery, followed by the positive cable.

9 Disconnect the battery hold-down strap and lift the battery out.

FXR models

10 Release the seat latch and tilt the seat up.

11 Disconnect the battery cables from the battery terminals (be sure to detach the negative cable first).

12 Remove the cable securing the seat to the battery hold-down bracket. Loosen the locknut, then remove the screw so the cable can be released.

13 Push the battery hold-down bracket toward the left rear side of the motorcycle. The bracket can then be separated from the slots in the frame.
14 Lift the battery out of the carrier after disconnecting the vent hose located near the negative terminal.

All models

15 Check the electrolyte level in each cell of the battery. If any of the cells are low, fill them to the proper level with distilled water. Do not use tap water (except in an emergency) and do not overfill the battery. The cell holes are quite small, so it might help to use a plastic squeeze bottle with a small spout to add the water.
16 Next, check the specific gravity of the electrolyte in each cell with a small hydrometer made especially for motorcycle batteries. They are available at most dealer parts departments and motorcycle accessory shops.
17 Remove the caps, draw some electrolyte from the first cell into the hydrometer and note the specific gravity. Compare the reading to the Specifications. Return the electrolyte to the appropriate cell and repeat the check for the remaining cells. When the check is complete, rinse the hydrometer thoroughly with clean water.
18 If the specific gravity of the electrolyte in each cell is as specified, the battery is in good condition and is apparently being charged by the machine's charging system.
19 If the specific gravity is low, the battery is not fully charged. This may be due to corroded battery terminals, a dirty battery case, a malfunctioning charging system, or loose or corroded wiring connections. On the other hand, it may be that the battery is worn out, especially if the machine is old, or infrequent use of the motorcycle prevents normal charging from taking place. If the specific gravity of any two cells varies by more than 50 points, replace the battery with a new one.
20 Be sure to correct any problems and charge the battery if necessary before reinstalling it in the machine. Refer to Sections 4 and 5 for additional battery maintenance and charging procedures.
21 Install the battery by reversing the removal procedure. Be very careful not to pinch or otherwise restrict the battery vent tube, as the battery may build up enough internal pressure during normal charging system operation to explode.

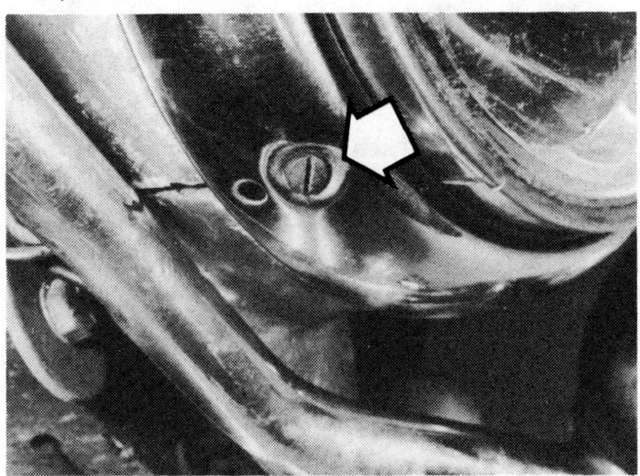

7.1 The chrome headlight trim ring is secured by a single screw

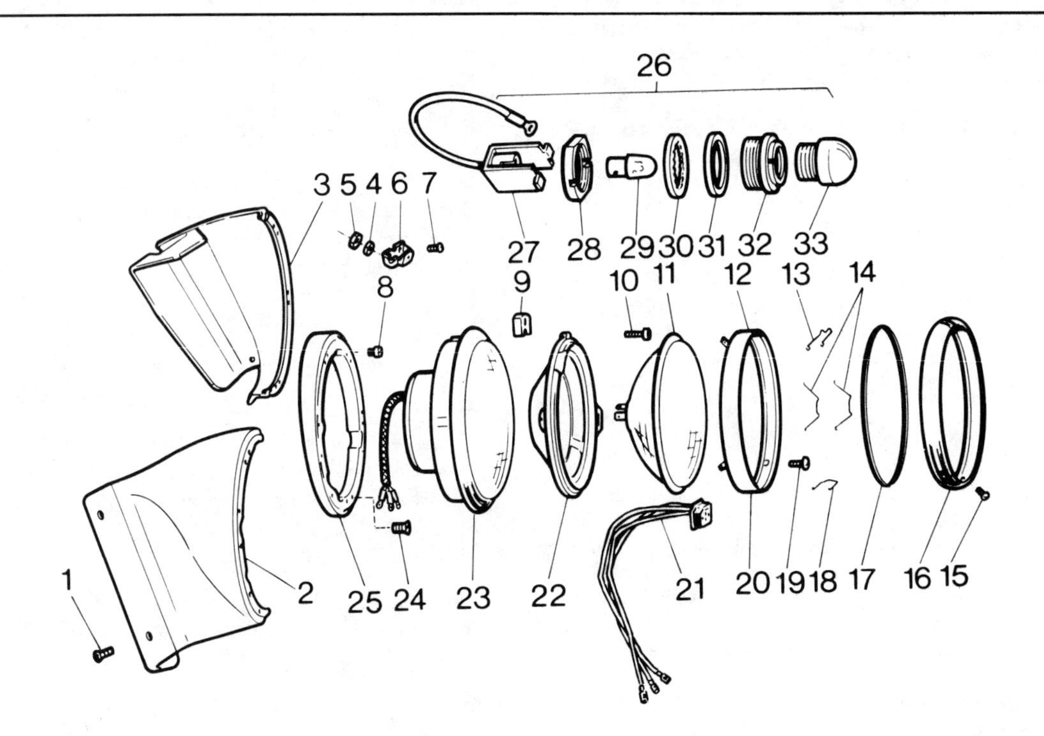

7.2 Typical headlight assembly components (FL models)

1 Screw
2 RH shell half
3 LH shell half
4 Fiber washer
5 Nut
6 Retaining catch
7 Screw
8 Rubber buffer
9 Threaded block
10 Adjusting screw
11 Sealed beam unit
12 Headlight retaining ring
13 Catch spring
14 Side retaining spring
15 Screw
16 Headlight rim
17 Rubber ring
18 Lower spring
19 Adjusting screw
20 Inner rim (if equipped)
21 Wire harness
22 Mounting ring
23 Headlight unit
24 Screw
25 Headlight housing
26 High beam warning light
27 Light body
28 Nut
29 Bulb
30 Serrated washer
31 Fiber washer
32 Light holder
33 Lens

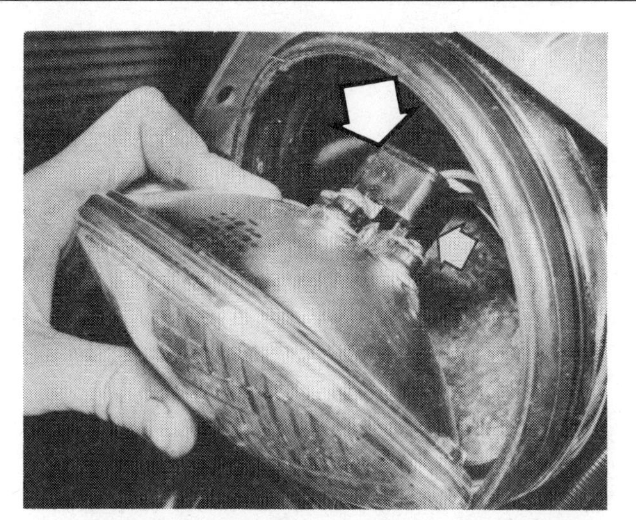

7.3 Unplug the wire harness connector (arrow) from the rear of the sealed beam

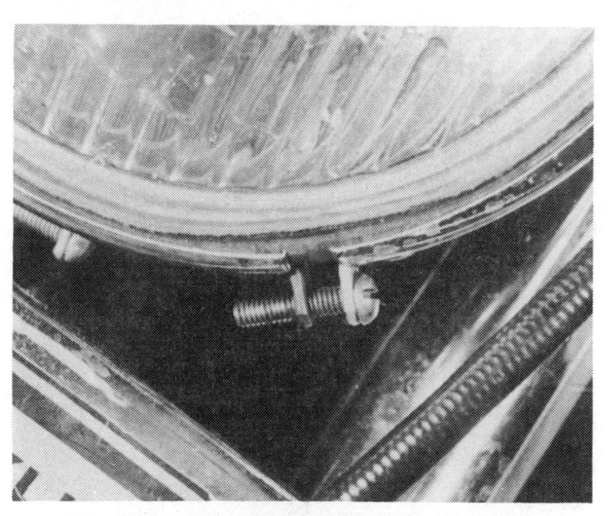

7.7 Remove the clamping ring screw to detach the headlight

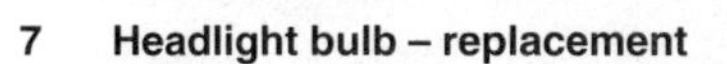

7 Headlight bulb – replacement

Note: *Either a sealed beam or quartz bulb headlight is fitted, depending on the model. The sealed beam unit comprises the reflector, glass and bulb as a single unit, whereas the quartz bulb can be replaced separately from the reflector unit.*

Caution: *Replaceable quartz headlight bulbs must be handled very carefully. Never touch the glass with your fingers – the oil from your skin will etch the glass and cause the bulb to fail. Wrap the bulb with paper or a clean, dry cloth when handling it. Also, wear safety glasses or a face shield – the bulb contains halogen gas under pressure and could cause injury if it breaks.*

1970 through 1980 FL models

Refer to illustrations 7.1, 7.2 and 7.3

1 Remove the screw that retains the chrome plated rim surrounding the headlight **(see illustration)**. Detach the rim.

2 Remove the three screws securing the headlight retaining ring. Do not turn the adjuster screws **(see illustration)**.

3 Pull the headlight out of the housing until the wire harness can be unplugged from the rear **(see illustration)**.

4 Attach the wire harness plug to the new headlight and position the new headlight in the housing. The headlight is usually marked in some way to indicate the top. Be sure the mark is at the top when the headlight is positioned in the housing.

5 Secure the headlight in the housing by attaching the retaining ring with the three screws.

6 Install the decorative chrome rim around the headlight. Adjustment shouldn't be necessary if the adjusting screws weren't turned.

All FX/Softail/Dyna models

Refer to illustrations 7.7 and 7.8

7 Remove the screw from the chrome plated clamp that surrounds the headlight **(see illustration)**. Take off the clamp.

8 Pry the sealed beam out of the rubber mount **(see illustration)**.

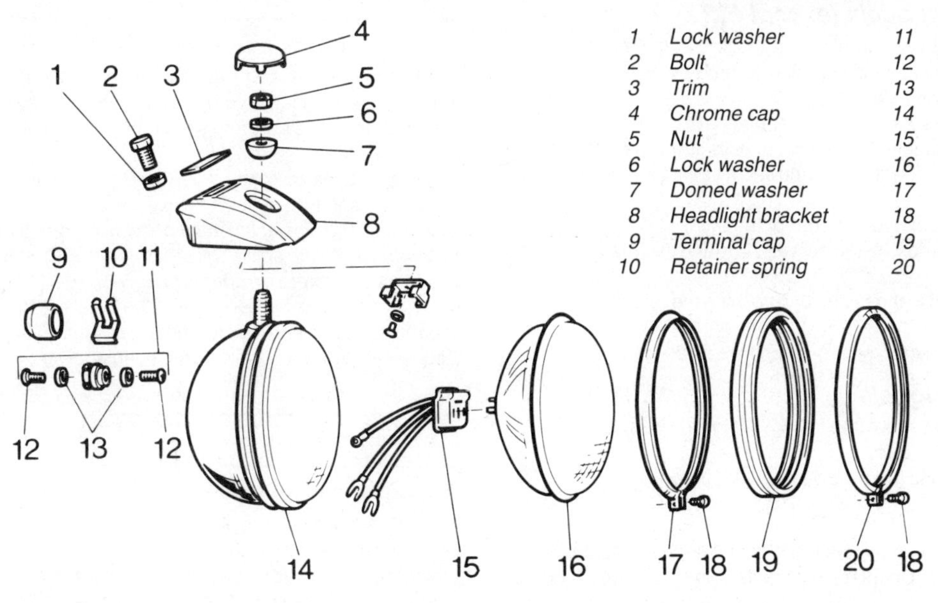

7.8 Typical headlight assembly components (FX models) (early models shown)

1	*Lock washer*	*11*	*Terminal unit*
2	*Bolt*	*12*	*Screw*
3	*Trim*	*13*	*Serrated washer*
4	*Chrome cap*	*14*	*Headlight*
5	*Nut*	*15*	*Wire harness*
6	*Lock washer*	*16*	*Sealed beam unit*
7	*Domed washer*	*17*	*Inner clamp*
8	*Headlight bracket*	*18*	*Screw*
9	*Terminal cap*	*19*	*Rubber mounting ring*
10	*Retainer spring*	*20*	*Outer clamp ring*

8.1a The taillight lens and . . .

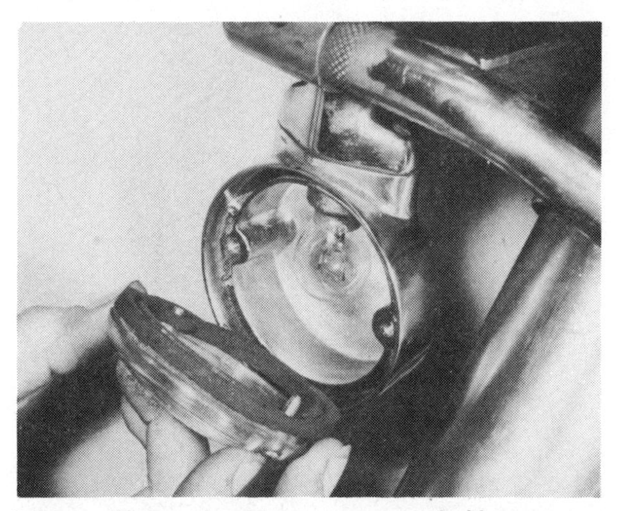

8.1b . . . the turn signal lenses are secured with two screws

9 Pull the sealed beam out of the housing until the wire connector at the rear of the unit can be unplugged. **Note:** *On models with a quartz bulb, the bulb can be removed from the reflector after the retainer clip (where fitted) is released (see Caution above).*

10 To install the sealed beam unit, attach the wire connector to the rear of the reflector and position it in the rubber mount. Where a quartz bulb is fitted, install the bulb without touching its glass envelope and secure with the retainer clip (where fitted), then attach the wire connector and rubber boot.

11 Install the clamp ring around the headlight and tighten the screw securely.

1980 and later FLT models

12 Remove the three screws and detach the cover plate.

13 Both sealed beams are secured with a retaining ring. Loosen the three screws securing the retaining ring that must be removed (the screws don't have to be removed). As soon as they're loose, the retaining ring can be turned and lifted out.

14 Withdraw the sealed beam until the wire harness plug can be disconnected from the rear of the defective headlight.

15 Connect the wire harness plug to the rear of the new sealed beam and position the headlight in the housing.

16 Rotate the retaining ring into position and tighten the screws securely.

17 Install the cover plate with the three screws.

1981 and later FL models (except FLT)

18 Remove the headlight from the fairing as described in Steps 1 through 3 above. On 1985 and earlier models, you may have to loosen and tip the fairing to remove the chrome rim.

19 Remove the rubber boot from the rear of the headlight.

20 Press the wire clip together and pull the bulb out of the reflector.

21 Install the new bulb in the reflector and secure it with the wire clip (see Caution above).

22 Attach the rubber boot over the rear of the bulb.

23 Connect the wire harness to the bulb and position the headlight in the fairing.

24 Secure the headlight in the fairing by attaching the retaining ring with the three screws.

25 Install the decorative chrome rim around the headlight.

FXRT and FXRD models

26 Working from the rear side of the fairing, unplug the wire harness connector from the rear of the bulb.

27 Follow the instructions in Steps 19 through 23 above.

All models

28 Adjustment should not have been altered while changing the headlight. If necessary, refer to Chapter 1 for the headlight adjustment procedure.

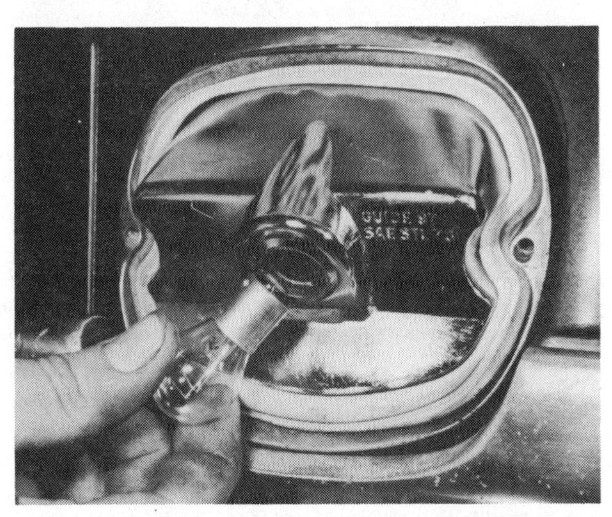

8.2 The two-filament taillight bulb has offset pins so it can only be installed one way in the socket

8 Taillight and turn signal bulbs – replacement

Refer to illustrations 8.1a, 8.1b and 8.2

1 Remove the screws securing the plastic lens cover to the taillight or the turn signal and detach the cover **(see illustrations)**.

2 Push in on the bulb and simultaneously turn it counterclockwise **(see illustration)**.

3 Replace the bulb with a new one of the same type by pushing it into the socket and turning it clockwise. **Note:** *The taillight bulb is a two filament bulb. The pins at the base of the bulb are offset so the bulb can only be inserted into the socket one way. If the bulb will not go into the socket, pull it out and rotate it 180-degrees, then reinsert it into the socket.*

4 Place the lens in position over the housing. Be sure the rubber seal is in good condition and makes contact all around the perimeter of the lens. Secure the lens with the two mounting screws.

9 Alternator – check

Refer to illustrations 9.4a and 9.4b

1 Checking the output of the alternator and other charging system components requires various electrical test devices which include a 0-50 amp ammeter, a 0-20 volt voltmeter, a variable resistor (rheostat) of 250 watt, 15 ohm resistance and an accurate ohmmeter. Unless the owner has the

9.4a Alternator output test instrument hook-up – 1970 through 1975 models

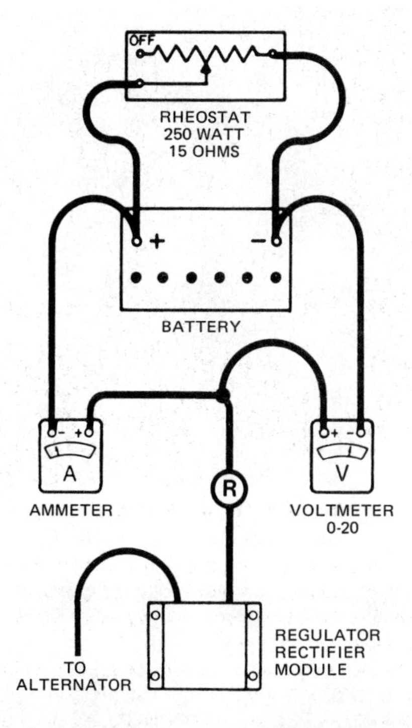

9.4b Alternator output test instrument hook-up – 1976 and later models

necessary equipment and experience required for the following procedures, the check should be done by a Harley-Davidson dealer service department.

2 If the tests are done, note the following:

a) Do not run the engine with the battery disconnected and do not reverse the battery cables.

b) Do not disconnect or reconnect the alternator-to-rectifier/regulator module wires while the engine is running.

3 Machines from 1976-on have a similar but modified electrical system compared to the one used on earlier models (particularly in the construction of the rectifier/regulator module). The test procedures and the results differ from one to the other – don't mix them up!

4 Connect the ammeter, voltmeter and variable resistor to the circuit as shown in the accompanying illustrations. Start and run the engine at 3600 rpm and adjust the variable resistor until the output is 3.5 amperes. If the air temperature measured at the rectifier/regulator module is 75-degrees F, the voltage reading should be between 13.8 and 15.0 volts. If this test is positive, reduce the engine speed to 2000 rpm and adjust the variable resistor until a constant 13.0 volts is obtained:

a) On 1975 and earlier models, the current (amperage) reading should be 10.5 amps minimum.

b) On 1976 through 1983 models, it should be 14 amps minimum.

c) On 1984 through 1988 models, it should be 19 amps minimum to 23 amps.

d) On 1989 and later models, it should be 29 amps minimum to 33 amps.

5 If the above test results are obtained, the alternator and rectifier/regulator module are in good condition. If not, the following test will determine whether the alternator or rectifier/regulator unit is bad. Disconnect the alternator at the socket.

6 On 1970 through 1975 models, run the engine at 2000 rpm. Connect a 0-150 volt AC voltmeter across the two white wire pins; the reading should be 50 to 100 volts AC. Make a similar connection across the blue and red wire pins; the reading should be 75 to 125 volts AC.

7 On 1976 and later models, connect the voltmeter across the two black (stator wire) pins. The reading should be:

a) On 1976 through 1988 models, 19 to 26 volts AC for every 1000 rpm.

b) On 1989 and later models, 16 to 20 volts AC for every 1000 rpm.

8 On all models, a satisfactory test result indicates the alternator is in good condition. The rectifier/regulator module should be checked as described in the next Section.

10 Rectifier/regulator module – check

1 The tests carried out in Steps 4 through 7 in the preceding Section will reveal whether the alternator or rectifier/regulator module is malfunctioning.

2 On 1970 through 1975 models, check the module resistance values as shown in the following table. Make one test with positive polarity and the second test with the polarity reversed. The module must be replaced with a new one if the values are not within the specified range .

Probe connection	Positive polarity	Negative polarity
White wire-to-module base	Infinity	3 to 15 ohms
Blue wire-to-black wire	3 to 15 ohms	Infinity
Red wire-to-module base	Infinity	Infinity

3 On 1976 and later models, no test procedures are possible. If you suspect the rectifier/regulator is malfunctioning, replace it with a new one. It may be a good idea to have your test results confirmed by a dealer service department before buying a new part.

11 Starter motor – removal, overhaul and installation

1 Three different types of starter motors are used on these machines, although their operation is similar. One is made by Prestolite (USA) and is used on machines through 1983, while another is made by Hitachi (Japan) and is used on machines from late 1978 through 1988. A third starter, with an integral solenoid and drive/overrunning clutch assembly is used on

11.3 The battery carrier may have to be removed to gain access to the starter motor

11.4 Detach the cable and remove the nut, then pull the clutch arm off the shaft

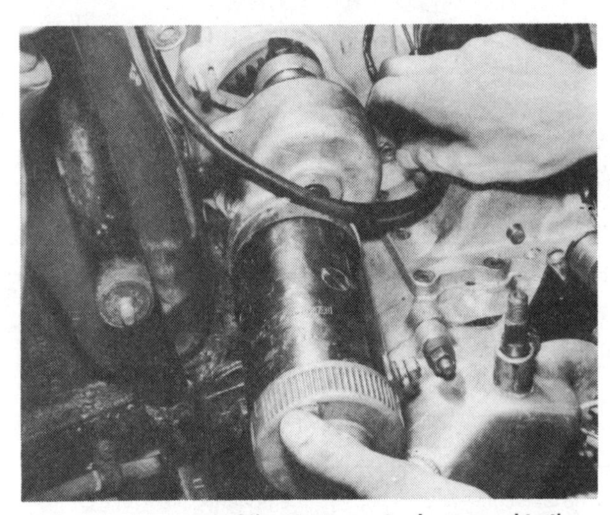

11.5 The drive end of the starter motor is secured to the primary chaincase by two nuts or bolts (1970 through 1983 models)

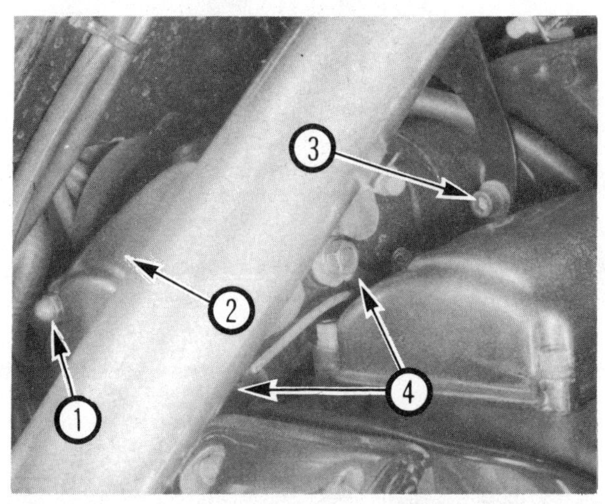

11.8 Starter motor mounting details – 1989 and later models

1 *Acorn nuts (one is hidden in this photo)*
2 *Chrome starter motor end cover*
3 *Mounting bolts (one is hidden in this photo)*
4 *Wires (remove the nut to detach the large wire)*

1989 and later models. Generally speaking, the starter motor requires very little attention during the normal service life of the machine, although periodic examination of the brushes and commutator is recommended.

2 If the starter motor refuses to function, first eliminate other possible components that may give the impression the starter motor needs attention. Check the wires for good connections and make sure the battery is fully charged. The handlebar switch must make good contact and the starter solenoid should engage with an audible click. Finally, the engine itself may be stiff if its been recently rebored and had new bearings installed, if it's very cold, or if excessively high viscosity oil has been used.

Removal

1970 through 1988

Refer to illustrations 11.3, 11.4 and 11.5

3 To remove the starter motor from the machine, first remove the battery cover and disconnect the battery cables (negative first, then positive). On some models, the battery and carrier must be removed in order to gain access to the motor **(see illustration)**. The carrier is attached by two rubber mounted studs and two or three bolts.

4 On 1983 and earlier models, disconnect the clutch cable at the operating arm, then remove the arm from the shaft projecting from the end cover of the transmission. The arm is retained by a single nut **(see illustration)**.

5 Disconnect the wires from the starter motor, then remove the outer support bracket, which is attached to the transmission end cover. The starter motor is secured to the rear of the primary chaincase by two nuts on studs. Remove the nuts (from the right side) and pull the starter motor out, leaving the driveshaft in position in the case **(see illustration)**. Pull the starter motor driven reduction gear out of the housing.

6 On 1984 and later models, disconnect the solenoid cable from the starter terminal (except FLT and FXR models).

a) On FXEF, FXSB and FXWG models – Remove the nuts that fasten the master cylinder reservoir bracket to the transmission. Remove the nut and detach the starter bracket. Remove the chrome end cover and detach the starter bracket from the through-bolt stud. Remove the through-bolts and hold the starter by both end covers to keep it from coming apart as you remove it.
b) On FXST/C and FLST/C models – Remove the nuts that hold the starter bracket to the transmission. Detach the chrome end cover and bracket, then remove the relay-to-starter ground wire. Remove the through-bolts and hold the starter by both end covers to keep it from coming apart as you remove it.

c) On FLT and FXR models – Remove the bolt that fastens the starter to the transmission side door. Remove the upper bolts (and the battery cable), then detach the starter and solenoid/drive gear as an assembly. Detach the solenoid cable from the starter motor terminal, then remove the through-bolts to separate the motor from the starter shaft housing.

1989-on

Refer to illustration 11.8

Note: *On FX/Softail models, remove the rear exhaust pipe. You may also have to loosen the oil tank mounts to provide room for starter removal.*

7 Remove the primary chaincase cover (see Chapter 2), then hold the starter drive pinion gear and remove the jackshaft bolt and lockplate (flatten the lockplate ear before removing the bolt).

8 Remove the two acorn nuts and detach the chrome end cover from the right side of the motor **(see illustration)**.

9 Remove the two Allen-head starter mounting bolts and pull the starter out, then detach the wires from the starter terminals and remove it from the right side. **Note:** *The jackshaft-to-starter coupling may come off with the starter or it may stay on the jackshaft. If it comes off with the starter, put it back on the jackshaft. On 1990 models, the coupling end with the counterbore must be on the jackshaft.*

Overhaul

Refer to illustrations 11.10a, 11.10b, 11.11a and 11.11b

10 Remove the two nuts/screws that retain the starter motor end cover, then remove the two long through-bolts (they were already removed on 1984 and later models). Detach the reduction gear housing (not all models). With the end cover removed, it will be possible to see the brushes, which must be replaced with new ones as a set if they have worn beyond the specified limits **(see illustrations)**.

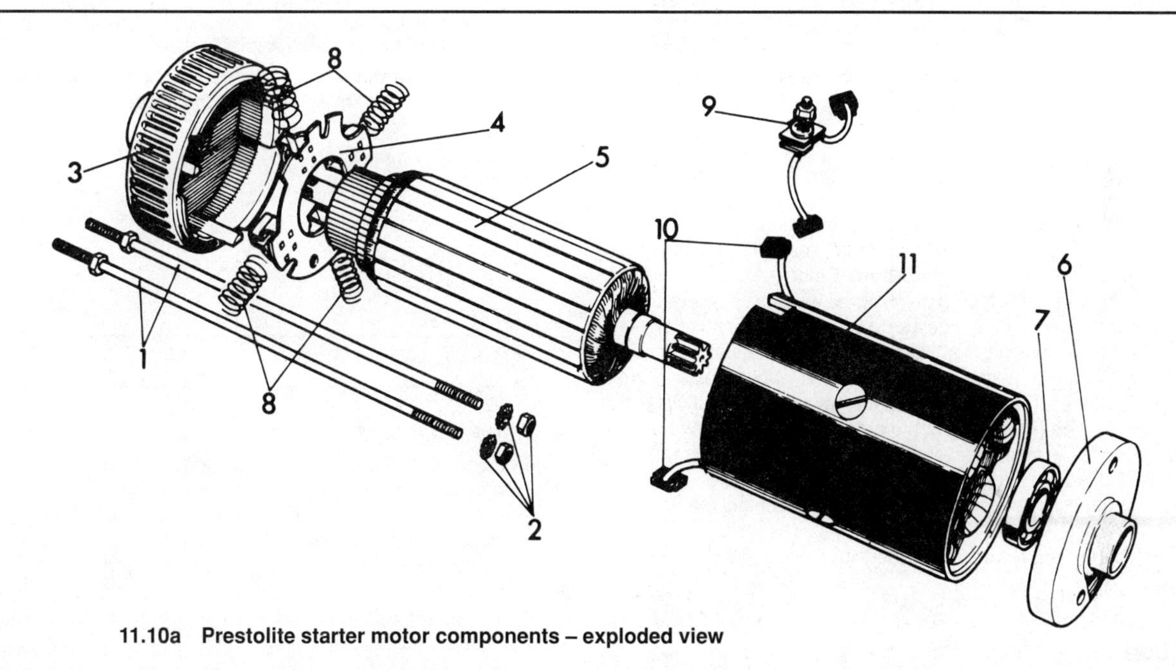

11.10a Prestolite starter motor components – exploded view

1 *Through-bolt*
2 *Washer and lock washer*
3 *Commutator end cover*
4 *Brush plate and holder assembly*
5 *Armature*
6 *Drive end cover*
7 *Drive end ball bearing*
8 *Brush spring*
9 *Terminal and brush assembly*
10 *Ground brush*
11 *Main body shell and field coil assembly*

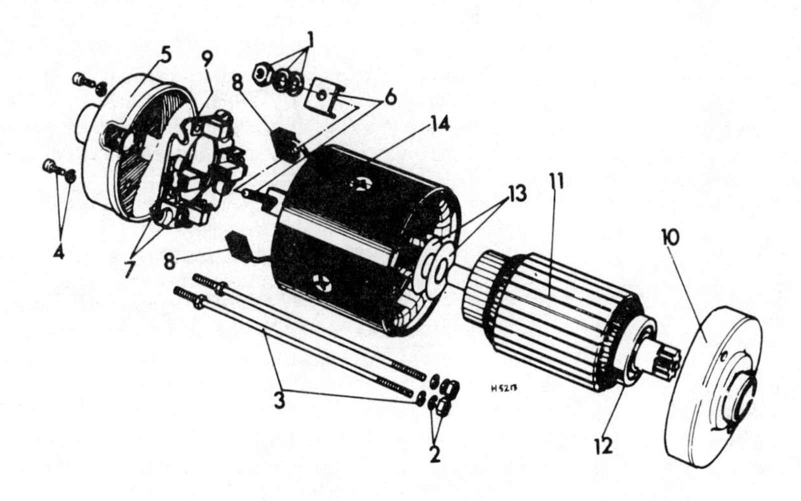

11.10b Hitachi starter motor components – exploded view

1 *Terminal nut, lock washer and washer*
2 *Through-bolt nut and lock washer*
3 *Through-bolt and lock washer*
4 *Rear cover screw and lock washer*
5 *Rear (commutator end) cover*
6 *Terminal and insulator*
7 *Negative brush*
8 *Positive brush*
9 *Brush holder assembly*
10 *Front (drive end) cover*
11 *Armature*
12 *Armature ball bearing*
13 *Thrust washer*
14 *Housing*

11.11a Check and clean the commutator and brushes . . .

11.11b . . . and measure the brush length after removing them from the holders

11 Make sure the brushes can move freely in the holders **(see illustration)**. If not, remove them so the holders can be cleaned with solvent to remove built-up carbon dust **(see illustration)**. Clean the commutator at the same time, using solvent – NOT a harsh abrasive such as emery cloth. If the starter motor has been in service a long time, the mica insulation between the individual copper segments of the commutator must be undercut by 1/32-inch (a task requiring dealer attention). When brushes are replaced with new ones, the original leads will have to be cut off close to the field coil connection (Prestolite) or unsoldered from the brush holder (Hitachi). Do not use too much heat or solder, otherwise the leads will become clogged with solder and lose their flexibility.
12 Before reassembly, make sure none of the leads contact the body of the motor or otherwise short out.
13 Other defects will require dealer attention or the installation of a new starter motor.

Installation

14 Installation is the reverse of removal. On 1989 and later models, make sure the starter jackshaft bolt lockplate tab fits into the keyway in the sleeve.

12 Starter solenoid – check

Refer to illustrations 12.2 and 12.4

1 Disconnect the negative cable from the battery to avoid an accidental short circuit.
2 Pry the rubber cap off the rear of the starter solenoid, exposing the wire terminals **(see illustration)**.
3 Label the wires and terminals, then disconnect the wires by removing the nuts.
4 Using jumper wires, connect a 12-volt battery to the terminals on the solenoid **(see illustration)**. One wire should go to the small terminal and the other wire should go to the shorter of the large terminals.
5 The solenoid should make a clicking sound. If a click or heavy spark at the terminal does not occur, the solenoid is defective and must be replaced with a new one.

12.2 On all models through 1988, the solenoid is attached to the primary chaincase – pull back the rubber cap to expose the terminals and wires

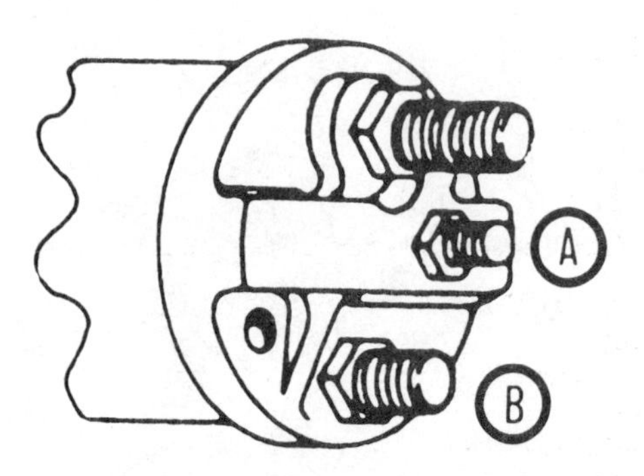

12.4 Check the solenoid by attaching the jumper wires to terminals A and B – the solenoid should click when battery voltage is applied to the terminals

13 Starter motor pinion shaft and engagement mechanism – disassembly, inspection and reassembly

1 The starter motor utilizes a reduction gear to drive the gear mounted on the clutch outer drum. On all models through 1988, the reduction driven gear turns a shaft on which a free-sliding pinion gear is mounted. When the starter button is operated, the starter solenoid moves the pinion gear into mesh with the clutch outer drum gear a moment before power is provided to the starter motor. The starter solenoid operates a pivot arm, which operates the pinion gear. On 1989 and later models, the pinion shaft is replaced by a jackshaft. A gear on the left end of the jackshaft is always engaged with the gear on the clutch outer drum and the starter drive connects and disconnects the motor from the jackshaft.

1970 through 1988

All models except FLT, FXR, FXB, FXSB, FXWG, FXST/C and FLST/C (belt-drive) models

Refer to illustrations 13.2, 13.3, 13.4, 13.5, 13.8a, 13.8b, 13.9 and 13.10

2 Remove the primary chaincase cover as described in Chapter 2 to gain access to the engagement mechanism. Any broken or worn parts should be obvious immediately **(see illustration)**.

3 To disassemble the components, pull the bronze thrust washer off the

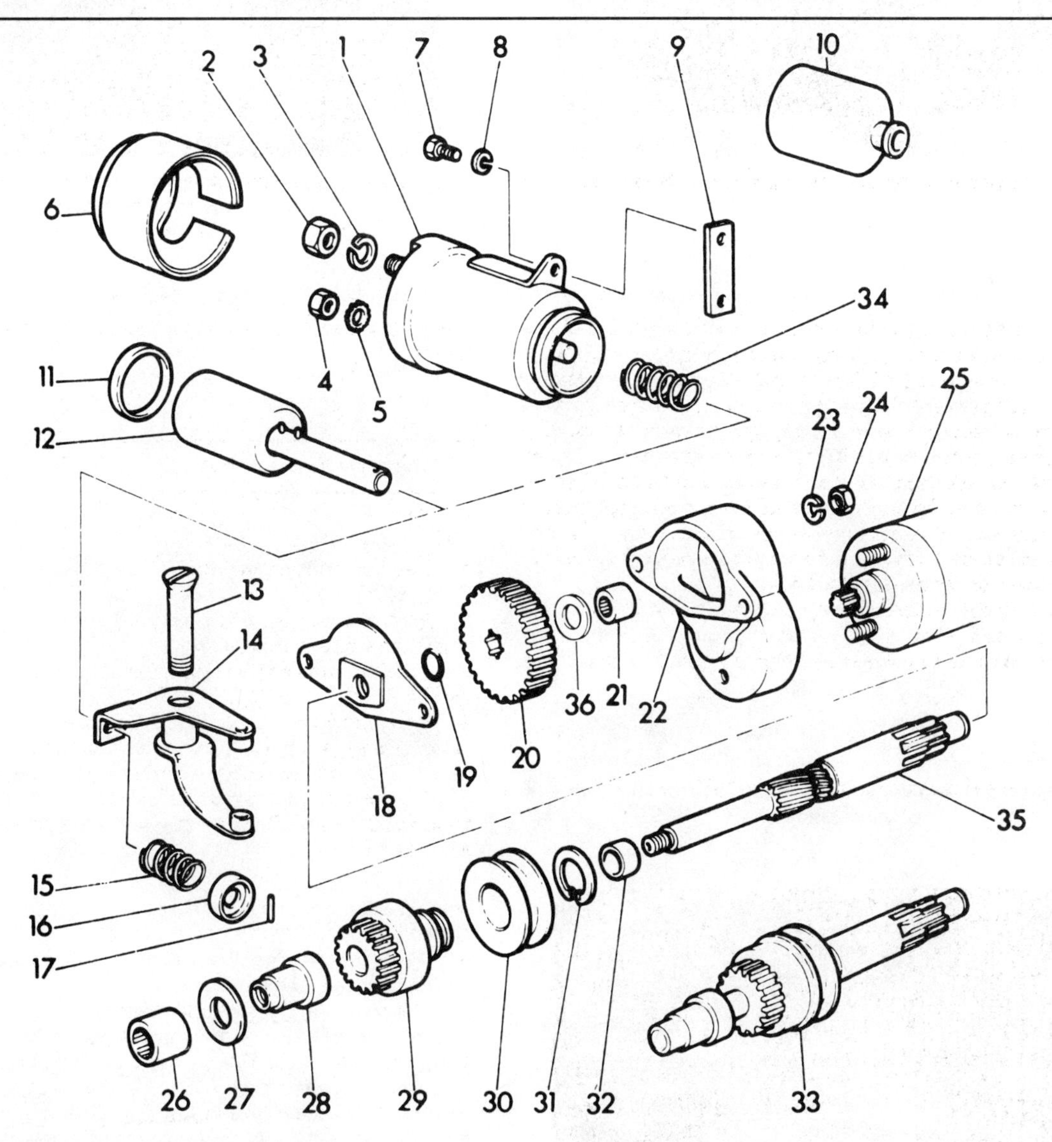

13.2 Typical starter solenoid and engagement mechanism components – chain-drive models (belt-drive models slightly different)

1 Solenoid
2 Nut
3 Lock washer
4 Nut
5 Serrated washer
6 Terminal boot
7 Bolt
8 Lock washer
9 Plate
10 Plunger boot
11 Sealing ring
12 Solenoid plunger
13 Pivot screw
14 Pivot arm
15 Spring
16 Collar
17 Pin
18 Oil seal carrier plate
19 O-ring
20 Reduction driven gear
21 Inner needle roller bearing
22 Reduction gear housing
23 Lock washer
24 Nut
25 Starter motor
26 Outer needle roller bearing
27 Bronze thrust washer
28 Pinion shaft collar
29 Pinion gear
30 Shifter collar
31 Snap-ring
32 Spacer
33 Pinion shaft
34 Spring
35 Pinion shaft assembly
36 Thrust washer

13.3 Slide the bronze thrust washer (arrow) off the pinion shaft

13.4 Remove the pin (arrow) to disengage the solenoid plunger from the pivot arm

13.5 Remove the pivot arm screw (arrow) from the top of the chaincase

13.8a Check the starter gear, . . .

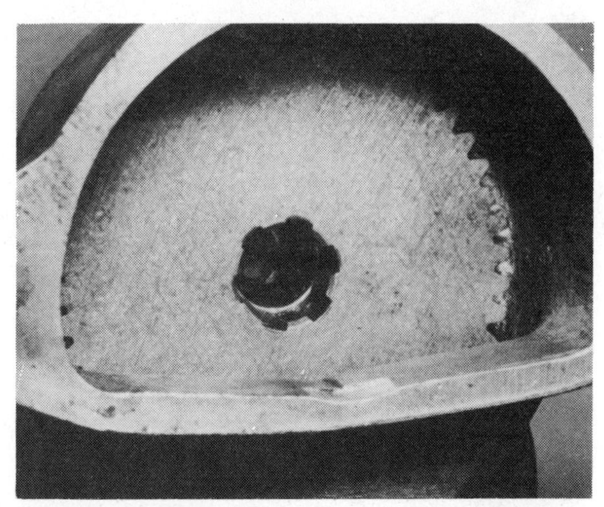

13.8b . . . and the gear teeth and splined hole of the reduction driven gear for wear and damage

shaft **(see illustration)** and remove the oil seal carrier plate from the opposite end.

4 Push the coil spring on the solenoid plunger back so the retaining pin can be removed **(see illustration)**.

5 The pivot arm rides on a large diameter countersunk screw which threads directly into the inner primary chaincase from above **(see illustration)**. The screw is usually very tight – be very careful to avoid damaging the head when removing it.

6 Manipulate the pivot arm so it leaves the solenoid plunger, then pull the complete shaft assembly out. The components on the shaft can be taken off after removing the end nut (left-hand threads) and pinion shaft collar.

7 The engagement components and pinion shaft can be installed by reversing the removal procedure.

8 Before installation, check the starter drive and reduction (driven) gears for wear and damage **(see illustrations)**. Make sure the splines in the reduction driven gear center hole that mate with the pinion shaft are in good condition (check the fit of the gear and shaft).

9 If either of the pinion shaft needle roller bearings are worn, they must be replaced with new ones. The bearing in the starter motor reduction gear housing is particularly prone to wear, since no means of lubrication is provided **(see illustration)**.

13.9 The needle roller bearing in the starter motor reduction gear housing should turn freely – apply grease to it before installation

13.10 Check the O-ring in the oil seal carrier plate – a new one should be installed each time the pinion shaft is removed

10 Make sure the O-ring in the oil seal carrier plate is in good condition. Replace it with a new one if it's cracked or otherwise deteriorated or damaged **(see illustration)**.

11 During reassembly, remember to install the bronze thrust washer on the left end of the pinion shaft. It's a loose fit and easily misplaced.

FXB, FXSB, FXWG, FXST/C and FLST/C (belt-drive) models

12 Remove the starter motor as described in Section 11, then remove the primary chaincase cover as described in Chapter 2.

13 Unscrew the three bolts securing the outer reduction driven gear housing and remove the housing and gasket.

14 Remove the reduction driven gear, then unscrew the two bolts securing the inner housing. Detach the inner housing and gasket.

15 Disengage the fingers of the shifter lever (pivot arm) from the shifter collar to remove the pinion shaft assembly and pinion gear from the clutch side.

16 Unscrew the solenoid mounting bolts and remove the solenoid, along with the spacer, felt gasket and spring.

17 Remove the screw securing the starter drive engagement lever and detach the lever. It may be necessary to remove the oil tank mounting hardware, the battery and the battery carrier to gain access to the lever retaining screw.

18 Check the gears and bearings for signs of wear and damage. If the reduction driven gear is worn, be sure to check the condition of the starter gear. Replace any defective components with new parts.

19 If the seal in the inner reduction driven gear housing requires replacement, carefully pry it out and press a new seal into the housing with the lip side facing the gear.

20 Use new gaskets during reassembly.

21 Reassembly is the reverse of disassembly. Lubricate the bearing in the outer housing, the starter drive engagement lever, the screw and bearing and the parts on the pinion shaft assembly, with high-temperature grease.

22 Align the groove in the shifter collar with the fingers of the starter drive engagement lever.

23 Lubricate the outer thrust washer and install it on the pinion shaft collar just before installing the primary chain cover.

FLT and FXR models

24 Remove the starter motor as described in Section 11, then remove the primary chaincase cover as described in Chapter 2.

25 Remove the thrust washer from the left end of the pinion shaft, then pull out the shaft.

26 Unscrew the solenoid mounting bolts and detach the solenoid, spring and felt gasket.

27 Remove the solenoid plunger from inside the primary chaincase. Pull the pin out, which will release the collar and the spring from the plunger. The plunger can then be withdrawn from the right side of the chaincase.

28 Unscrew the plug from the top of the chaincase, then remove the shaft securing the pivot arm from the opening in the chaincase.

29 Lift the pivot arm out.

30 Check the components as described in Steps 8 and 9 above.

31 Reassembly can be done by reversing the disassembly procedure. Be sure the lip on the pinion shaft collar fits against the pinion gear during reassembly.

32 Slide the shifter collar onto the pinion gear and secure it with the snap-ring. Install the assembly on the pinion shaft with the snap-ring facing the spacer.

1989 and later

33 Remove the primary chaincase cover and the clutch (see Chapter 2).

34 Flatten the lockplate ear. Hold the starter drive pinion gear and loosen the jackshaft bolt. Pull out the bolt, lockplate, thrust washer (1990 and later) and O-ring (1989 through 1992).

35 Pull the jackshaft assembly out of the inner chaincase as an assembly.

36 On 1989 models, remove the sleeve from the shaft. The Woodruff key may come out with it.

37 Pull off the pinion gear. Remove the coupling and spring (the spring and a retaining ring are inside the coupling).

38 If the starter coupling didn't come off with the jackshaft assembly, remove it from the starter shaft. Make sure the retaining ring is in place in the groove in the coupling.

39 On FLT and FXR models:

a) Check the bushing, seal and scraper in the back side of the primary chaincase. If the bushing is worn or damaged, remove it and install a new one – it must be flush to 0.010-inch (0.25 mm) below the face of the case.
b) If the seal is leaking, drive the seal and scraper out and discard them. Install a new seal and make sure the inner side is 0.110 to 0.120-inch (2.79 to 3.0 mm) from the end of the case bore. The scraper is no longer needed.
c) Check the bushing in the chaincase cover. If it's worn or damaged, remove it and install a new one – it must be flush to 0.030-inch (0.76 mm) below the case boss.
d) During reassembly, make sure the large coupling is installed with the retaining ring end facing the starter. If it's reversed it will contact the inner chaincase and the pinion gear won't engage the clutch ring gear.

15.4 The ignition switch on FL models (arrow) is retained by four screws

40 Install the large coupling on the jackshaft and place the spring inside the coupling.
41 Install the pinion gear, sleeve (1989 only) and Woodruff key.
42 Slide the lockplate, thrust washer (1990 and later) and O-ring (1989 through 1992) onto the bolt, then install the bolt in the jackshaft. Make sure the lockplate tab is in the pinion gear or sleeve cutaway (on 1990 and later models, it must also pass through the thrust washer slot).
43 Place the starter shaft coupling on the inner end of the jackshaft, then slide the jackshaft into the chaincase. **Note:** *On 1990 and later models, make sure the end of the starter coupling with the counterbore faces away from the starter shaft.*
44 Thread the jackshaft bolt into the end of the starter shaft and tighten it to 7 to 9 ft-lbs (9 to 12 Nm). Bend up the lockplate ear against the bolt head. On 1990 and later models, make sure it doesn't extend past the outside diameter of the pinion gear journal. The jackshaft will jam and the chaincase cover bushing will be damaged if the lockplate ear touches it when the starter drive is engaged.
45 Install the clutch and primary chaincase cover.

14 Handlebar switches – removal and installation

1 Generally speaking, the handlebar switches are trouble-free, but, if necessary, they can be detached by removing the screws and separating the halves, which form a split clamp around the handlebar.
2 To prevent the possibility of a short circuit, disconnect the battery before removing the switches.
3 Most problems are caused by dirty contacts, which can be cleaned with an aerosol contact cleaner specially formulated for this purpose.
4 Switch repair is generally impractical. If a switch breaks or wears out, replace it with a new one.

15 Ignition and light switch – removal and installation

Caution: *Disconnect the battery negative cable before working on the ignition switch.*

1 The main switch that controls both the ignition system and the lights is mounted in one of four places: in the control console on the fuel tank, in the base of the instrument panel, on a bracket on the left side of the motorcycle just below the fuel tank, or just below the seat on the left side of the motorcycle. Later FLT models also incorporate a fork lock in the switch.
2 If the switch malfunctions, replace it with a new one – repair is not possible. If the switch is replaced, the ignition key will have to be replaced also.
3 The switch doesn't normally require attention and it should never be oiled. If the switch is oiled, there is a risk of oil reaching the electrical contacts and acting as an insulator.

Tank console-mounted switch

Refer to illustration 15.4

4 The switch must be in the OFF position and the key removed. Detach the instrument console from the tank as described in Chapter 3. Label all wires at the bottom of the switch, then disconnect them **(see illustration).**
5 Remove the four screws retaining the switch and detach it from the console bracket or console underside (as applicable).
6 Install the switch in the reverse order of removal.

Instrument panel-mounted switch

7 Access to the switch necessitates a considerable amount of dismantling and the use of a service tool to set up the fork lock mechanism on reassembly; it is recommended that this operation is carried out by an authorized Harley-Davidson dealer.

Frame-mounted switch

8 The switch is attached to the mounting plate with a threaded decorative ring around the exterior. When the ring is unscrewed from the switch, it can be removed from the rear of the mounting plate.
9 Unplug the wires from the rear of the switch.
10 Install the new switch by reversing the removal procedure.

16 Horn – adjustment

1 The horn is equipped with an adjusting screw on the back of the horn body so the volume can be varied.
2 To adjust the tone or volume, turn the screw 1/2-turn in either direction and test the sound. If the sound is weaker or lost altogether, turn the screw in the opposite direction. Continue adjusting until the desired tone and volume is achieved.
3 If the horn malfunctions completely and cannot be restored by adjustment, install a new one. Repair of the horn isn't possible, because the assembly is riveted together.

17 Spotlight – bulb replacement

Refer to illustrations 17.3 and 17.4

1 Many FL models are equipped with two spotlights. These lights are attached to a common mounting bar, one on each side of the headlight. As with the headlight, the bulb is a sealed beam and must be replaced as a complete unit.
2 Disconnect the negative cable from the battery.
3 Loosen the screw located at the base of the clamp ring and remove the ring **(see illustration)**.

17.3 The spotlight sealed beam is retained by a clamp ring

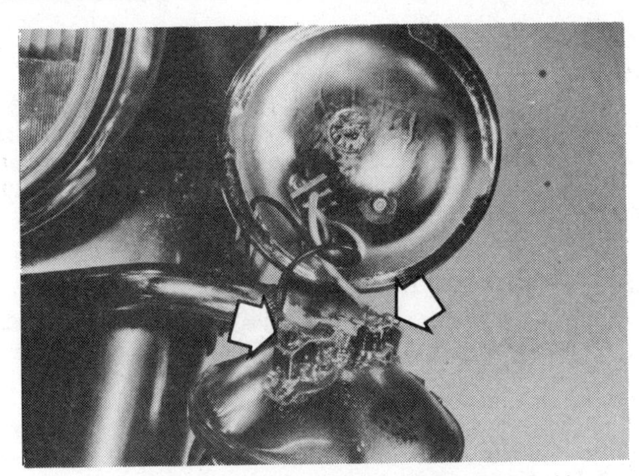

17.4 The wires are attached to the spotlight sealed beam unit with small screws

4 Withdraw the bulb from the housing and disconnect the wire harness from the rear of the sealed beam **(see illustration)**.
5 Attach the wire harness to the new sealed beam and place the light in position in the housing.
6 Attach the clamp ring and tighten the retaining screw securely.
7 Connect the negative cable to the battery.

18 Evaporative emission control system – solenoid test (1992 and later California models)

1 The solenoid is clamped to the rear of the air cleaner baseplate and operates a butterfly valve located in the bottom of the air cleaner housing, a mechanical linkage connects the two components. If operating normally, it will shut the valve when the engine is stopped (ignition switch in OFF position), open it when the starter circuit is operated via the pull-in winding, and keep it open while the engine is running via its hold-in winding.
2 To test the two windings trace its wiring up to the 4-pin connector and separate it at this point. Making the tests on the solenoid side of the connector, use an ohmmeter to measure the resistance between the black/red and gray/black wires (pull-in winding) – a reading of 4 to 6 ohms should be obtained. Take another reading between the white and black wires of the connector (hold-in winding) – a figure of 21 to 27 ohms should be obtained.
3 If either resistance reading is widely different from that specified, the solenoid is confirmed faulty and must be renewed.
4 If the windings prove sound, yet the butterfly valve still fails to operate normally, make continuity checks along the supply and ground circuits to isolate the fault – it will most likely be due to a corroded connector or broken wire.

19 Wiring diagrams – general information

Since it isn't possible to include all wiring diagrams for every year covered by this manual, the following diagrams are typical.

Prior to troubleshooting a circuit, check the fuse and circuit breaker (if equipped) to make sure they're in good condition. Make sure the battery is fully charged and check the cable connections (Sections 4, 5 and 6).

When checking a circuit, make sure all connectors are clean, with no broken or loose terminals or wires. When unplugging a connector, do not pull on the wires. Pull only on the connector housings themselves.

Refer to the accompanying table for the wire color codes.

Wiring diagram color code key

BE	BLUE
B or BK	BLACK
BN	BROWN
GN	GREEN
GY	GRAY
O	ORANGE
PK	PINK
R	RED
V	VIOLET
W	WHITE
Y	YELLOW
T or TN	TAN

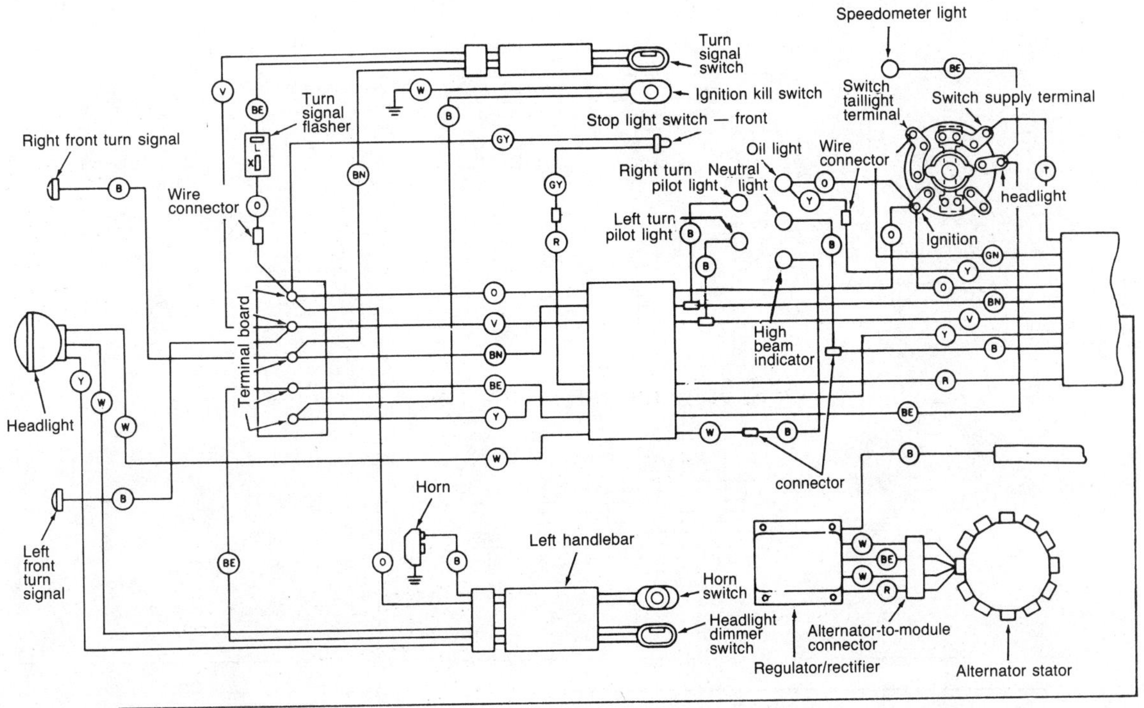

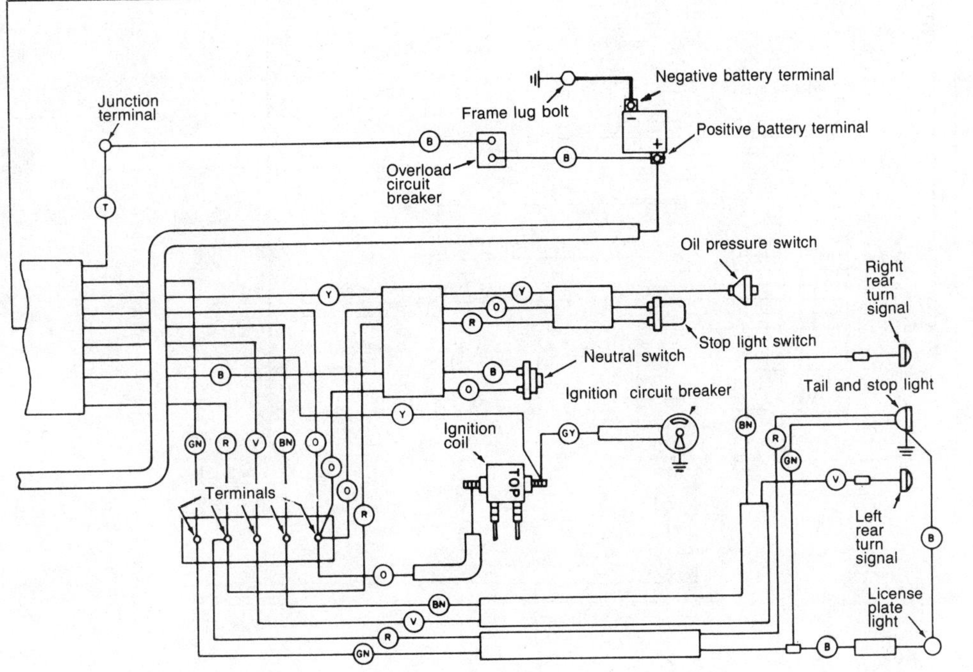

Typical wiring diagram for 1970 through 1972 FX models (1971 FX/FXE shown)

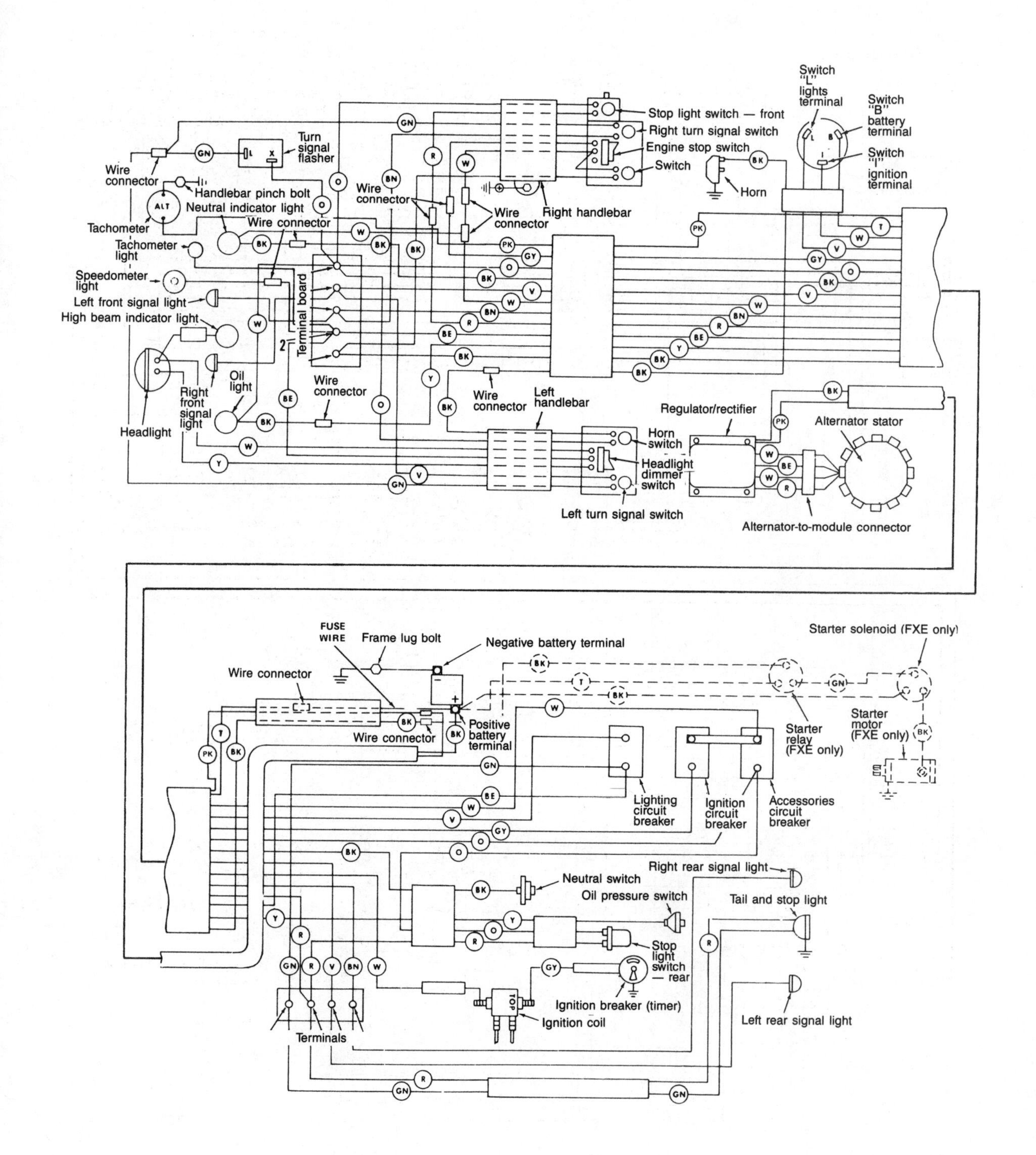

Typical wiring diagram for 1973 and 1974 FX models (1974 FX/FXE shown)

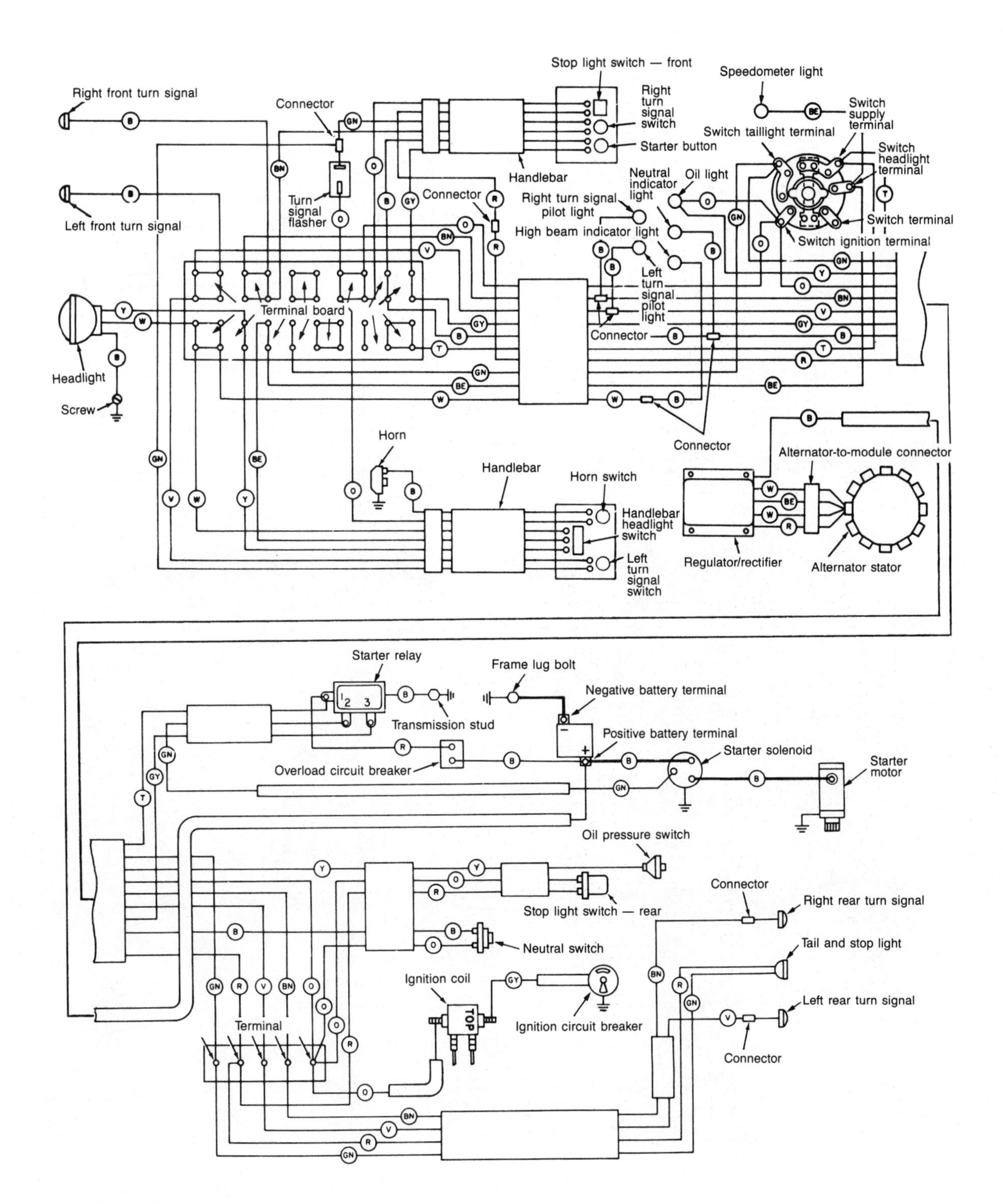

Typical wiring diagram for 1970 through 1972 FL models (1972 FL/FLH shown)

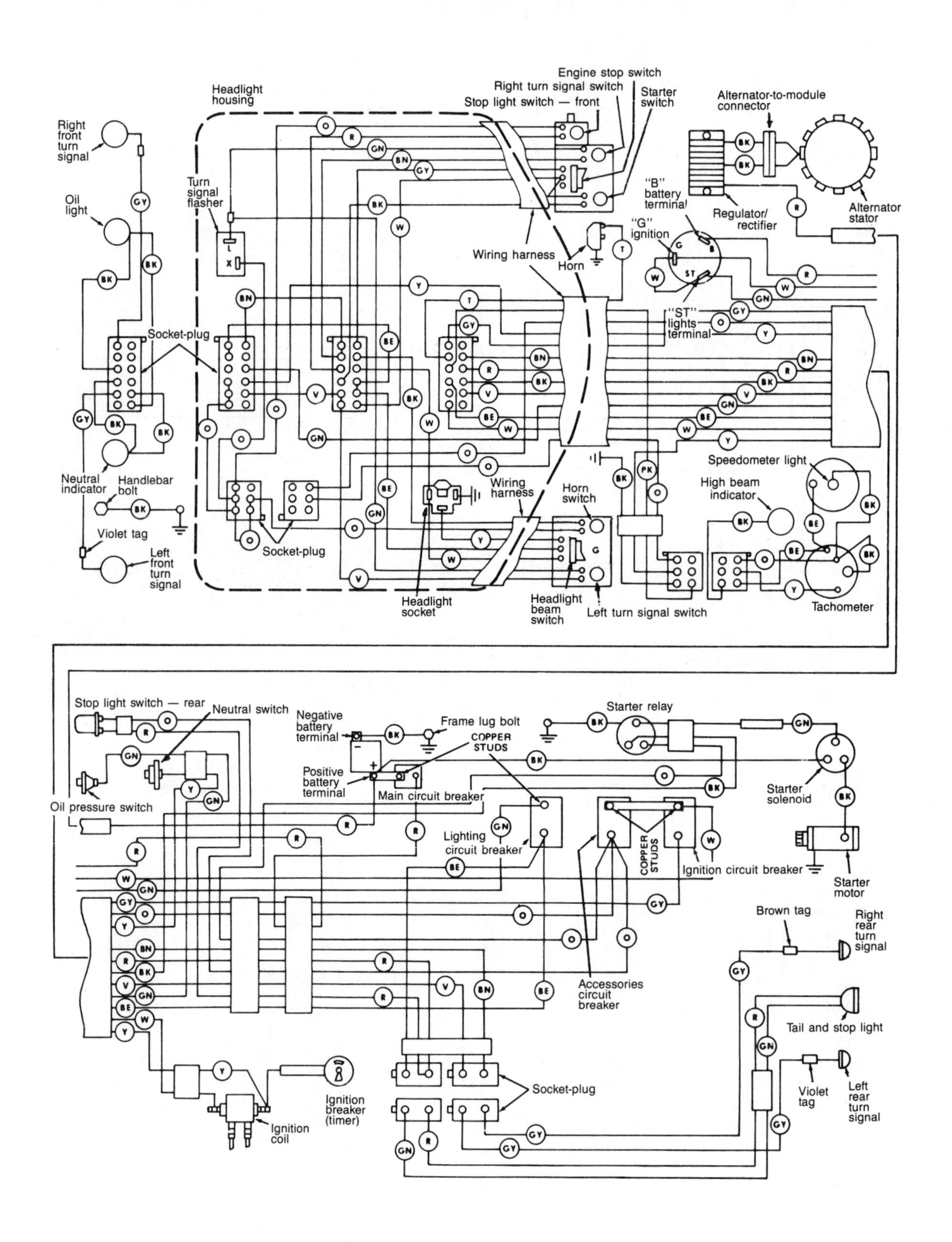

Typical wiring diagram for 1975 through early 1978 FX models (1978 FXS shown)

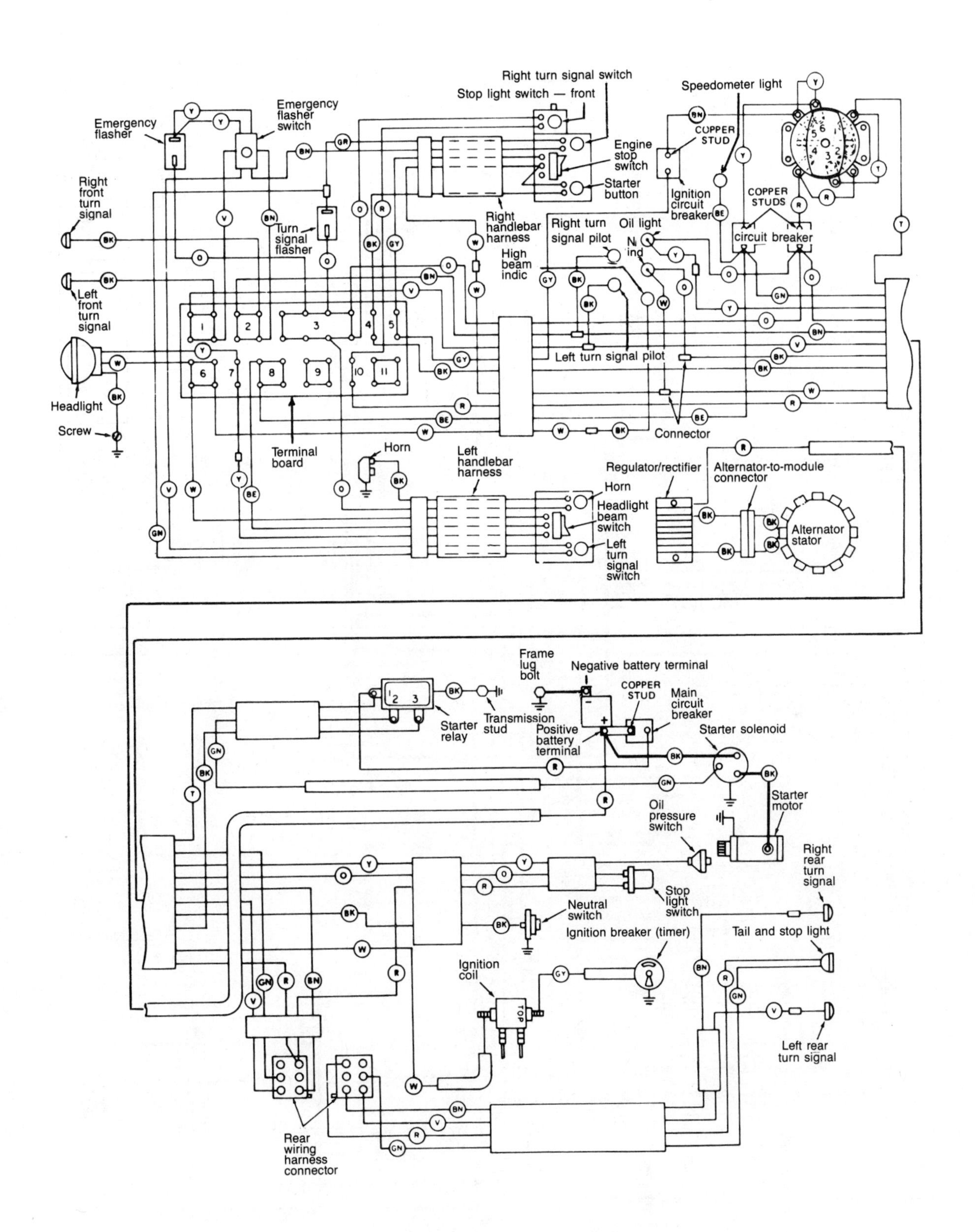

Typical wiring diagram for 1973 through 1978 FL models (1978 FL/FLH shown)

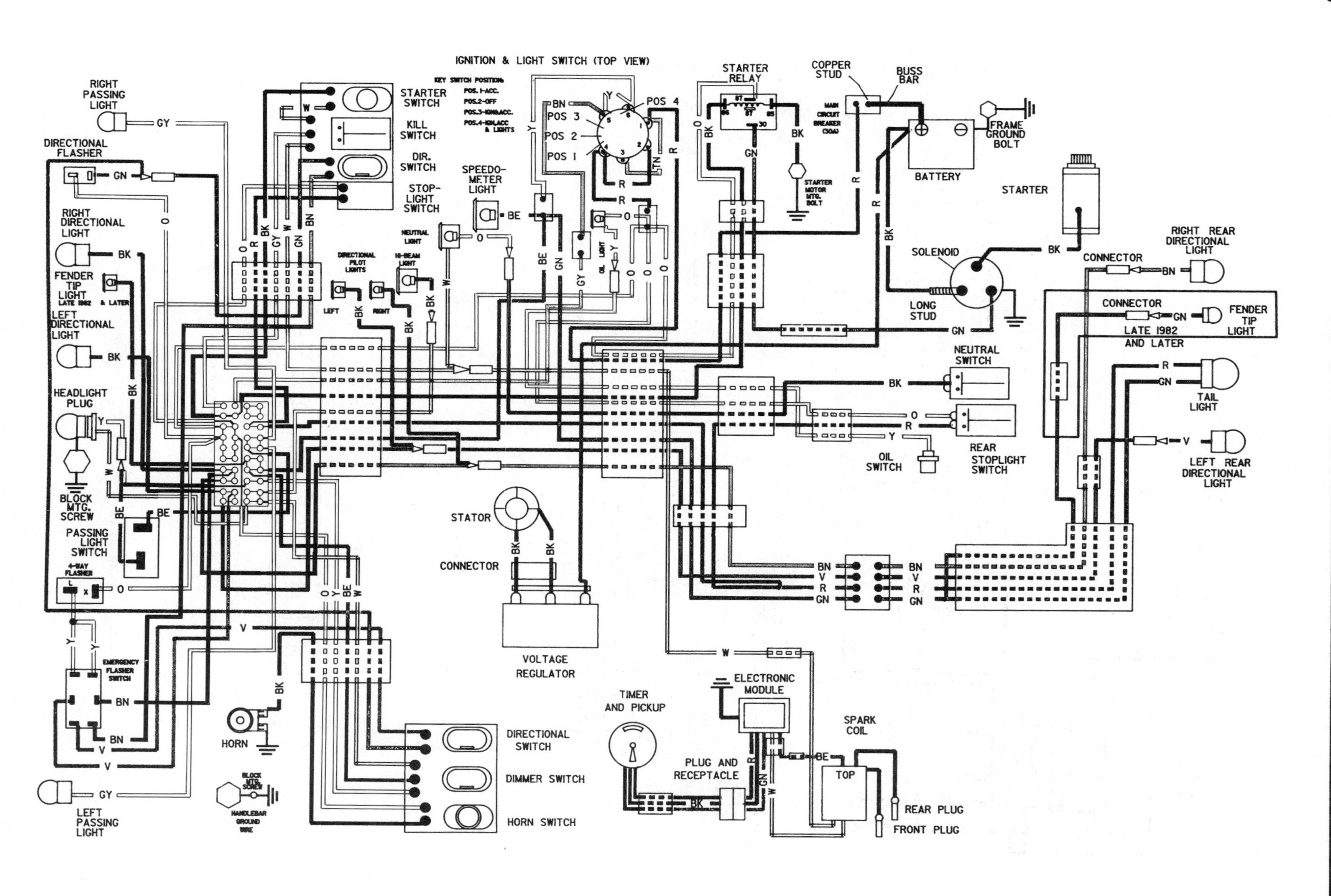

Typical wiring diagram for late 1978 through 1983 FL models (1980 through 1983 FL/FLH shown) – later models similar

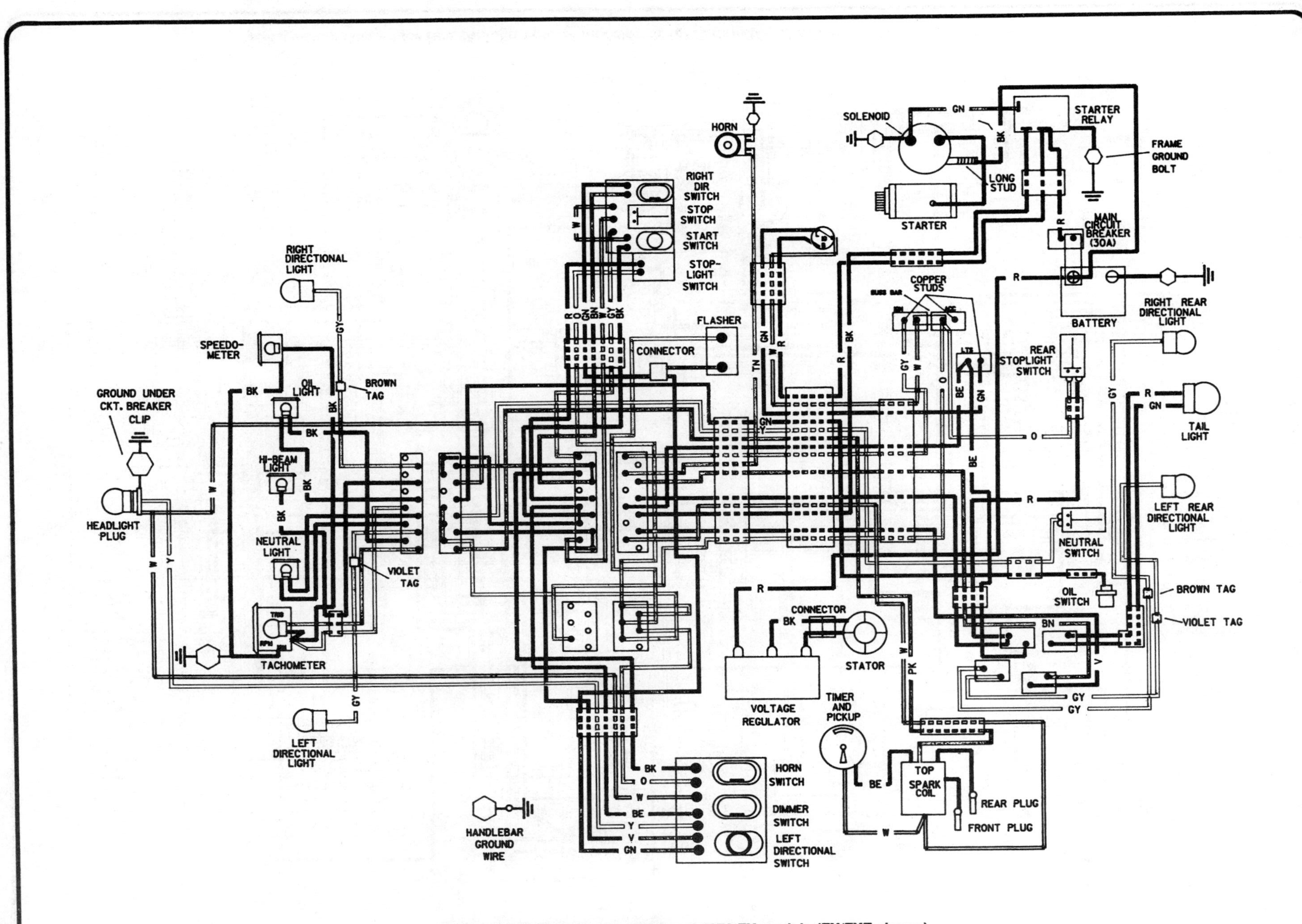

Typical wiring diagram for late 1978 and 1979 FX models (FX/FXE shown)

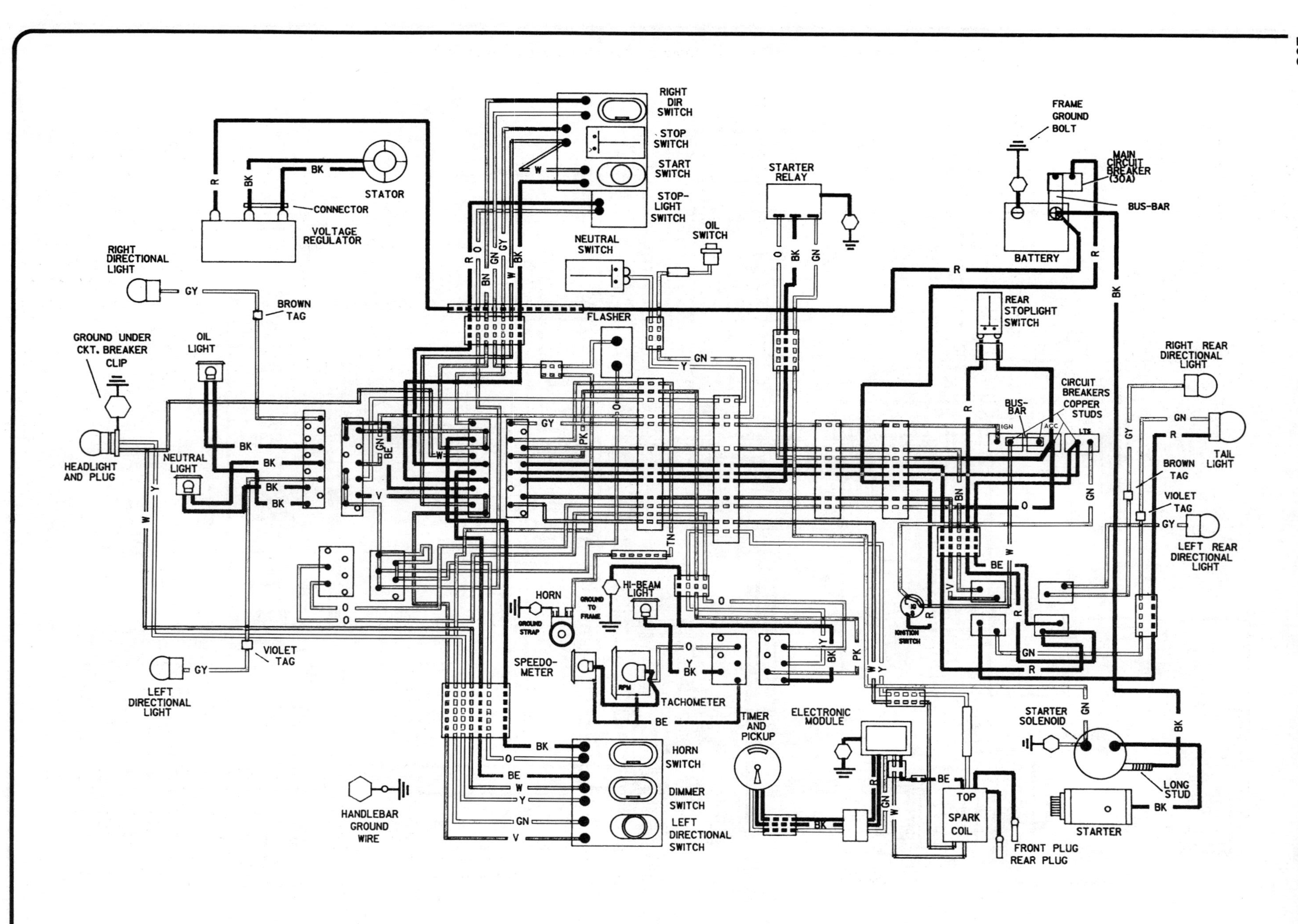

Typical wiring diagram for 1980 through 1983 FX models, except FXR (FXB/FXS/FXSB/FXEF shown) – later models similar

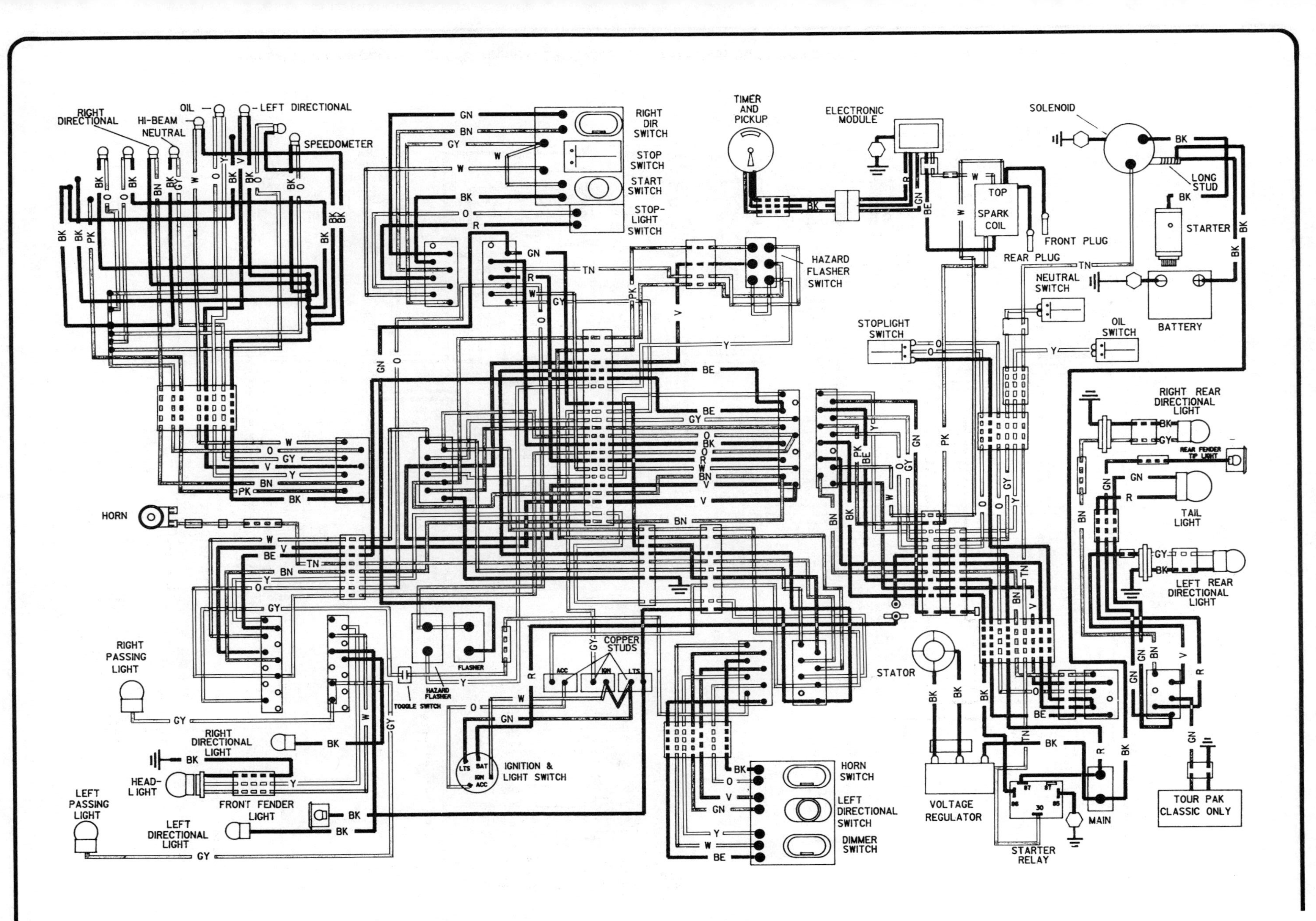

Typical wiring diagram for 1980 through 1983 FLT models (1983 FLHT shown) – later models similar

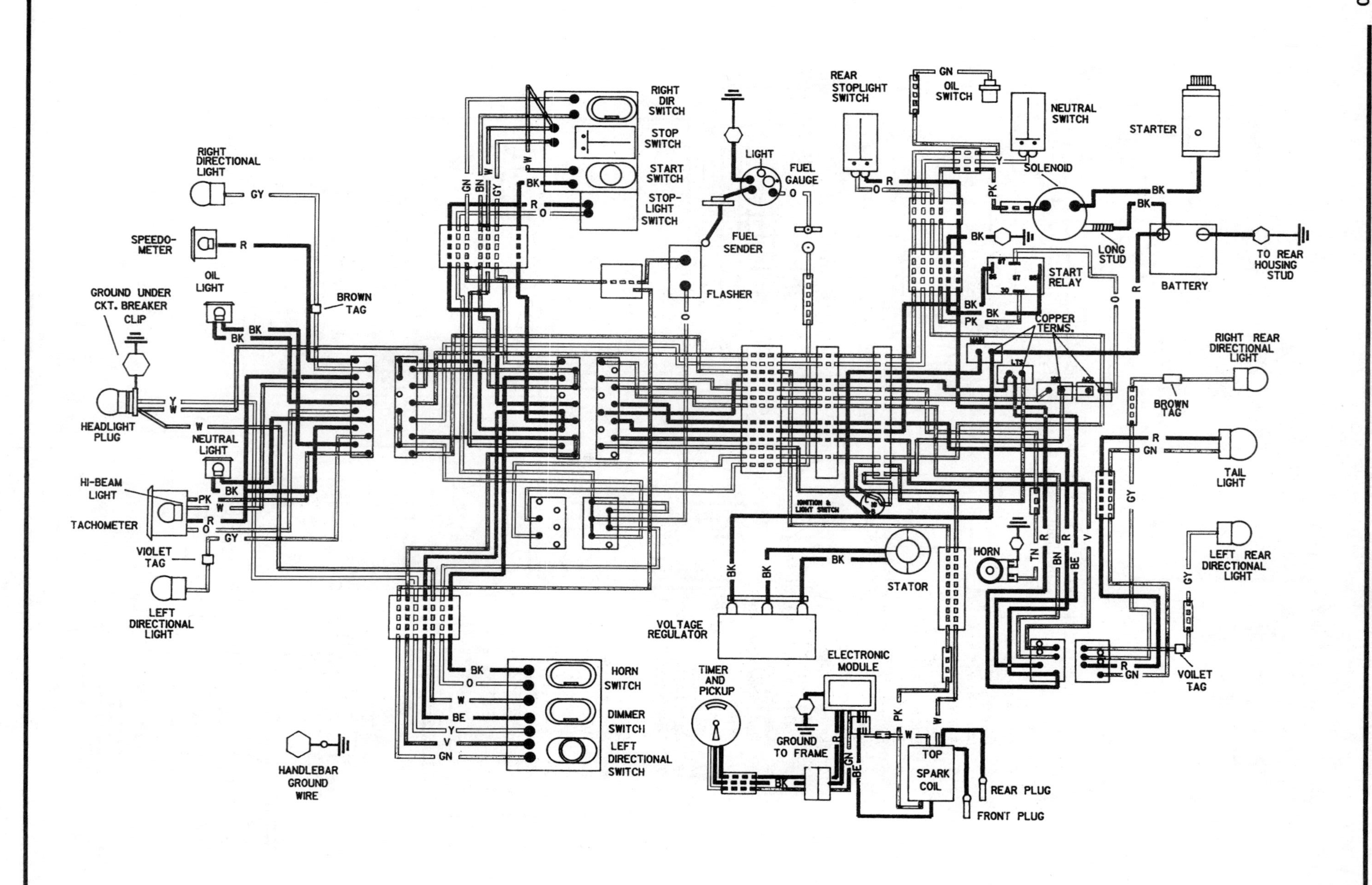

Typical wiring diagram for 1980 through 1983 FXR models (1982 and 1983 FXR/FXRS shown) – later models similar

Component key for FLHS and FLSTC wiring diagrams for diagrams on pages 212 and 213

1 Passing lights switch
2 Connection to turn signal cancel unit (later models)
3 Anti-dive solenoid
4 Right turn signal switch
5 Engine stop switch
6 Start switch
7 Front brake stop switch
8 Ignition coil
9 Ignition module
10 Ignition sensor
11 Oil pressure switch
12 Vacuum operated electric switch
13 Neutral switch
14 Rear brake stop switch
15 Horn
16 Rear fender/mudguard tip light
17 Stop/tail light
18 Rear right turn signal
18 Rear left turn signal
20 Fuses
21 Starter relay
22 Starter motor
23 Battery
24 Alternator
25 Voltage regulator
26 Fuel gauge sender
27 Ignition/light switch
28 Horn switch
29 Headlight dimmer switch
30 Left turn signal switch
31 Emission control solenoid (later California models only)
32 Right passing light
33 Front right turn signal
34 Headlight
35 Front fender/mudguard tip light
36 Front left turn signal
37 Left passing light
38 Speedometer
39 Tachometer
40 Instrument cluster front view
41 Instrument illumination
42 Left turn signal indicator light
43 Right turn signal indicator light
44 Oil pressure warning light
45 Neutral indicator light
46 High beam indicator light
47 Cruise indicator light
48 Fuel gauge
49 Main fuse
50 Ignition/light switch top view
51 Turn signal cancel unit (later models)
52 Handlebar ground/earth wire
53 To frame ground/earth tab
54 Reed switch

Color key

BL Blue
BK Black
BR Brown
GN Green
GY Gray
O Orange
P Pink
R Red
T Tan
V Violet
W White
Y Yellow

Typical wiring diagram for 1988 through 1993 FLHS

See page 211 for key

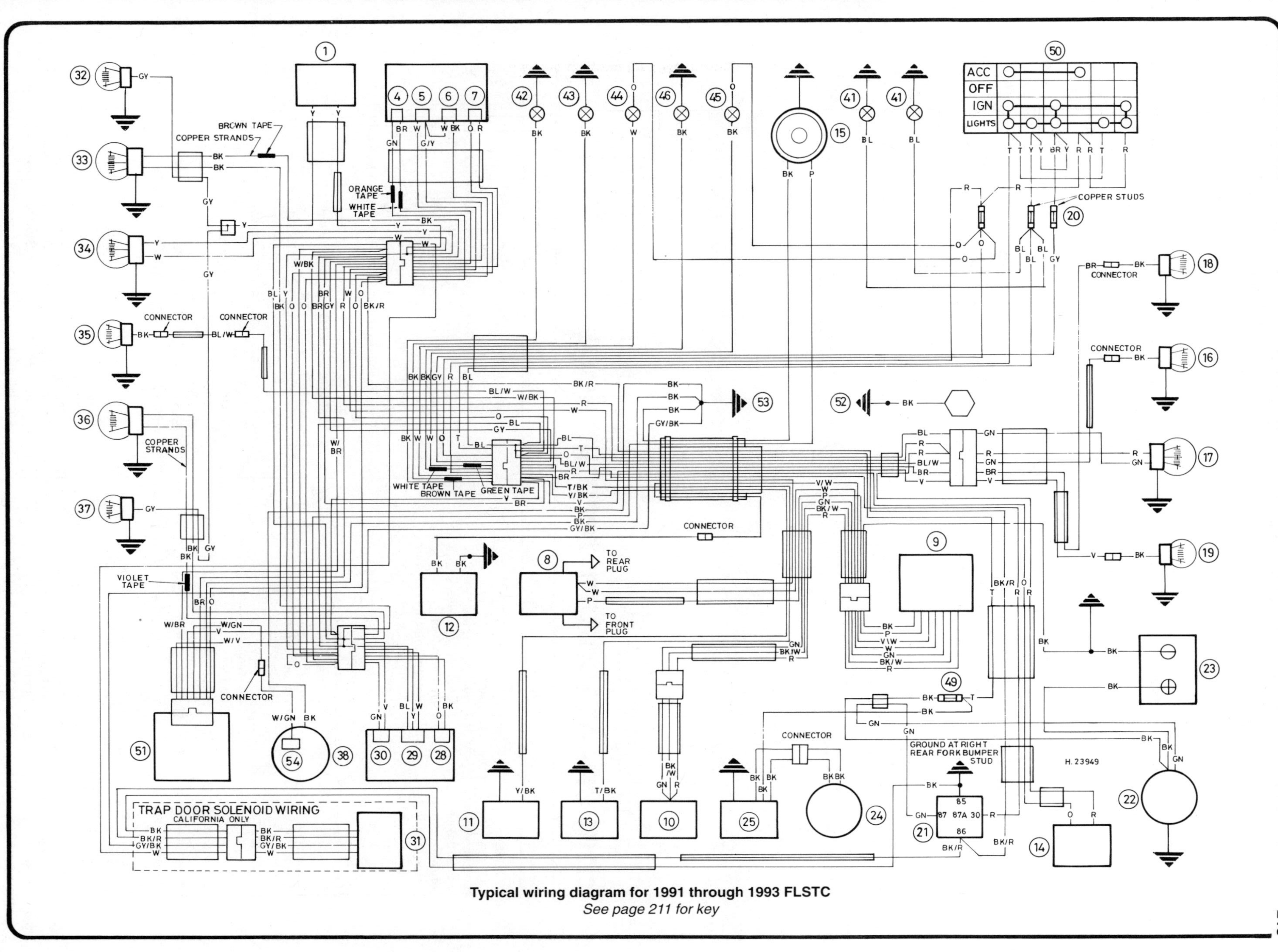

Typical wiring diagram for 1991 through 1993 FLSTC

See page 211 for key

English/American terminology

English	American
Air filter	Air cleaner
Alignment (headlamp)	Aim
Allen screw/key	Socket screw/wrench
Anticlockwise	Counterclockwise
Bottom/top gear	Low/high gear
Bottom/top yoke	Bottom/top triple clamp
Bush	Bushing
Carburettor	Carburetor
Catch	Latch
Circlip	Snap-ring
Clutch drum	Clutch housing
Dip switch	Dimmer switch
Disulphide	Disulfide
Dynamo	DC generator
Earth	Ground
End float	End play
Engineer's blue	Machinist's dye
Exhaust pipe	Header
Fault diagnosis	Troubleshooting
Float chamber	Float bowl
Footrest	Footpeg
Fuel/petrol tap	Petcock
Gaiter	Boot
Gearbox	Transmission
Gearchange	Shift
Gudgeon pin	Wrist/piston pin
Indicator	Turn signal
Inlet	Intake
Input shaft or mainshaft	Mainshaft
Kickstart	Kickstarter
Lower leg	Slider
Mudguard	Fender
Number plate	License plate
Output or layshaft	Countershaft
Panniers	Side cases
Paraffin	Kerosene
Petrol	Gasoline
Petrol/fuel tank	Gas tank
Pinking	Pinging
Rear suspension unit	Rear shock absorber
Rocker cover	Valve cover
Selector	Shifter
Self-locking pliers	Vise-grips
Side or parking lamp	Parking or auxiliary light
Side or prop stand	Kickstand
Silencer	Muffler
Spanner	Wrench
Split pin	Cotter pin
Stanchion	Tube
Sulphuric	Sulfuric
Sump	Oil pan
Swing arm	Swingarm
Tab washer	Lock washer
Top box	Trunk
Two/four stroke	Two/four cycle
Tyre	Tire
Valve collar	Valve retainer
Valve collets	Valve keepers
Vice	Vise
Wheel spindle	Axle
White spirit	Stoddard solvent
Windscreen	Windshield

Conversion factors

Length (distance)

Inches (in)	25.4	= Millimetres (mm)	X 0.0394	= Inches (in)
Feet (ft)	0.305	= Metres (m)	X 3.281	= Feet (ft)
Miles	1.609	= Kilometres (km)	X 0.621	= Miles

Volume (capacity)

Cubic inches (cu in; in³)	X 16.387	= Cubic centimetres (cc; cm³)	X 0.061	= Cubic inches (cu in; in³)
Imperial pints (Imp pt)	X 0.568	= Litres (l)	X 1.76	= Imperial pints (Imp pt)
Imperial quarts (Imp qt)	X 1.137	= Litres (l)	X 0.88	= Imperial quarts (Imp qt)
Imperial quarts (Imp qt)	X 1.201	= US quarts (US qt)	X 0.833	= Imperial quarts (Imp qt)
US quarts (US qt)	X 0.946	= Litres (l)	X 1.057	= US quarts (US qt)
Imperial gallons (Imp gal)	X 4.546	= Litres (l)	X 0.22	= Imperial gallons (Imp gal)
Imperial gallons (Imp gal)	X 1.201	= US gallons (US gal)	X 0.833	= Imperial gallons (Imp gal)
US gallons (US gal)	X 3.785	= Litres (l)	X 0.264	= US gallons (US gal)

Mass (weight)

Ounces (oz)	X 28.35	= Grams (g)	X 0.035	= Ounces (oz)
Pounds (lb)	X 0.454	= Kilograms (kg)	X 2.205	= Pounds (lb)

Force

Ounces-force (ozf; oz)	X 0.278	= Newtons (N)	X 3.6	= Ounces-force (ozf; oz)
Pounds-force (lbf; lb)	X 4.448	= Newtons (N)	X 0.225	= Pounds-force (lbf; lb)
Newtons (N)	X 0.1	= Kilograms-force (kgf; kg)	X 9.81	= Newtons (N)

Pressure

Pounds-force per square inch (psi; lbf/in²; lb/in²)	X 0.070	= Kilograms-force per square centimetre (kgf/cm²; kg/cm²)	X 14.223	= Pounds-force per square inch (psi; lbf/in²; lb/in²)
Pounds-force per square inch (psi; lbf/in²; lb/in²)	X 0.068	= Atmospheres (atm)	X 14.696	= Pounds-force per square inch (psi; lbf/in²; lb/in²)
Pounds-force per square inch (psi; lbf/in²; lb/in²)	X 0.069	= Bars	X 14.5	= Pounds-force per square inch (psi; lbf/in²; lb/in²)
Pounds-force per square inch (psi; lbf/in²; lb/in²)	X 6.895	= Kilopascals (kPa)	X 0.145	= Pounds-force per square inch (psi; lbf/in²; lb/in²)
Kilopascals (kPa)	X 0.01	= Kilograms-force per square centimetre (kgf/cm²; kg/cm²)	X 98.1	= Kilopascals (kPa)
Millibar (mbar)	X 100	= Pascals (Pa)	X 0.01	= Millibar (mbar)
Millibar (mbar)	X 0.0145	= Pounds-force per square inch (psi; lbf/in²; lb/in²)	X 68.947	= Millibar (mbar)
Millibar (mbar)	X 0.75	= Millimetres of mercury (mmHg)	X 1.333	= Millibar (mbar)
Millibar (mbar)	X 0.401	= Inches of water (inH_2O)	X 2.491	= Millibar (mbar)
Millimetres of mercury (mmHg)	X 0.535	= Inches of water (inH_2O)	X 1.868	= Millimetres of mercury (mmHg)
Inches of water (inH_2O)	X 0.036	= Pounds-force per square inch (psi; lbf/in²; lb/in²)	X 27.68	= Inches of water (inH_2O)

Torque (moment of force)

Pounds-force inches (lbf in; lb in)	X 1.152	= Kilograms-force centimetre (kgf cm; kg cm)	X 0.868	= Pounds-force inches (lbf in; lb in)
Pounds-force inches (lbf in; lb in)	X 0.113	= Newton metres (Nm)	X 8.85	= Pounds-force inches (lbf in; lb in)
Pounds-force inches (lbf in; lb in)	X 0.083	= Pounds-force feet (lbf ft; lb ft)	X 12	= Pounds-force inches (lbf in; lb in)
Pounds-force feet (lbf ft; lb ft)	X 0.138	= Kilograms-force metres (kgf m; kg m)	X 7.233	= Pounds-force feet (lbf ft; lb ft)
Pounds-force feet (lbf ft; lb ft)	X 1.356	= Newton metres (Nm)	X 0.738	= Pounds-force feet (lbf ft; lb ft)
Newton metres (Nm)	X 0.102	= Kilograms-force metres (kgf m; kg m)	X 9.804	= Newton metres (Nm)

Power

Horsepower (hp)	X 745.7	= Watts (W)	X 0.0013	= Horsepower (hp)

Velocity (speed)

Miles per hour (miles/hr; mph)	X 1.609	= Kilometres per hour (km/hr; kph)	X 0.621	= Miles per hour (miles/hr; mph)

*Fuel consumption**

Miles per gallon, Imperial (mpg)	X 0.354	= Kilometres per litre (km/l)	X 2.825	= Miles per gallon, Imperial (mpg)
Miles per gallon, US (mpg)	X 0.425	= Kilometres per litre (km/l)	X 2.352	= Miles per gallon, US (mpg)

Temperature

Degrees Fahrenheit = (°C x 1.8) + 32

Degrees Celsius (Degrees Centigrade; °C) = (°F - 32) x 0.56

** It is common practice to convert from miles per gallon (mpg) to litres/100 kilometres (l/100km), where mpg (Imperial) x l/100 km = 282 and mpg (US) x l/100 km = 235*

Index

C

D

E

F